Management for Professionals

The Springer series "Management for Professionals" comprises high-level business and management books for executives, MBA students, and practice-oriented business researchers. The topics cover all themes relevant to businesses and the business ecosystem. The authors are experienced business professionals and renowned professors who combine scientific backgrounds, best practices, and entrepreneurial vision to provide powerful insights into achieving business excellence.

The Series is SCOPUS-indexed.

Pengfei Sun

A Guidebook for 5GtoB and 6G Vision for Deep Convergence

Springer

Pengfei Sun
Huawei Technologies
Shenzhen, China

ISSN 2192-8096 ISSN 2192-810X (electronic)
Management for Professionals
ISBN 978-981-99-4023-3 ISBN 978-981-99-4024-0 (eBook)
https://doi.org/10.1007/978-981-99-4024-0

Jointly published with Posts & Telecom Press
The print edition is not for sale in China mainland. Customers from China (mainland) please order the print book from: Posts & Telecom Press.

Cover designer (key designer): eStudio Calamar, Berlin/Figueres

This Springer imprint is published by the registered company Springer Nature Singapore Pte Ltd.
The registered company address is: 152 Beach Road, #21-01/04 Gateway East, Singapore 189721, Singapore

Foreword

John Hoffman

In 2021, the world continued to witness unprecedented challenges, but responses to the pandemic have again highlighted the importance of mobile technologies. Throughout the pandemic, the mobile industry has been a steady hand, strongly supporting socio-economic recovery and sustainable development. Mobile technology is playing a critical role in the process of digital transformation and the expansion of the digital economy, with far-reaching impact in many different industries. As we look ahead to the post-pandemic era, innovative mindsets will be key to responding to the ever-changing challenges, while ensuring that we continue to enable the mobile industry to seek greater welfare for people and the planet. This also means that we must continue to break boundaries and explore new possibilities to continuously innovate and widen implementations in the 5G era.

We are in the era of intelligent connectivity. The combination of 5G, AI, Internet of Things, and Big Data makes the mobile industry one of the most powerful enablement platforms in the world. Despite the pandemic, the momentum of 5G deployment remains strong. The GSMA's "Mobile Economy Report 2021" forecasts that by 2025, China will account for nearly half the total number of 5G connections in the world. But this will depend on the industry's ongoing commitment to innovation and sustainable investment. Between now and 2025, the mobile industry will invest more than USD900 billion in mobile networks, of which 80% will be in 5G networks. Importantly, this investment will enable us to better care for and empower global sustainable development.

2021 was a crucial year for 5G to achieve initial success and move towards mature development. Around the world, 5G is advancing steadily and has become the fastest-growing mobile technology on the commercial scale. China leads in the number of 5G base stations and connections. By the end of 2021, the total number of 5G connections in China reached 480 million, and, according to GSMA Intelligence, by the end of 2022, it is estimated that there will be 640 million connections, accounting for more than 60% of the total global mobile connections.

The mission of 5G is threefold—to empower vertical industries, promote digital transformation across industries, and become the new infrastructure for societies and economies. The success of 5G for verticals is a critical element to realising the full value of 5G. The ability to seize new opportunities in enterprise business is one of the most important indicators for the transformation of mobile operators, and will also impact the success of digital transformation across the globe. 2021 was the first year in which 5G was commercialised for verticals, and also the year in which we saw the large-scale development of 5G for vertical application taking shape. The core issue that the industry now faces is how 5G for verticals can move from "1" to "*N*" and realise scaled commercial development. Driving greater economy of scale for industry applications is the only way for the success of 5G for verticals, and is also an important path towards new 5G business models.

This book comes at a critical juncture. In this new edition, it explores an in-depth analysis of 5G commercial use cases on the developmental trends and challenges faced during scaled replication, while bringing in the perspectives of key industries that can be empowered by 5G, such as Smart Manufacturing, Smart Ports, Smart Mining, Smart Steel, Smart Healthcare, and Smart Agriculture. It also provides a detailed review of practices and solutions, followed by suggestions to the development of 5G industry standards, cooperation, innovation, and scaled adoption.

We are pleased to participate in this exploration again, and to work hand in hand with many industry partners from mobile operators, Huawei and other vendors, vertical industries, academic and research institutions, to jointly contribute to the book. I hope our joint efforts will provide insightful best practices and suggestions to promote more scaled deployment of 5G for vertical applications.

We believe this book will greatly help many to better utilise 5G to empower various vertical industries and accelerate the commercial success of 5G. It will also become an important reference for best practices of 5G for verticals and large-scale replication.

John Hoffman
CEO, GSMA Ltd.
London, UK

Zhang Ping

2022 has been a milestone year for China. Amidst a challenging environment and the COVID-19 pandemic, China managed to host the Beijing Winter Olympics in February 2022, an event I was proud to attend as a participant in the torch relay. The Beijing Games were also the first-ever Winter Olympics to feature full 5G coverage. Millions of spectators and attendees marvelled at the possibilities brought by 5G, 4K/8K, cloud computing, IoT, blockchain, and IPv6+, and China once again demonstrated its leading role in 5G applications.

The 5G applications at the Olympics are just a microcosm for the revolutionary technologies that are sweeping the globe. Since its commercial rollout, 5G has made its mark in every sector, from academia to industry and beyond, driving the development of a seamlessly digital economy. At the macro level, 5G has served as an engine for economic growth, in China and around the world. At the micro level, 5G has facilitated the development of ToC and ToB applications, stimulating consumer demand and spurring a fresh wave of business investment. For example, in the intelligent transportation field, 5G has made the vehicle-to-everything (V2X) possible, which makes safe autonomous driving more feasible than ever; in the smart healthcare field, 5G has played an active role in COVID-19 prevention, control, and remote consultation; likewise with smart education, 5G has made education fairer and more accessible across the board.

Nonetheless, major challenges still stand in the way of scaling up 5G adoption within industries, including network construction, application convergence, industry supplies, and convergence ecosystems. Matching capabilities and requirements, implementing 5G+vertical industries, and disseminating 5G to Business (5GtoB) on a large scale, are other key concerns within the ICT industry.

This book comes at an opportune moment, as it addresses 5G innovation and exploration, as well as 5GtoB large-scale replication on a systematic level. With its keen insights on global 5G development, it identifies challenges related to large-scale replication, proposes new ideas and methods, and offers an in-depth analysis from industry-specific perspectives to help overcome these challenges. I believe that this book will resonate strongly with industry leaders and professionals.

When exploring 5GtoB large-scale replication, one can't help but notice that 6G research is also in full swing. This book details the evolution of 5G and the development trends behind 6G technologies. 6G will extend from the real to virtual worlds, and will connect people, machines, and things, as well as "genies", harnessing the continually improving disciplines of artificial intelligence, big data, material science, brain-computer interaction, and emotional cognition. "Genies" will prove capable of expanding the scope of emotional intelligence, to encompass intuition, emotion, reasoning, perception, analysis, and more, creating new forms of expression and new paradigms for cognitive development. This will help lead to the harmonious coexistence of artificial intelligence and human intelligence, ushering in a new era in which everything is seamlessly connected and effortlessly smart.

Zhang Ping
Academician of the Chinese Academy of Engineering
Professor of Beijing University of Posts and Telecommunications
Beijing, China

Rahim Tafazolli

Every generation of mobile system aims to address the societal needs with economical benefits.

New features in every generation also include previous generations' features and 5G is no exception to this rule.

The first four generations supported connectivity between people. In addition to that, 5G is taking a giant step in support of connectivity between machines. Machine connectivity is the most transformative aspect of 5G and hugely differentiates it from previous generations. For this reason, 5G can be considered as "Communication and Automation" supported in one network technology. 5G is the first network technology after Internet that supports mix of different services but with important differences to internet in the form of by-design mechanisms for; managing quality of service for each service class through network slicing, robust security and offers huge flexibility through its unique service-based architecture (SBA). 5G is a flexible system that can be deployed at any frequency band in the form of public and non-public or their hybrid network.

Automation capability, enabled by 5G and Artificial Intelligence (AI), transforms through modernisation many industries and can create new business opportunities. Automation when adopted leads to enhancement in efficiency, productivity and reduces costs. Unique capabilities offered by 5G for automation are "guaranteed and reliable low latency" and ability to support "massive number of machines". Low latency and high reliability of connectivity enable automation of time and mission critical tasks. Some of the industries that would hugely benefit from 5G automation are: Manufacturing, Health, Transportation/Logistics, Financial services, Retails, Digital Creativity and Information services. 5G will also provide huge societal benefits in Education, Telemedicine/Care, Smart Homes and Smart Cities, and many more. As a representative of advance economy country, an economic modelling of 5G study identified contribution to the UK GDP by 2030 of 1.5% from mobile broadband whereas automation contribution is expected to be more than 4%.

We are only at the beginning of 5G revolution that started with deployment of mobile broadband to address capacity crunch of 4G. The most impactful aspects of 5G through automation and connected machines has just started. 5G with its continuous evolution in capabilities and its inherent flexibility is far-reaching beyond terrestrial networks into non-terrestrial networks. The global ecosystem behind 5G ensures continuous innovation and its evolution with new capabilities to serve the world for the foreseeable future. Its transformative role has started to be understood and appreciated by non-telecom businesses.

5G standards are progressing with speed as new use cases are emerging. In December 2021, ambitious 3GPP Release 18, that is known as 5G-Advanced, was agreed. These standards are targeting even higher capacities for both up and downlinks, higher energy and spectral efficiencies, high-accuracy geo-location, mix of high-quality virtual and augmented reality (XR) all to be supported through unicast, multi-cast, broadcast services, and many more rich services and features.

This book covers, with rigour and sufficient depth, 5G capability with good examples of its use in different representative industries.

It is the best source of information for all in telecom and non-telecom sectors for purposes of research, staff training, and 5G adoption in their respective organisations.

I strongly recommend this book for all stakeholders from industry, business, academia, research institute, and government departments and organisations.

Rahim Tafazolli
Regius Professor, FREng, FIET, Fellow of WWRF
Director of Institute for Communication Systems (ICS)
Founder and Director of 5GIC and 6GIC
The University of Surrey
UK

Wen Ku

As our society embraces digitalisation, a new form of dynamism has emerged. Digital technologies, namely 5G, cloud computing, artificial intelligence (AI), and big data, have developed at a rapid pace, inspiring a new technological revolution akin to the industrial revolutions of the past. This has enormous implications at a societal level, from the economy to production, to culture and daily life. Digitalisation, networking, and ubiquitous intelligence will be the key attributes of this paradigm shift.

5G industry applications are on the verge of becoming widespread, during the ramp up from 0 to 1, as the supply capabilities of 5G networks, such as 5G virtual private networks, continue to improve, and 5G is deeply integrated into business and operations models. Some successful 5G applications will be replicated on a large scale, from 1 to *N*, a process that is unprecedented and will be fraught with challenges.

5G applications exist within extremely complex ecosystems, in which projects and solutions are implemented differentially depending on the industry value chain. The use of different methods and standards will lead to redundant construction and low efficiency, ultimately making it impossible to form closed loops. Therefore, standardisation is a prerequisite of large-scale 5G deployment; it requires buy-in throughout the industry value chain.

A "Set Sail" Action Plan for 5G Applications (2021–2023) was issued by ten government agencies in July 2021, including the Ministry of Industry and Information Technology (MIIT), with the goal of guiding the standardisation of 5G applications within China. This book makes a strong case for the need to enhance protocol standard interoperability across industries, formulate a comprehensive 5G standard system, with more than 30 standards for key industries, and implement these standards.

The 5G application standard system in China has continued to progress under the leadership of the China Communications Standards Association (CCSA), providing important technical specifications and guarantees for 5G application development.

A total of 447 industry standards have been released thus far, laying the groundwork for a comprehensive framework. Meanwhile, the key industry standards have been completed, which facilitate the dissemination of 5G applications on a large scale.

CCSA will continue to commit itself to openness and mutually beneficial global partnerships, by planning and deploying 6G research, studying trends for fusion technologies, such as communication + sensing and communication + AI, and actively carrying out 6G technology exchanges and industry partnerships on a global scale.

I'm glad to see that this book discusses the major challenges and development paths for 5G on a large scale, as well as the evolution towards intelligent 6G, on both a theoretical and practical level. This book also explores 5G application standards in general fields and key industries. Its release is timely, as large-scale applications for 5G lay just around the corner, and related standards continue to be unveiled around the world, giving us the winds to "set sail on a great voyage".

Wen Ku
Vice Chairman of the Council and Secretary-General of China Communications Standards Association (CCSA)
Beijing, China

Jiang Wei

The iron and steel industry is fundamental to China's economy, and a key contributor to modernisation and an important sector for green, low-carbon development. Iron and steel touch on a vast number of industries, starting from mining and refining, and extending all the way to automobile, shipbuilding, home appliance, machinery, and rail transportation.

China's iron and steel sector is now among the world's most formidable, on the heels of decades of reform, opening up, innovation, and automation. Under the 13th Five-Year Plan, enterprise equipment has become world-class, with a dramatic improvement in informatisation. To stimulate high-quality development, the Ministry of Industry and Information Technology (MIIT) recently issued a set of guidelines that aims to put the sector on the path towards sustainability and global competitiveness by 2025.

Despite with such tangible progress, China is still at the initial stage of smart manufacturing, and faces challenges related to software and information security capabilities, a lack of industry standards, and key technology bottlenecks. While some leading players have reached Industry 3.0 and are moving towards Industry 4.0, a multitude of enterprises are still at the Industry 2.0 stage. Smart manufacturing is not widely deployed, whereas remote automatic production is urgently needed, to improve overall living standards and working conditions.

The iron and steel industry still needs to be digitalised, and 5G can serve as the driving force for this paradigm shift. 5G is the core of new-generation ICT, and it continues to be integrated with edge computing, cloud computing, big data, artificial intelligence, industrial Internet, digital twin, and other technologies, providing a strong foundation for industrial digitalisation.

5G applications in vertical industries have evolved from the exploration and incubation phase, to large-scale development, in which experiences and models are lacking. This book couldn't have arrived at a more opportune time. It provides a solid theoretical analysis and summarises the best practices we gained to help readers understand the major challenges and common capabilities that determine the success of large-scale development.

I believe that 5G, and even 6G, will pave the way towards digitalisation and smart manufacturing in the iron and steel industry, and countless other sectors, unleashing game-changing innovations and novel ways of doing business.

Jiang Wei
Vice President of China Iron and Steel Association
Beijing, China

Acknowledgments

5G industry applications are being replicated on a large scale, from 1 to N, as 5G is deeply integrated into industries. In this process, 5G collaborates with other innovative information technologies, paving the way for the future development of industry applications. Beyond that, 6G is emerging as a new research focus, with its vision and requirements being defined. This book comes at this critical juncture. It incorporates the collective wisdom of the industrial chain and the latest research results from multiple disciplines. We thank all experts and scholars for their great work and support.

Our heartfelt appreciation also goes to the translation team for their high level of professionalism. Without their contributions, this book would not have been possible. They are Li Xiaochun, Feng Qiangqiang, Chen Mengyuan, Wang Xiaofen, Wang Ying, Chen Gong, Chen Xiexia, Liu Yong, Wang Lili, Yang Xiaojing, Zhang Shuangshuang, Nathanael Schneider, Kyle Melieste, CHRISTOVA EKATERINA ROUMENOVA, Vladyslav Malska, Kang Dan, and Hou Junyan.

Abstract

5G has the potential to revolutionize industry as we know it, and unleash a new wave of digitalization, made possible by its fusion with artificial intelligence, big data, cloud computing, and other technologies.

This book explores the roadmap for widespread implementation of 5G to Business (5GtoB) on a large scale, including business models and standards. We also address how 5G, and even 6G, can be applied on an industry-specific basis.

First, we provide an overview of how 5G stimulates industrial transformation in the wake of global digitalization, involving commercial applications, policy support, and global trends.

Second, we summarize the main challenges facing large-scale 5GtoB adoption from a theoretical side, and provide insight on the development path for XtoB, as well as future paradigms.

Third, we shed light on common applications for 5G in key industries based on best practices and key capabilities, and detail the progress in 5GtoB convergence standards.

Fourth, we present a vision for 6G and innovative ToB enabling technologies, by describing such technologies in depth, including semantic communication, on-purpose network, and cell-free cooperative massive multiple-input multiple-output (MIMO).

Lastly, we forecast typical industry applications of 6G, such as those involving the metaverse, human-machine interaction, and hyper-connected future cities.

5G is being integrated into a wide range of industries, a path that 6G is expected to follow in the future. We hope that this book serves as a beacon on the journey toward industry-wide digitalization.

Editorial Board

- Adrian Scrase, Head of the 3GPP Mobile Competence Center and ETSI CTO
- Cai Kang, Vice President of the China Telecom Research Institute
- Cao Lei, Deputy General Manager of the China Telecom Government and Enterprise Business Group
- Fan Ji'an, Chief Scientist in Big Data for China Unicom
- Huang Yuhong, General Secretary of GTI and President of the China Mobile Research Institute
- Si Han, President of GSMA Greater China
- Wang Zhiqin, Vice President of the China Academy of Information and Communications Technology
- Wei Bing, Deputy General Manager of the China Mobile Government & Enterprise Business Department
- Wen Ku, Vice Chairman of the Council and Secretary-General of the China Communications Standards Association (CCSA)
- Ye Xiaoyu, Vice President of the China Unicom Research Institute

- **Secretary of the Editorial Board**: Lin Li
- **Writing Team** (in alphabetical order)

Leaders		
Chen Dan	Deng Wei	Li Jiamin
Lin Li	Liu Guangyi	Liu Hong
Lu Lu	Pan Feng	Tan Hua
Wang Xiaoqi	Xiao Shanpeng	Xu Xiaodong
Yang Jin	Zhang Xueli	Zhao Shizhuo
Deputy Leaders		
Chen Li	Dong Chen	Du Jiadong
Li Ying	Liao Yunfa	Pan Guixin
Peter Jarich	Tan Zhenlong	Wan Ming
Wang Weiguo	Xiao Yue	Yang Xinjie
Yu Jiang	Zheng Shaowen	
Members		
Cheng Jinxia	Cong Jianxun	Dong Jing
Du Bin	Fu Chenglong	Gao Peng
Guan Jialing	Guan Qinghe	Guo Keqiang
Guo Xin	Han Shujun	Li Jian
Li Zejie	Lian Yue	Liang Yongming
Liu Jiawei	Liu Jing	Liu Shuai
Liu Yajian	Lou Mengting	Lu Yuhao
Ma Hongbin	Ma Shuai	Meng Xiangwei
Qi Xu	Shi Lei	Shi Nanxiang
Si Zhe	Tan Ziwei	Wang Bizhu

Wang Rui	Wang Yonghui	Wei Liurong
Wu Cheng	Xiao Xie	Xu Furong
Xu Shu	Yang Bohan	Yang Dewu
Yang Peng	Yang Xinhua	Yang Yi
Zeng Kaiyue	Zhang Haitao	Zhang Jian
Zhang Long	Zhang Lulu	Zhang Qi
Zhang Tianjing	Zhang Yan	Zhang Yu
Zhao Ying	Zheng Shiying	Zhou Lisha
Zuo Yun		

Contents

Abbreviations

3GPP	3rd Generation Partnership Project
4IR	Fourth Industrial Revolution
5G PPP	5G Infrastructure Public-Private Partnership
5G S-Module	5G Superior Universal Module
5GAIA	5G Applications Industry Array
5GCroCo	Fifth Generation Cross-Border Control
5GS	5G System
5GtoB	5G to Business
5GTT	5G Testbeds and Trails Programme
AC	Application Client
ACAI	Animal Crossing Artificial Intelligence
ADAS	Advanced Driver Assistance System
ADSL	Asymmetric Digital Subscriber Line
AEF	API Exposing Function
AEHF	Advanced Extremely High Frequency
AF	Application Function
AGV	Automated Guided Vehicle
AI	Artificial Intelligence
AIDS	Acquired Immunodeficiency Syndrome
AIoT	Artificial Intelligence of Things
AKMA	Authentication and Key Management for Applications
AMR	Automated Mobile Robot
ANFR	French National Frequency Agency
AP	Access Point
API	Application Programming Interface
APNOMS	Asia-Pacific Network Operations and Management Symposium
AR	Augmented Reality
ARMG	Automated Rail-Mounted Gantry
ARPU	Average Revenue Per User
ART	Artificial Intelligence Robot of Transportation
ATO	Automatic Train Operation

ATP	Automatic Train Protection
AWS	Amazon Web Service
BAF	Basic-Advanced-Flexible
CA	Carrier Aggregation
CAC	Cyberspace Administration of China
CAGR	Compound Annual Growth Rate
CAICT	China Academy of Information and Communications Technology
CAM	Connected and Automated Mobility
CAPIF	Common API Framework
CBCF	Cell Broadcast Center Function
CCF	CAPIF Core Function
CCFD	Co-frequency Co-time Full Duplex
CCSA	China Communications Standards Association
CF-RAN	Cell-Free Radio Access Network
CLI	Cross-Link Interference
CMOS	Complementary Metal-Oxide-Semiconductor
CN	Core Network
CNN	Convolutional Neural Network
CPC	Communist Party of China
CPE	Customer-Premises Equipment
CPHA	China Ports and Harbors Association
CPU	Central Processing Unit
CRTA	China Road Transport Associations
CSI	Channel State Information
CSMF	Communication Service Management Function
CT	Communication Technology
CTISC	Computer Technology, Information Science and Communications
C-V2X	Cellular Vehicle-to-Everything
DAS	Distributed Antenna System
DDoS	Distributed Denial of Service
DN	Data Network
DoD	Department of Defense
DoS	Denial of Service
DOS	Disk Operating Systems
DOU	Data of Usage
DQN	Deep Q-Network
DRL	Deep Reinforcement Learning
DSA	Digital Subtraction Angiography
DSRC	Dedicated Short-Range Communications
DSS	Dynamic Spectrum Sharing
DTU	Data Transfer Unit
E2E	End-to-End
EAS	Edge Application Server
ECS	Edge Configuration Server
EDGEAPP	Edge Application

EDN	Edge Data Network
EEC	Edge Enabler Client
EECC	European Electronic Communications Code
EES	Edge Enabler Server
EHIH	Emory Healthcare Innovation Hub
EIR	Equipment Interchange Receipt
eMBB	Enhanced Mobile Broadband
EMR	Electronic Medical Record
ERP	Enterprise Resource Planning
E-RTG	Electric Rubber-tired Gantry
EU	European Union
F5G	Fifth Generation Fixed Network
FAE	FF Application Enabler
FCC	Federal Communications Commission
FDD	Frequency Division Duplex
FF	Factories of the Future
FL	Federated Learning
FlexE	Flexible Ethernet
FP	Favorable Propagation
FRL	Federated Reinforcement Learning
FRMCS	Future Railway Mobile Communication System
FWA	Fixed Wireless Access
GCN	Graph Convolutional Network
GDP	Gross Domestic Product
GEO	Geostationary Earth Orbit
GIS	Geographic Information System
GLONASS	Global Navigation Satellite System
GNSS	Global Navigation Satellite System
GPS	Global Positioning System
GPT	General-Purpose Technology
GPU	Graphical Processing Unit
GSMA	Global System for Mobile Communications Association
GSM-R	Global System for Mobile Communications-Railway
HARI	Human Augmented Robotics Intelligence
HD	High-Definition
HDMI	High-Definition Multimedia Interface
HSR	High-Speed Railway
IAB	Integrated Access and Backhaul
ICECA	International Conference on Electronics, Communication and Aerospace Technology
ICS	Institute for Communication Systems
ICT	Information and Communications Technology
ICU	Intensive Care Unit
IEEE	Institute of Electrical and Electronics Engineers
IGV	Intelligent Guided Vehicle

IoE	Internet of Everything
IoS	Internet of Ships
IoT	Internet of Things
IP RAN	IP Radio Access Network
IRS	Intelligent Reflecting Surface
ISAC	Integrated Sensing and Communication
IT	Information Technology
ITOS	Intelligent Terminal Operation System
ITU	International Telecommunication Union
IUR	Industry-University-Research
KOSF	Korea Smart Factory Association
KPI	Key Performance Indicator
KPN	Koninklijke PTT Nederland
KRRI	Korea Railway Research Institute
KTX	Korea Train Express
LAN	Local Area Network
LEO	Low-Earth Orbit
LoRa	Long Range Radio
LOS	Line of Sight
LPHAP	Low-Power High-Accuracy Positioning
LPIPS	Learned Perceptual Image Patch Similarity
LTE	Long Term Evolution
LTE Cat.1	Long Term Evolution Category 1
LTE-U	LTE-Unlicensed
M2M	Machine to Machine
MaaP	Messaging as a Platform
MAB	Multi-armed Bandit
MADDPG	Multi-agent Deep Deterministic Policy Gradient
MARL	Multi-agent Reinforcement Learning
MCPTT	Mission-Critical Push-to-Talk
MCU	Micro Control Unit
MDT	Minimization of Drive Tests
MEC	Multi-access Edge Computing
MEO	Multi-access Edge Orchestrator
MEP	Multi-access Edge Platform
MES	Manufacturing Execution System
MGI	McKinsey Global Institute
MIB	Master Information Block
MIIT	Ministry of Industry and Information Technology
MIMO	Multiple-Input Multiple-Output
MIoT	Massive Internet of Things
mMTC	Massive Machine-Type Communications
mmWave	Millimeter Wave
MOT	Ministry of Transport
MPA	Maritime and Port Authority of Singapore

MR	Mixed Reality
MRI	Magnetic Resonance Imaging
MSGin5G	MIoT over the 5G System
MSIT	Ministry of Science and Information Communications Technology
MSMEs	Micro, Small, and Medium-sized Enterprises
MUOS	Mobile User Objective System
mURLLC	Massive URLLC
MVNO	Mobile Virtual Network Operator
MxD	Manufacturing Times Digital
NAFD	Network-assisted Full Duplex
NASA	National Aeronautics and Space Administration
NB-IoT	Narrowband Internet of Things (NB-IoT)
NDRC	National Development and Reform Commission
NE	Network Element
NEA	National Energy Administration
NERCEL	National Engineering Research Center for E-Learning
NFC	Near Field Communication
NFV	Network Functions Virtualization
NGP	Next Generation Port
NHC	National Health Commission
NLOS	Non-Line-of-Sight
NOMA	Non-Orthogonal Multiple Access
NPIF	National Productivity Investment Fund
NPN	Non-Public Network
NSA	Non-Standalone
NSCE	Network Slice Capability Enablement
NSMF	Network Slice Management Function
NSSAI	Network Slice Selection Assistance Information
NSSMF	Network Slice Subnet Management Function
NTT	Nippon Telegraph and Telephone Corporation
OB	Outside Broadcasting
OEM	Original Equipment Manufacturing
OMA	Orthogonal Multiple Access
OT	Operation Technology
PBCH	Physical Broadcast Channel
PCF	Policy Control Function
PDA	Personal Digital Assistant
PIN	Personal IoT Network
PINAPP	Personal IoT Network Application
PLC	Programmable Logic Controller
PLMN	Public Land Mobile Network
PMU	Phasor Measurement Unit
PNI-NPN	Public Network Integrated Non-Public Network
PNT	Position Navigation and Timing
PoE	Power over Ethernet

PON	Passive Optical Network
PPP	Public Private Partnership
PSS	Primary Synchronization Signal
PV	Photovoltaics
QoS	Quality of Service
RaaS	Robot as a Service
RAN	Radio Access Network
RAT	Radio Access Technology
RAU	Remote Antenna Unit
RB	Resource Block
RedCap	Reduced Capability
RFID	Radio Frequency Identification
RIS	Reconfigurable Intelligent Surface
RMG	Rail-mounted Gantry
RMSI	Remaining Minimum SI
RNN	Recurrent Neural Network
ROI	Return on Investment
RSS	Received Signal Strength
RSU	Road Side Unit
RTG	Rubber-tired Gantry
RTK	Real-time Kinematic
RTT	Round-Trip Time
SA	Standalone
SASAC	State-owned Assets Supervision and Administration Commission
SAWAPs	Small-Area Wireless Access Points
SBA	Service-based Architecture
SC	Superimposed Coding
SCaN	Space Communications and Navigation
SDN	Software-defined Network
SD-RAN	Software-defined RAN
SEAL	Service Enabler Architecture Layer
Seb	Semantic Base
SIC	Successive Interference Cancellation
SIM	Subscriber Identity Module
SLA	Service Level Agreement
SLAM	Simultaneous Localization and Mapping
SMC	Samsung Medical Center
SME	Small- and Medium-sized Enterprise
SMF	Session Management Function
SNPN	Standalone Non-Public Network
SNR	Signal-to-Noise Ratio
SON	Self-Organizing Network
SSB	Synchronization Signal and PBCH Block
SSS	Secondary Synchronization Signal
STS	Ship-to-Shore

TCP	Transmission Control Protocol
TDD	Time Division Duplex
TEU	Twenty-Foot Equivalent Unit
THz	Terahertz
TN	Transport Network
TOS	Terminal Operating System
TSAT	Transformational Satellite Communications System
TSN	Time Sensitive Networking
UAS	Uncrewed Aerial System
UAV	Unmanned Aerial Vehicle
UCET	UK-China Emerging Technologies
UDP	User Datagram Protocol
UE	User Equipment
UGC	User-generated Content
UHD	Ultra-High-Definition
UNCTAD	United Nations Conference on Trade and Development
UPF	User Plane Function
URLLC	Ultra-Reliable Low-Latency Communication
USIM	Universal Subscriber Identity Module
U-TDOA	Uplink Time Difference Of Arrival
UTM	Unmanned Aerial System Traffic Management
UW	Ultrasonic Wave
UWB	Ultra-Wideband
V2I	Vehicle to Infrastructure
V2P	Vehicle to Pedestrian
V2V	Vehicle to Vehicle
V2X	Vehicle-to-Everything
VAL	Vertical Application Layer
VDSL	Very-High-Bit-Rate Digital Subscriber Loop
VLAN	Virtual Local Area Network
VPN	Virtual Private Network
VR	Virtual Reality
VXLAN	Virtual Extensible LAN
WAN	Wide Area Network
WEF	World Economic Forum
WLAN	Wireless Local Area Network
WM5G	West Midlands 5G
XISC	Xiangtan Iron and Steel Company
xPON	x Passive Optical Network
XR	Extended Reality
YoY	Year-on-Year

List of Figures

List of Tables

Chapter 1
5G Drives New Developments in Global Digital Transformation

1.1 Communication Technologies Lead Continuous Social Progress

Information exchange is fundamental to the existence and development of human society. The means of information exchange have been continuously evolving alongside improvements in productivity, in response to the growing demand for immediate access to information. Communication technologies have developed from pigeon delivery in ancient times, to telegraph and telephone in the modern era, and to mobile communication today, profoundly changing our lifestyles and production modes. It is no wonder that studies on scientific development, technological progress, production innovation, and economic growth defined information communication as a general purpose technology (GPT)—a technology that has revolutionary impact on the progress of society, along with materials, transportation, and power supply systems.

In ancient times, people used to exchange information through verbal communication and body gestures. Written words, bamboo slips, and paper books were later invented in response to the demand for information recording. Afterwards, beacon fires, horses, and pigeons, among others, were creatively used as means of communication for more efficient societal management. In 1044, ceramic movable type printing was created by Bi Sheng, a Chinese inventor, and in 1450, metal movable type printing was created by Gutenberg, a German inventor. These technological innovations hugely accelerated the spread of knowledge. The invention of metal movable type printing, in particular, ushered in a massive wave of printing in Europe. Being set against the backdrop of the Renaissance, this invention helped break up the church's monopoly on knowledge, which further catalyzed the religious reform and laid the foundation for the Industrial Revolution.

During the first Industrial Revolution, modes of transportation such as ships and railways developed rapidly, extending people's connections and the scope of trade. This in turn gave rise to the need for real-time communication, which could not be met by the previous communication methods such as letters. In 1835, Samuel

P. Sun, *A Guidebook for 5GtoB and 6G Vision for Deep Convergence*,
Management for Professionals, https://doi.org/10.1007/978-981-99-4024-0_1

F.B. Morse, an American painter and inventor, developed the world's first telegraph system, marking the beginning of the electronic communication era. The year of 1876 witnessed the world's first telephone invented by American inventor Alexander Graham Bell. In 1878, he successfully made a call from Boston to New York, which were 300 km apart. It was the first long-distance telephone call in human history. Telegraph and telephone technologies transformed communication because instead of relying on human- and object-based physical means, people were able to use electronic communication, which shattered the existing temporal and spatial barriers. In conjunction with the advances in transportation, these innovations changed the ways of production and trade, and contributed significantly to global economic integration and industrial development.

The advent of the ARPANET in 1969 revolutionized communication for another time. Internet technologies, along with computer technologies, software technologies, digital communication technologies, etc., have taken human society from the industrial age to a brand-new information age. The Internet has made it significantly cheaper and more convenient to transmit large amounts of information over long distances, substantially changing the mode of information dissemination. It has further made global collaborative production a reality, and in turn spurred the third wave of globalization, which is reshaping the world economy and the social and geopolitical landscape.

From the mid-1970s to the 1980s, the first generation of cellular mobile communication technology was developed and put into commercial use. Since 2000, mobile communication technologies have found ways to integrate with Internet technologies, helping popularize the Internet, empower new mobile Internet products, service forms, and models such as smartphones, mobile payment, mobile commerce, and the sharing economy, and nurture the consumer digital economy. This has all had an immeasurable impact on people's ways of life.

1.2 The World Is Embracing a New Wave of Digital Transformation

The world economy is gradually shifting from old economic engines to new ones, as the revolution of science and technology has been the catalyst for another wave of industrial revolution. The next-generation information and communications technologies (ICTs) are promoting the digital transformation of economies, societies, production methods, and people's daily lives all around the world. Digitalization, connectivity, and intelligence are becoming the most outstanding features of the new information era.

With the focus on boosting total factor productivity (TFP), the previous industrial revolutions centered on innovation and application of tangible technologies, such as steam and electric power, along with transformation in management and organizational restructuring. By contrast, the current era is characterized by accelerated overproduction and highly unstable demand fluctuations, because demands are increasingly personalized, diversified, and dynamic. In this new phase of development, the defining factors of success are no longer the scale of economy, the abundance of materials, or the strength of social relationships, but rather innovativeness, adaptability, swiftness, resilience, and the ability to learn. The inflexibility nature of material investments means that alone they are often not enough to secure a competitive edge. Data, information, and knowledge must be strategically explored. Powered by digital technologies, digital transformation can build data-driven loops for optimization through in-depth integration and interaction between the digital space and the physical world. This will drive reconfiguration of material investment and restructuring of organizational management, and as a result, will inspire more agility, vitality, and resilience across the industry. Digital transformation has forged a new path to developing long-term competitive advantage in the current highly uncertain climate.

In this new phase of accelerated development, digital transformation is playing a more prominent role in enterprises, industries, and economies. Digital technologies were first used in the 1950s and 1960s in industries for efficiency improvements, and their application in the service industry originated in the 1990s. In those times, however, the applications were mostly standalone, in non-critical processes, and for transaction purposes, limited by the low-level demands and underdeveloped technical systems, among other factors. Presently, new ICTs, such as mobile communication, the Internet of Things (IoT), big data, cloud computing, and artificial intelligence (AI), are rapidly developing while constantly converging with each other, creating new paths for industrial and economic digital transformation. These technologies will greatly reduce the costs of digital transformation and lower the technical barriers, and will eventually accelerate digital transformation on a larger scale and in greater depth. Since the outbreak of the COVID-19 pandemic, digital transformation has gone from being a "nice-to-have" to a "must-have". The next 10 to 15 years will witness a remarkable acceleration of digital transformation across society. Whether an enterprise seizes the landmark opportunity of digital transformation will determine its future comprehensive strengths and competitive position. This is also true for a country.

For this reason, countries around the world are placing great importance on digital transformation, with hopes to build a strategic digitalization framework and formulate policies aligned with digital transformation, in order to expedite digital transformation, boost the digital economy, and promote sustainable and comprehensive economic development. China is among the most proactive in this regard, having listed accelerating digitalization-based development as a strategic national priority in the *Outline of the People's Republic of China 14th Five-Year Plan (2021-2025) for National Economic and Social Development and Long-Range Objectives for 2035* (hereinafter the *Outline of the 14th Five-Year Plan (2021-2025)*). On October 18, 2021, President Xi Jinping called for grasping the trend and law of digital economic development and pushing forward the sound development of the digital economy in the country. In a word, digital transformation is now a global consensus and a general trend.

1.3 Digital Transformation Brings Big Opportunities

Digital transformation is rapidly becoming a stable development strategy in a variety of industries, and this is a fitting indication of the potential scale of 5GtoB development. How well this opportunity is grasped will directly determine the trajectory of 5GtoB. For example, China's digital economy, growing at a compound annual growth rate (CAGR) of over 10%, is projected to be worth CNY65 trillion in 2025, accounting for over 50% of the country's gross domestic product (GDP). Therefore, to achieve fast-paced 5GtoB expansion, it is critical to take an accurate pulse of the new missions of digital transformation, new expectations for digital transformation, and new trends in digital transformation, and plan accordingly.

1.3.1 Digital Transformation: New Missions

As the global digital economy booms, a new round of technological revolution and industrial transformation featuring digitalization, connectivity, and intelligence will change the global pattern of innovation and reshape the global economy.

Digital Transformation as a Strategic Choice for Building a Nation's Competitive Advantage

Society is entering a new historic stage which will be defined by digital productivity. Many countries around the world, in an attempt to dominate the new round of development, are proactively seizing the opportunities offered by digital technological innovations and attaching importance to digitalization in their economic development and technological innovations. The ability to adapt to and lead digital development will be pertinent to the future of a country.

Digital Transformation as Essential Support for the New Development Blueprint

China's economic development, driven by new momentum from the evolving digital economy, has shifted from speed and scale to high quality. Digital development accelerates the formation of a "dual circulation" development pattern, in which the domestic and international markets support and reinforce each other, which will help China more effectively respond to the increasingly complex international environment and maintain its economic security and stability. Digital development is fundamental to providing better-quality public services, which will lead to a narrowed digital divide. Leveraging modern information technologies and progressive governance is key to boosting the efficiency of governance and fostering a stronger sense of fulfillment in hundreds of millions of people when sharing the digital development achievements.

1.3.2 Digital Transformation: New Expectations

China has prioritized digitalization in its journey to building a digital economy and society. This is reflected in many of its national strategic plans, including the newly unveiled *Outline of the 14th Five-Year Plan (2021-2025).* This plan stresses the importance of harnessing the massive amounts of data and its wide range of applications, and promoting an in-depth adoption of digital technologies into the real economy, in order to empower traditional industries to transform and upgrade, incubate new industries, service forms, and business models, and identify and strengthen new engines of economic development. The plan has a dedicated section titled "Accelerate digitalization-based development and construct a digital China", in which the government proposes that China embraces the digital era, unleashes the potential of data elements, advances the buildup of a cyber powerhouse, accelerates the shaping of a digital economy, digital society, and digital government, and drives overall changes in production methods, lifestyles, and governance with digital transformation.

Initiatives have been taken to implement the strategies outlined in the plan. The Ministry of Industry and Information Technology (MIIT) of China has launched the *Integrated Computing Network*, *5G Application "Set Sail" Action Plan*, *Dual Gigabit Network Coordinated Development*, and *East-to-West Computing channeling* initiatives. The State-owned Assets Supervision and Administration Commission (SASAC) of the State Council has outlined the steps to take in order for state-owned enterprises to take a lead in digital transformation. The National Development and Reform Commission (NDRC) has put in place the *Cloud-based Big Data and AI* initiative to promote the digital transformation of micro, small and medium-sized enterprises (MSMEs). Local authorities have introduced policies to boost the digital economy. Nearly 200 cities have set up digital economy management offices, developed data resource systems, and built data infrastructure to promote the overall development of the digital economy.

1.3.3 Digital Transformation: New Trends

It is now historically significant time when China has triumphed on a new journey toward the second centenary goal of fully building a modern socialist country. The next 10 to 15 years will be a strategic period for China's digital transformation, for which the new trends of digital transformation must be accurately analyzed.

Data Is at the Center of High-Quality Economic Development.

Data is a new, replicable, and shareable factor of production, and is thus a fundamental resource that embodies great potential value and strategic importance. The *Guideline on Improving the Market-based Allocation Mechanism of Production Factors* proposes the acceleration of data element market cultivation. As the digital economy advances, data elements will be rapidly marketized. This means that in the course of digital transformation, focus should be placed on stimulating the efficiency of data elements, exploring the value of data, revitalizing data assets, and magnifying the effect of data on quality improvement, efficiency, and service innovation.

Digital Infrastructure Empowers Economic and Social Digital Transformation at an Unprecedented Speed.

New technologies such as 5G, industrial Internet, and AI are exerting an even more significant impact when used together in creative ways, giving rise to new use cases, business models and service forms. The systematic deployment and moderately advanced construction of new types of digital infrastructure can lay a new cornerstone for economic and social development, while highlighting the pioneering role of digital infrastructure and driving digital transformation forward in all fields.

Digital Industrialization Is Trending Toward a Comprehensive Digital Ecosystem.

The digital industry, featuring 5G, e-manufacturing, software, Internet, IoT, big data, and AI, among others, will develop even rapidly. Digital technologies will be more widely used by industries for digital penetration, integration, and restructuring. This will gradually lead to the formation of a digital ecosystem featuring diverse resources, industry convergence, advanced technologies, and quality services.

Industrial Digitalization Will Evolve to Be AI-driven, Differentiated in Scenarios, and Managed on Platforms.

Digital technologies such as 5G, big data, cloud computing, and AI will drive industries to become more specialized in terms of digital production, and will fuel producer services with more specialization while pushing them toward the higher end of the value chain. The innovations and applications of digital technologies will lead digital transformation into a new phase which is Internet of Everything (IoE)-enabled, data-driven, platform-based, software-defined, and intelligence-led.

Digitalization has become an industry consensus. It is seen as the inevitable path to transformation for enterprises. Using digital technologies as means of production can boost the quality and efficiency of traditional modes of production. Using them as management tools can enable business decision-making to be powered by data, computing, and algorithms for service innovation, capability development, governance, and value creation in enterprises.

1.4 5G Unlocks the Potential of Industry Transformation

2019 saw the commercial adoption of 5G technologies around the globe. The coincidence of the mobile communication technology revolution with the new round of industrial revolution is of historic significance. 5G technologies will lay a new foundation for economic and social development in the new round of industrial revolution, and create new ways, methods, and paths for the digital transformation of industries, which will unleash productivity and improve standards of living, catalyzing a new round of industrial revolution.

5G provides key technical support for data integration in various industries. Effective data collection is a prerequisite for the digitalization, connectivity, and intelligentization of any industry. 5G can provide connectivity for devices anytime and anywhere. With its unique features of massive connectivity, low latency, and high bandwidth, 5G can adequately meet the requirements of real-time transmission and distribution of tremendous amounts of production, service, and management data among various devices. When used in tandem with other technologies such as big data and AI, normalization, identification, processing, computing, and retrieval of data can be realized. 5G is becoming a key element bridging the digital and physical worlds.

5G paves the path for unmanned, intelligent production, as the only route leading to future industrial upgrade and advancement is building digital, manpower-saving, and intelligent enterprises. 5G can provide multifaceted support for enterprises in the innovation of their production modes. Firstly, 5G supports multi-channel and high-speed video transfer, and thus can enable high-precision, real-time monitoring, analysis, and control of production processes, when combined with machine vision technologies. Secondly, 5G is a key force in the fulfilment of cost-effective, long-distance, and large-scale remote control of mobile devices. Featuring high transmission rates and high reliability, 5G can ensure secure connections between remotely-controlled devices and the console while reducing installation, commissioning, and maintenance costs. Thirdly, 5G increases production flexibility. By enabling production equipment to be wirelessly connected via the cloud, 5G can help realize timely updates and flexible adjustments of equipment functions. Fourthly, 5G facilitates unmanned inspections and remote monitoring, enabling enterprises to perform real-time monitoring in new and effective ways.

5G opens up new space for the development of products and services. In the production field, 5G will be deeply integrated with technologies such as edge computing, cloud computing, and AI to nurture remotely controllable or automated products and solutions, such as 5G UAVs, 5G engineering vehicles, 5G+unmanned agricultural machinery, and 5G+autonomous driving. With regard to people's lives, 5G will promote the development of consumer-oriented devices and applications with three of its capabilities: cloud-edge-device integration, multi-device collaboration, and extensive connectivity. 5G's cloud-edge-device integration capability will bolster technologies such as ultra-high-definition (UHD), virtual reality (VR), and augmented reality (AR), which are expected to lead people into a new virtual world and revolutionize people's lives, entertainment, and education. And "5G+extensive connectivity" will hopefully lead to the realization of new use cases like smart home and smart fitness.

Looking ahead, 5G will be adopted even more widely in all aspects of society, gradually reshaping the development model of traditional industries. 5G will also nurture new demands, new services, and new business models, fully unlocking the potential of digital transformation and stimulating new momentum for economic and social transformation.

Chapter 2
5G Development Around the Globe

The commercial adoption of 5G continues to surge around the globe, with more than 30% of countries and regions having launched 5G networks. 5GtoC applications are still in their early stages. Extended reality (XR), high-definition (HD), and immersive experience are the buzzwords and telecom operators are mainly offering entertainment applications, such as live streaming of games, sports, and performances and immersive interaction during these events, and virtual reality (VR)/augmented reality (AR). 5GtoB applications are now in the extensive verification and demonstration phase. There have been implementations in industry, ports, healthcare, and autonomous driving, among others. However, replicating the success to scale up these applications still requires further study.

Governments and regulators in major countries and regions have realized that 5G offers great economic and social development opportunities, and have stepped up efforts to create inclusive, favorable environments for 5G application innovations in ways suited to their local conditions, for example by organizing strategic planning and project launches, with hopes to promote the application of 5G in vertical industries, foster 5G application ecosystems, and drive the digital transformation of the economy and society. South Korea has made a top-level strategic plan for 5G development, and the government reviews the progress regularly and adjusts the plan as required for accelerated application of 5G. The U.S. has been focusing on developing technical advantages in 5G and has released the *5G FAST Plan*, a comprehensive strategy providing guidance for the industry, in a bid to promote 5G network buildout. In Europe, endeavors have been made both at the national level and at various industry levels. Specifically, policies have been formulated and a series of projects has been rolled out to forge ahead with the application of 5G. A clear pathway to integrating 5G into vertical industries has been laid out. Japan has worked out an overarching plan for 5G applications. With the vision to realize "Society 5.0", Japan is actively pushing for the convergence of 5G with other advanced technologies, and strongly encouraging the large-scale, early deployment and expanded application of 5G in key sectors.

P. Sun, *A Guidebook for 5GtoB and 6G Vision for Deep Convergence*,
Management for Professionals, https://doi.org/10.1007/978-981-99-4024-0_2

China has been one of the countries pioneering the commercial adoption of 5G. It has been three years since June 2019 when 5G commercial licenses were issued. Now 5G network coverage has expanded from urban areas to counties and towns, with 70% of new phones being 5G capable and 5G user penetration exceeding 25%. The economic benefits of 5G commercialization are starting to show in the financial reports of telecom operators. Another highlight is that 5G commercialization started from the business (ToB) sector. 5GtoB, after four years of development, has gained scaled replication in some industries, and a system favoring further 5GtoB growth has begun to take shape. This shows that 5GtoB is already in a positive cycle featuring strong momentum in application innovation and ecosystem fostering. President Xi Jinping stressed that the development of the digital economy represents a strategic choice to seize the opportunities presented by the latest round of scientific and technological revolution and industrial transformation. Being a key enabler of the digital economy, 5G is considered a key success factor to digital transformation in both economic and societal domains in the current industrial revolution. The China Academy of Information and Communications Technology (CAICT) made a forecast that, in 2021, 5G would generate a direct economic output of CNY1.3 trillion, a direct added output of CNY300 billion, an indirect economic output of CNY3.38 trillion, and an indirect added output of CNY1.23 trillion, each of which representing an increase of over 30% over 2020. Looking ahead, as 5G is increasingly adopted across industries, its role in promoting the transformation of the real economy and creating a better life for people will become more prominent.

2.1 5G Development in China

2.1.1 5G Commercial Adoption in China

For China, mobile communications is one of the few world-leading fields that are of fundamental and strategic importance to the economic and social development. China's mobile communications industry, which is worth over CNY1 trillion, has been steadily growing in size and strength since its early days. From being a follower in 2G, to a strong competitor in 3G, and a forerunner in 4G, China has achieved major breakthroughs in the core mobile communications technologies. The overall level of the industry is significantly higher, with explosive growth in information consumption and vigorous development in the digital economy. China is steadily progressing to be a leader in 5G coverage, technology, and adoption.

The economic benefits of 5G commercialization are more tangible than ever before, bringing a phenomenal increase to the telecom operators' revenue. From January to August 2021, the telecom services of the telecom industry registered a total revenue of CNY991.9 billion, growing 8.4% year on year to reach a new high since 2014. Also, 5G has strongly promoted the development of industrial digitalization services and mobile data services. Known for its strong convergence capability, 5G has been used together with cloud computing, edge computing, big data, Internet of Things (IoT), artificial intelligence (AI), and other technologies to provide digital transformation services for enterprises. As its financial report for the first half of 2021 shows, China Mobile Communications Group Co., Ltd. (hereinafter China Mobile) added CNY6 billion to its data and ICT (DICT) revenue thanks to 5G. China United Network Communications Group Co., Ltd. (hereinafter China Unicom) signed contracts on 5GtoB applications worth over CNY1.3 billion in the first half of 2021. On the mobile data services side, the increase in the 5G user penetration rate boosted the average revenue per user (ARPU) for mobile services. As shown in their financial reports for the first half of 2021, China Mobile's 5G ARPU was CNY88.9, far higher than its overall mobile ARPU of CNY52.2, and the increase in its 5G user penetration rate reversed the downturn of its mobile Average Revenue Per User (ARPU) since 2018. China Telecom's 5G ARPU climbed to CNY57.4 and its 4G-to-5G upgrade user ARPU increased by 10%. With new data plans launched, China Unicom saw its 5G user penetration rate increase to 37% and its mobile ARPU increase to CNY44.4, registering a year-on year growth of 8.5%.

The 5G standards have been constantly evolving to support enhanced capabilities. 3GPP announced 5G Release 17 finalized in June 2022. The primary aim of Release 17 is to expand 5G connections to a wider range of services while enhancing the basic communications capabilities. The launch of this release also marks the readiness of the 5G reduced capability (RedCap) specification, which is meant for medium- and high-speed connections. Work on Release 18 has also started and the content was approved in December 2021. As Release 17 meets the network coverage requirements, China has aimed to further differentiate the current 5GtoB applications and foster new applications in more vertical industries. To this end, China has set several strategic directions for network development. The first is to support more mid-band and high-band spectra, realize wider network coverage by using technologies such as coverage enhancement and non-terrestrial communication, and expand multi-input multi-output (MIMO) to more spectra such as frequency division duplex (FDD). The second is to provide capabilities for supporting a wider range of services, such as IoT with medium- or high-speed transmission and low power consumption, submeter-level positioning, radio slicing enhancement, AR/VR enhancement, and sidelink. The third is to go intelligent to some degree, for example by achieving continuous enhancement of data collection, wireless self-organizing network (SON)/minimization of drive tests (MDT), and network automation.

China is leading in building high-quality and well-targeted 5G networks. To promote the innovative development of 5G applications and the digital economy, China has taken a forward-thinking approach and has been steadily advancing the construction of refined 5G networks. Results have been encouraging so far in terms of network quality and cost efficiency. Firstly, the 5G network coverage has been extended to counties and towns. As of the end of August 2021, over 1.037 million 5G base stations had been deployed in China, accounting for over 70% of the world's total and equivalent to 18% of the number of 4G stations deployed in the country. 5G networks covered all prefecture-level cities, over 97% of the central areas of county towns, and over 40% of rural townships in China. Globally, 5G networks were able to cover 19.6% of the world population by the end of June 2021. Secondly, standalone (SA) networking became mainstream. With an SA 5G core network in operation, the three telecom operators in China have all deployed large-scale SA 5G networks and all newly registered 5G devices are supposed to support SA networking starting from May 17, 2021. According to CAICT's 5G cloud test platform, SA connections accounted for 74.7% of 5G network tests in China in the second quarter of 2021, a 22.8% increase over the previous quarter. SA outperforms non-standalone (NSA) in addressing the needs of vertical industries since it can realize all features and functions 5G has to offer while providing extra support for network slicing, low latency, and other network capabilities. Thirdly, network co-building and sharing has become the new normal. Such collaboration has been established between China Telecom and China Unicom, and between China Mobile and China Broadcasting Network Group Corporation Ltd. (hereinafter China Broadcasting Network), forming a preliminary "2+2" landscape. By the end of September 2021, China Telecom and China Unicom had co-built and shared more than 500,000 5G base stations, and by doing so had saved up to CNY100 billion on network construction. In addition, over 12 billion kWh of electricity can be spared each year. China Broadcasting Network and China Mobile have agreed to co-build and share a 700 MHz 5G network. Under this agreement, they pledged to make concerted efforts in network construction and operation, "192" number segment service provisioning, network interconnection, and more areas, in order to expedite the formulation of network specifications, speed up frequency refarming, and accelerate the construction of a 700 MHz "network + content" ecosystem. In addition, the 5G access network roaming standards and the related test specifications have been worked out, and lab tests are being conducted by the four telecom operators.

As 5G has been increasingly adopted in various industries and society, new links emerged in the traditional 5G system, forming a 5G platform consisting of five chains: the device chain, network chain, platform chain, application solution chain, and security chain. The device chain has new products with distinctive industry features, but work has to be done to lower the costs and break through the technological bottlenecks. Moreover, market fragmentation remains an issue to be solved. The network chain has some new devices featuring technological convergence and some lightweight devices, while customization, O&M, and scalability are still challenges that need to be addressed. The platform chain has been enriched with the operations platform and the edge platform, though efforts are still required to build a universal, multi-purpose platform and cultivate a 5G application ecosystem. The application solution chain is added with applications featuring the convergence of 5G into industries, and yet continuous innovation is required to nurture more use cases and accelerate the convergence. For the security chain, which now also covers industry security, substantial advances have to be made in terms of technology and product design to form a cross-industrial, certified 5G application security system.

As more telecom operators in the world, and in particular the Chinese ones, started to scale up 5G networks, the 5G infrastructure supply market also flourished. China's mobile communications industry has been growing rapidly thanks to the large-scale deployment and early commercial adoption of 5G in the country. This has also helped China earn and maintain its position as a leader in communications equipment and 5G mobile phones. According to Omdia, the global 5G radio access network (RAN) equipment market grew to USD13.3 billion in the first half of 2021, up 43% compared with a year earlier. Omdia forecast that China would contribute as much as 44% to the global market. Chinese providers have taken over half of the global market, but are faced with challenges when expanding outside China. The Omdia data also showed that Huawei Technologies Co., Ltd. (hereinafter Huawei) topped the industry with 35.2% of the global market share in the first half of 2021, followed by Ericsson Inc. (hereinafter Ericsson), Zhongxing Telecommunications Equipment Corporation (hereinafter ZTE), Nokia Corporation (hereinafter Nokia), and Samsung Group (hereinafter Samsung), which had market shares of 21.5%, 16.4%, 12.1%, and 9.1%, respectively. As for the mobile network infrastructure, Huawei kept its leading position in the first half of 2021, with Ericsson, Nokia, ZTE, and Samsung ranking second to fifth, respectively. China shipped 52.8% (128 million [1] out of 239.5 million [2]) of the world's 5G mobile phones in the first half of 2021. In the same period, China's global share of smartphone shipments was 27.8% [3]. This means that China's global share of 5G mobile phone shipments was 25% higher than its global share of smartphone shipments. Canalys statistics showed that nearly 70% of all Android-based 5G phones shipped worldwide were made by Chinese manufacturers, which launched 110 models of 5G mobile phones in the first half of 2021, accounting for 67% of all 164 new models launched globally in that period of time.

2.1.2 5G Promotion Policies in China

The 14th Five-Year Plan period (2021–2025) covers the first five years during which China began its march towards the goal of building a modern socialist country in all respects. It is also a critical period for the large-scale deployment of 5G networks in China. China has proposed to accelerate the rollout of 5G networks and increase its user penetration rate to 56% during the 2021-2025 period. It has identified 10 key sectors for 5GtoB applications, covering smart transportation, smart energy, smart manufacturing, smart agriculture and water conservation, smart education, smart healthcare, and smart cultural tourism, in order to develop 5G-based use cases and 5G industry ecosystems.

Initiatives have been taken at all levels to implement the government's guiding principles for 5G development. The Ministry of Industry and Information Technology (MIIT) has prioritized 5G development in its "1+2+9" planning and has formulated the 14th Five-Year Plan (2021–2025) for the infocommunications industry, which outlined the key tasks and objectives for 5G development. The MIIT has also strengthened policy guidance and support with regard to network construction, use case creation, industrial development, and more. The specific strategies and objectives are covered in the *Notice on Accelerating the Development of 5G*, the *Dual Gigabit*

Network Coordinated Development Action Plan (2021–2023), and the *Promotion Plan for the "5G+Industrial Internet" 512 Project* it has issued.

The MIIT issued a further action plan in July 2021, namely the *5G Application "Set Sail" Action Plan (2021–2023)* (hereinafter the *Action Plan*), in conjunction with nine other central government authorities. The *Action Plan* unveiled 32 tasks included in eight initiatives to guide the adoption of 5G in the consumer (ToC), business (ToB), and public (ToG) sectors over the next three years. The plan identified 15 key areas for 5GtoB applications, including information consumption, industrial manufacturing, energy, transportation, agriculture, healthcare, education, cultural tourism, and smart city. According to this plan, regular summaries and promotion would be held nationwide to share the success experiences and channel more support for 5G growth from local governments. An example of this is that the MIIT, together with the Ministry of Education and the National Health Commission (NHC), organized the collection of "5G+Smart Education" and "5G+Healthcare" projects, aiming to promote innovations in adopting 5G in various sectors. In addition, local governments, competent authorities, and telecom enterprises in the country have been asked to coordinate their efforts to innovate and accelerate the buildout of high-quality 5G networks.

In the meantime, local governments in China have been offering various policy support for 5G development. They have set 5G development objectives based on their local economic and industrial conditions and have made endeavors to create favorable environments to spur high-quality 5G construction. According to CAICT's statistics, 569 5G policies had been released by the end of August 2021, including 67 at the provincial level, 259 at the municipal level, and 243 at the district level. Most of the policies have called for faster 5G network construction. The release of these favorable policies cleared up many persistent problems; however, sustained efforts are still required to ensure the accelerated and accurate implementation of the policies.

2.1.3 5G Application in China

Presently, 5GtoB applications are moving from trial use to the start of widespread rollout. Unlike 4G and even earlier mobile technologies, 5G is mainly used in production. Being a fundamental technology defined by high transmission rates, low latency, and massive connectivity, 5G delivers better mobility, more flexible networking, and easier deployment in extreme environments. This makes 5G well suited for the industrial manufacturing and port sectors. 3GPP Release 15 supports an end-to-end (E2E) latency of less than 30 ms with 99.99% reliability, which suffices to enable digitalization in most scenarios. Subsequent releases can support industrial applications requiring even lower latency and higher reliability. With its aggregation capabilities, 5G can also be deeply integrated with emerging technologies such as cloud computing, big data, AI, and VR/AR, to create cloud-network-edge-device synergies, build intelligence-enabled infrastructure, and accelerate the convergence of communications, information, and control technologies. In addition, with this convergence feature, 5G can be integrated with industry-specific knowledge and experiences through industrial big data and AI to address the needs of industries and foster new applications.

5GtoB applications have achieved "0 to 1" breakthroughs and are now embarking on the fast track. The number of 5G use cases has nearly doubled since the end of 2020, and the market demand is continuing to expand. The 4th "Bloom Cup" 5G Application Competition held in 2022 received over 12,000 5G application projects from across China, which are world-leading in terms of both quantity and level of innovation. Of these projects, over half are related to five sectors: industrial Internet, smart campus, smart city, information consumption, and smart healthcare. Telecom operators, equipment vendors, and industry enterprises have initiated joint innovations to level up 5G services, create scenario-based standard solutions, and integrate the solutions into production, in order to bring digital and intelligent solutions to various industries.

Breakthroughs were made first in the leading industries in China, where 5G applications have already produced notable results. To date, 5G has been employed in 39 of the total 97 economic categories of 15 out of the total 20 sectors in China. Telecom operators and equipment vendors have cultivated quite many pilot and demonstration 5GtoB applications and have refined a series of showcase projects in multiple leading industries, like manufacturing, energy, and mining, to drive forward the digital and intelligent transformation of industries. Industrial manufacturing is the main area where there has been significant advancement of 5G applications. A host of typical use cases have been defined, including remote device control, flexible production and manufacturing, onsite assembly assistance, machine vision aided quality inspection, and unattended intelligent inspection. For example, Sany Heavy Industry has built a lighthouse factory with full 5G connectivity, covering its entire production process, which has yielded significant cost reduction, quality enhancement, and efficiency improvement. 5G has made mining smart both above ground and underground. By remotely controlling the tunnel boring machines, coal shearers, and other devices, 5G has transformed the way of mining, reducing the safety risks of underground workers, boosting the mining efficiency, and facilitating unmanned and less-manned mining. In the ports sector, 5G applications such as remote container crane control, unmanned container trucks, and intelligent stocktaking have been widely used in China's major ports, leading to improved loading efficiency, decreased numbers of onsite personnel, and reduced safety risks, and bringing the vision of smart ports closer to reality. 5G is also playing a bigger role in various transport scenarios, from highways to urban roads, railways, and subways. 5G applications for smart education, smart healthcare, and smart city are booming, too. For example, the China-Japan Friendship Hospital has taken advantage of 5G's large capacity and low latency features on top of cloud computing and IoT, to launch applications such as emergency rescue, teleconsultation, remote diagnosis, and dynamic monitoring, which has effectively elevated the diagnosis and treatment service level and raised the management efficiency.

Upstream and downstream players on the industry chain have started to jointly explore new business models for 5G, but many are still a work in progress. China

Mobile, for example, has released a multi-dimensional business model that allows the basic and advanced network functions to be flexibly combined. This is made possible due to the deployment of multi-access edge computing (MEC) and dedicated slicing technologies on top of the basic network. China Unicom has adopted an innovative cash flow model, breaking down the maintenance expenses in both local and remote modes into smaller pieces. In addition, leading industry players have taken the initiative to streamline their internal processes in an attempt to future proof their operations against the uncertainty of upcoming policy decisions and mitigate the risks of future 5G rollout.

2.2 5G Development in South Korea

2.2.1 5G Commercial Adoption in South Korea

South Korea has been a leader in 5G commercialization. 5G was put into trial commercial use at the 2018 Pyeongchang Winter Olympics, which laid out a blueprint for how it would be deployed. In June 2018, South Korea closed the auction for 3.5 GHz and 28 GHz spectra. On December 1, 2018, the three South Korean telecom operators, that is, South Korea Telecom (SK Telecom), Korea Telecom (KT), and LG U+, launched 5G services for enterprises. On April 3, 2019, the three operators rolled out 5G services for mobile phone users in 17 cities and provinces.

According to the Ministry of Science and ICT (MSIT) of South Korea, as of June 2021, a total of 183,000 5G base stations had been constructed in South Korea, a ratio of 35.4 base stations for every 10,000 people, providing 5G coverage for the Seoul metropolitan area, the major surrounding cities, and major transportation routes, which lifted the 5G population coverage rate to over 95% (Fig. 2.1).

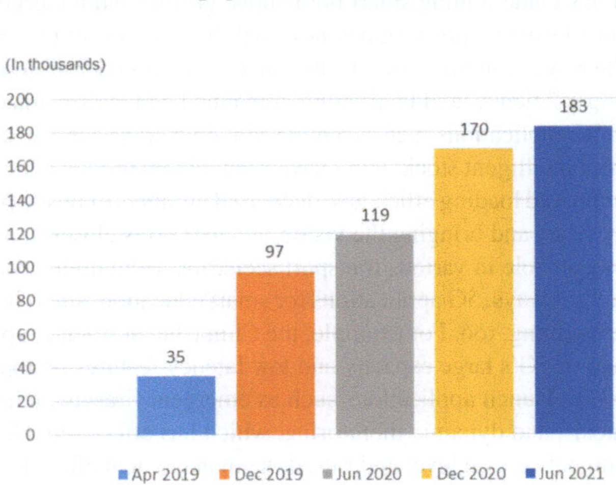

Fig. 2.1 Number of 5G base stations in South Korea. *Source* MSIT

As the 5G network coverage expands, the networks have also become faster and more accessible. MSIT began regularly assessing the 5G network service quality in the second half of 2020. According to the Ministry's third report, which was released in August 2021, full 5G coverage was then available in seven major cities, as well as densely populated areas in other 78 smaller cities. The report also noted that the average 5G coverage for the three operators was 6,271.12 km^2, up from 5,409.3 km^2 in 2020. The three telecom operators' average 5G download speed sat at 808.45 Mbit/s in the first half of 2021, a rise of 117.98 Mbit/s over 690.47 Mbit/s in 2020. Their average 5G upload speed was 83.93 Mbit/s, a rise of 20.61 Mbit/s compared with 63.32 Mbit/s in 2020. The stability of 5G networks has also improved. The average rate of switching to 4G Long Term Evolution (LTE) networks during 5G downloads was 1.22% in the first half of 2021, down from 5.49% in 2020, while that during uploads was 1.25%, previously 5.29% in 2020.

To accelerate the commercialization of 5G, South Korea adopted a 5G and 4G co-networking model, which has substantially shortened the time required for 5G network deployment. Its negative impact on the 4G network speed, however, has become apparent as the 5G user base expands. In addition, while the operators pooled their efforts into promoting 5G, they were not able to maintain the existing infrastructure such as 4G base stations in rural and fishing villages, in a timely manner, which dragged down the 4G network speed in those areas. Now, the operators have started to build up SA 5G networks, to better serve the mass market and facilitate the provision of 5G services for vertical industries. Due to limited investment, however, it may take quite a while before SA 5G networks are put into large-scale commercial use.

By the end of June 2021, there were 16.46 million 5G subscriptions in South Korea, representing 23% of the country's total mobile subscriptions, which had brought the 5G user penetration rate up to 31.6%. The share of 5G subscriptions for SK Telecom, KT, and LG U+ was 46.6%, 30.4%, and 22.7%, respectively, and there were an additional 36,949 5G subscriptions to mobile virtual network operators (MVNOs). In February 2021, the MSIT made a regulatory amendment, allowing MVNOs to independently launch 5G data packages. Prior to that, MVNOs had to launch mobile data packages in conjunction with a telecom operator. The amendment was meant to incentivize MVNOs to roll out affordable 5G plans, and by so doing to spur competition in the 5G market and attract more users to 5G. Backed by this amendment, 10 MVNOs, including Sejong Telecom, started to launch 5G data packages in April 2021, and in the following month their 5G user base grew significantly, with 29,273 new subscriptions recorded.

The monthly new subscriptions kept increasing in the five months following 5G's initial commercial rollout. In August 2019, with Samsung's launch of its Note 10 mobile phone, the monthly new 5G subscriptions recorded an initial peak of 880,000. Following that, new subscriptions slowed down. From August 2020 until the present, the average monthly new subscriptions neared 800,000, with the launch of new devices and packages as key drivers (Fig. 2.2).

The second month of 5G's commercial rollout saw 4G subscriptions in South Korea decline and the total mobile subscriptions increase slightly. This trend was

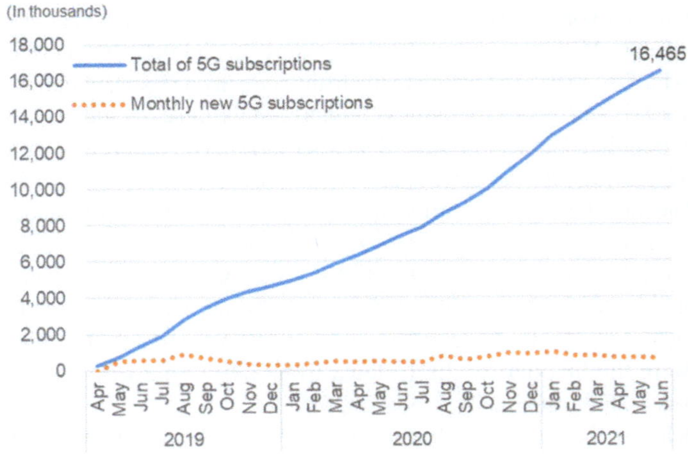

Fig. 2.2 5G subscription trends in South Korea. *Source* MSIT

kept for quite some time. By June 2021, the total mobile subscriptions stood at about 71.62 million, among which 71% were 4G subscriptions (Fig. 2.3).

South Korea has made 5G a high priority, with a focus on consumer-targeted 5G applications from an early stage. Pursuing the "5G+Entertainment" strategy on the consumer front, SK Telecom, KT, and LG U+ have channeled their efforts into cultivating industries featuring quality content, such as VR/AR, cloud gaming, and 4K HD video, and have rolled out traffic-intensive applications based around sports and idols, which have spurred the 5G traffic.

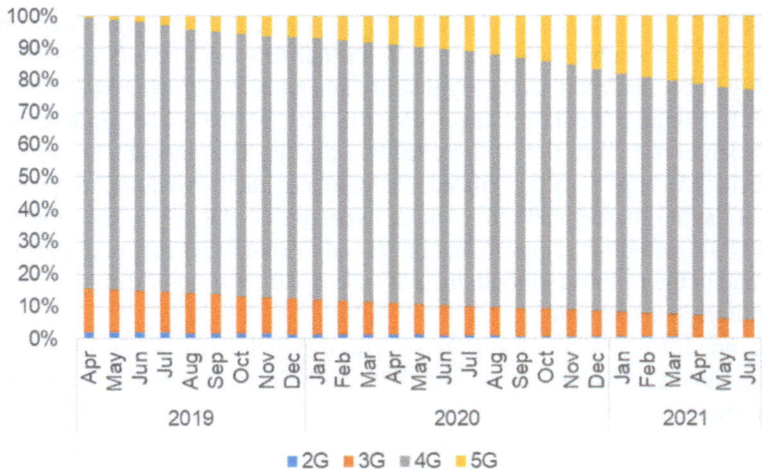

Fig. 2.3 Percentage of mobile subscriptions in South Korea. *Source* MSIT

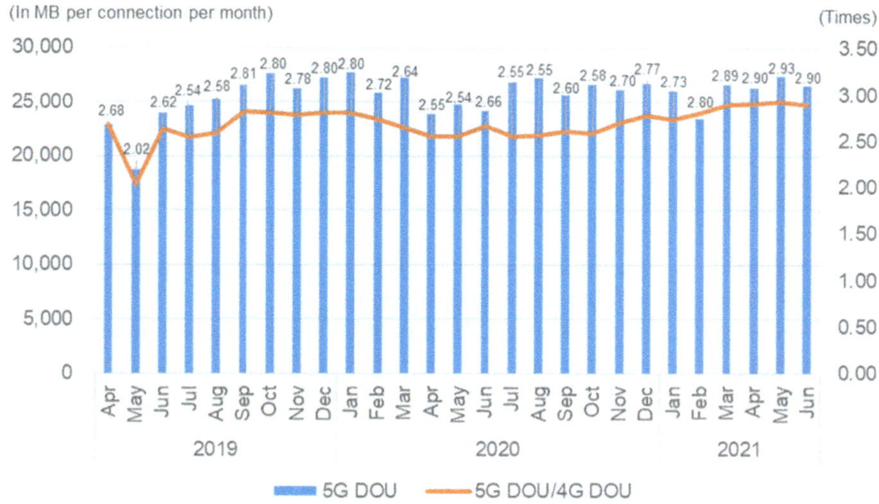

Fig. 2.4 5G DOU and its comparison to 4G DOU in South Korea. *Source* MSIT

In terms of data of usage (DOU) per connection, 5G has been leading 4G by a large gap. In June 2021, the average monthly 5G DOU reached 26,580 MB per connection, 2.90 times the 4G level, which was 9,177 MB per connection (Fig. 2.4).

2.2.2 *5G Promotion Policies in South Korea*

South Korea is among the first countries to have mapped out a national strategy for 5G. To accelerate the development of 5G, it has pursued a comprehensive strategy, boosted investment, intensified R&D efforts, promoted co-building and sharing, and released preferential tax policies, among others. These initiatives have worked to put South Korea's 5G commercialization one year ahead of schedule. After its initial 5G commercialization, South Korea released a second 5G strategy to beef up its 5G industry ecosystems and cultivate top-notch 5G+ converged ecosystems. The strategy included policies on accelerating 5G network deployment, enhancing service quality, and expanding the use of 5G into more industries.

2.2.2.1 Promotion Policies for 5G Commercial Adoption

In November 2013, South Korea launched the "Giga Korea" plan for ICT investment, with the goal of providing individual users with access to Gbps-grade mobile services by 2020. According to this plan, the investment in 5G infrastructure was projected to total KRW359.8 billion from 2013 to 2020, focusing on developing

Table 2.1 Main contents of the 5G+ strategy

Ten Core Industries	Network equipment, next-generation smartphones, VR/AR devices, wearable devices, intelligent CCTV, UAVs, connected robots, 5G-V2X (vehicle to everything) communication, information security, and edge computing
Five Core Services	Immersive content, smart manufacturing, autonomous vehicles, smart city, and digital healthcare

the core technologies for millimeter wave (10 GHz to 40 GHz) broadband mobile communications and building the practical systems.

In December 2013, South Korea's Ministry of Science, ICT and Future Planning (MSIP)[1] held a hearing on "Creative 5G Mobile Strategy", announcing its ambition to commercialize 5G services starting in 2020. It estimated that by 2026, the 5G equipment market would be worth KRW476 trillion and the 5G consumption market would be worth KRW94 trillion. The MSIP planned to invest KRW500 billion into R&D, standardization, infrastructure construction, and other key 5G sectors in 2014–2020, in addition to setting up a 5G industry-university-research (IUR) collaboration promotion workgroup, in order to develop the 5G industry.

In April 2018, the MSIT announced a series of measures aiming to put 5G into commercial use ahead of schedule, including amending regulations so that telecom operators could have access to government-managed infrastructure, such as street lamps and transportation facilities, for the purpose of constructing 5G networks. The operators also pledged to share their existing infrastructure for initial 5G deployment, such as pipes and utility poles, and to jointly build the new infrastructure required for launching 5G services, such as sandpoint wells and pipes.

2.2.2.2 Promotion Policies for 5G Converged Services

Riding the initial wave of 5G commercialization, the MSIT released the *5G+ Strategy to Realize Innovative Growth* in April 2019, seeking to comprehensively promote 5G+ converged services nationwide, nurture 5G applications in more industries, and to establish South Korea as a leader in the global market.

The 5G+ strategy pledged to build the world's best 5G ecosystem, with objectives of building a nationwide 5G network by 2022 through over KRW30 trillion of public and private investment, creating 600,000 quality jobs, and achieving KRW180 trillion in gross domestic product (GDP) and USD73 billion of exports by 2026. Table 2.1 lists the main contents of the 5G+ strategy.

The 5G+ strategy includes the establishment of a private-public partnership (PPP) to synergize the growth of upstream and downstream industries. In the PPP framework, the main responsibilities of the government include allocating frequency bands early, improving regulations (including tax revisions), standardizing testing

[1] Ministry of Science, ICT and Future Planning (MSIP) was renamed Ministry of Science and ICT (MSIT) in 2017.

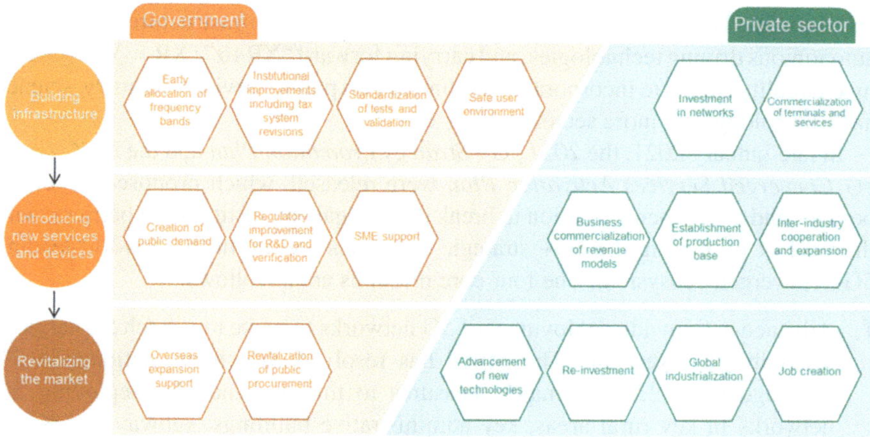

Fig. 2.5 The PPP framework. *Source* MSIT

and validation, establishing a secure user environment, stimulating public demand, improving R&D and verification supervision, supporting small- and medium-sized enterprises (SMEs), supporting overseas expansion, and revitalizing public procurement. On the private sector's side, the main responsibilities include investing in network construction, launching terminals and services, commercializing revenue models, establishing a production base, implementing inter-industry cooperation and expansion, advancing new technologies, reinvestment, global industrialization, and creating jobs (Fig. 2.5).

To ensure a thorough implementation of the 5G+ strategy, South Korea established the 5G+ Strategy Committee and 5G+ Working Groups, which are positioned to oversee the execution of the package of measures defined.

Moreover, the government has mapped out top-level designs to push the 5G+ strategy forward and strengthen support for 5G development. It released 5G+ annual promotion plans in both 2020 and 2021, which enacted measures such as sponsoring of innovation projects, tax reductions, and formulating of policies related to private 5G networks, all of which were meant to create a favorable policy environment and steer the development of the 5G market.

In April 2020, the 5G+ Strategy Committee released the *5G+ Strategy Development Status and Draft Future Plan*, announcing the government's commitment to invest approximately KRW650 billion (USD500 million) for the cultivation of the strategic 5G+ industries, and exploration and promotion of 5G converged services. The government also vowed to accelerate regulatory innovation, promote the commercialization of achievements, and establish a regular assessment mechanism. Specific measures include: launching 5G+ innovation projects in multiple sectors (including setting up 200 5G smart factories each year), carrying out pilot projects for digital healthcare (in order to lay the foundation for a 5G+ AI emergency healthcare system), completing the demonstration of smart city services, developing 5G-V2X infrastructure (accreditation service and testbed establishment) and the related technologies, developing vehicle-to-cloud infrastructure and the core autonomous driving technologies, and carrying forward "XR+α" (XR = VR + AR + mixed reality (MR)) to incorporate XR content into public services, industry, science and technology, and more sectors.

In late January 2021, the *2021 5G+ Strategy Promotion Plan* and the *MEC-based 5G Converged Services Activation Plan* were released, which proposed to review policies and strengthen execution to break down legal and institutional barriers, with the goal of promoting the 5G+ strategic industries and nurturing a world-leading 5G+ converged ecosystem. The four core missions are as follows:

1. Advance nationwide deployment of 5G networks to make top-notch 5G services accessible to everyone. The MSIT has resolved to achieve nationwide 5G coverage by 2022. The major measures to this end include: deploying 5G networks in key rural areas, key administrative buildings, subway and Korea Train Express (KTX) stations, and 4,000 multi-purpose facilities across 85 cities, extending 5G to the rural areas by enabling roaming and network sharing among the three telecom operators, increasing the tax credit for 5G investment from the record high of 2% in 2020 to 3% in 2021, strengthening 5G quality assessment, and reducing registration and licensing taxes by 50%.
2. Develop 5G converged services and the 5G equipment industry to ensure sustained competitiveness in 5G. Centered on the 5G+ core services, an investment of KRW165.5 billion would be made in 2021 into 5G+ innovation projects (see Table 2.2),providing strong support for the advancement, verification, and commercialization of 5G technologies. Other measures include enhancing inter-ministry collaboration, launching MEC-based 5G converged services, and rolling out private networks to expand the 5G market. In addition, end-to-end support (from module and device development to infrastructure construction and service activation, popularization, and development) would be strengthened to develop the 5G equipment industry.
3. Lead the world in 5G ecosystem development and pursue global expansion. With this mission, collaboration would be reinforced in key markets, while strong support would be provided for Korean enterprises to tap into the major and emerging markets abroad.
4. Strengthen the foundation for sustained 5G+ development, by allocating spectra necessary for 5G+ strategic industries, enhancing the management system, and offering talent training, etc.

In August 2021, the MSIT released the 5G+ Converged Services Promotion Strategy (Plan) to strive towards an all-round adoption of 5G+ converged services. The strategy mapped out the key tasks, timetable, and responsible bodies for the promotion of 5G+ converged services. The main measures include adopting 5G technologies in the key sectors, such as distance education, industrial security, and disaster response, to produce results easy for people to perceive, deploying private 5G networks, and using 5G services in government-sponsored projects. The strategy also set quantitative goals with regard to the use cases, engaged companies, and the technical level, in order to scale up 5G+ converged services and encourage proactive expansion into

Table 2.2 5G+ innovation projects in 2021

Sector	Project focus	2021 budget
Immersive content	VR/AR projects (XR converged projects)	KRW20 billion
	Developing next-generation immersive content	KRW25 billion
Autonomous driving	Developing innovative technologies for autonomous driving	KRW88.4 billion
Smart factory	Developing 5G-based food production & safety technologies	KRW6.2 billion
	Developing 5G-based core technologies for smart production	(Added in 2022)
Smart city	Promoting smart information services (5G-based digital twin in the public domain as priority)	KRW16 billion
	Developing centimeter-precision marine positioning and positioning, navigation, and timing (PNT) technologies (developing location information systems of high precision and high reliability)	KRW3.9 billion
Digital healthcare	Building up the foundation for the precision medicine industry (developing 5G-based emergency healthcare systems)	KRW6 billion

overseas markets. The specific goals are as follows: adopting 5G+ converged services in 195 locations in 2021, 630 by 2023, and 3,200 by 2026; raising the number of new 5G+ converged services that address social problems from 1 in 2021 to 5 by 2023, and 11 by 2026; lifting the number of 5G-specialized companies from 94 in 2021 to 330 by 2023, and 1,800 by 2026; boosting the 5G+ technical level from 84.5% in 2021 to 88% by 2023, and 95% by 2026.

2.2.3 5G Application in South Korea

2.2.3.1 5GtoC Application in South Korea

South Korea has given high priority to the development of 5G and focused on consumer (ToC) services from an early stage. The country has implemented a range of measures, including innovating the service models, optimizing the service packages, and enriching the service benefits, in a bid to cultivate killer applications based on its established content industry and promote such applications.

Taking advantage of the developed culture & entertainment, sports, and gaming industries in the country, South Korean operators have been exploring user demands and points-of-interest, and deploying consumer-oriented content services, such as HD video, VR/AR, and cloud gaming, to give full play to their networks.

VR/AR has been widely used during live streaming of sport events, performances, gaming & fitness events, shopping, social networking, and in libraries. SK Telecom has launched the Jump VR (touring esports stadiums in the League of Legends Park on mobile phone screens), LCK VR Live Broadcasting (watching esports games from close distance through 360-degree VR cameras), VR Replay (taking a 360-degree view of the battlegrounds from the perspective of game characters), and other VR services, to provide esports players and viewers with true-to-life immersive experiences. SK Telecom has also launched a 5G VR social networking service called "Virtual Social World". LG U+ has rolled out a great number of new VR/AR applications, including VR live streaming, AR user-generated content (UGC), AR navigation, and AR library, and created thousands of new XR experiences, ranging from drama performances, shopping, and fitness, to education and social networking. In addition, it has built a technical platform and ecosystem featuring "MR+AI" to create ubiquitous smart life experiences that bridge the virtual and physical worlds. For example, the U+ Idol Live app allows viewers to watch live performances of Korean idol artists with unique close-up angles, and the U+ 5G Professional Baseball app provides live baseball streams with new viewing angles, including a sweeping 360-degree view.

Cloud gaming is of critical significance to the promotion of 5G, as it is an area in which the users are likely to pay high prices for high speed and low latency. South Korean operators have partnered with specialized game companies to offer 5G-based cloud gaming services and offline hardware-free game experiences, which has greatly expanded the 5G user base. KT has collaborated with Ubitus, a cloud gaming technology company, to establish a 5G game streaming platform; SK Telecom has joined hands with Microsoft to launch cloud game streaming services over 5G networks; LG U+ has provided 5G users with cloud gaming services based on the NVIDIA GeForce Now platform, and has set up cloud gaming experience zones in 100 of its direct-sale stores in South Korea.

South Korean operators place a high value on the building of a content ecosystem. They have provided differentiated 5G services such as live streaming of sporting events and exclusive VR/AR games, bundled with value-added content from content providers, to attract new subscribers and encourage 4G subscribers to upgrade to 5G. A common approach they have taken is to enter into partnership with the country's professional baseball leagues, golf events, and esports events to offer live, HD, and multi-view event streaming services over 5G networks. Meanwhile, they have continuously invested in making quality content such as VR/AR products and cloud-based games, both separately and in collaboration with content producers, to expand the global content ecosystem. LG U+ has partnered with companies abroad, including the American startup Spatial, the AR product developer Nreal, and Qualcomm, to develop 5G-based AR solutions; SK Telecom has collaborated with Microsoft to launch Jump Studio, Asia's first MR capture studio, which uses the advanced volumetric video capture technologies to produce MR content, such as 3D holographic videos, quickly and cost-effectively. While enriching its own VR/AR content library, SK Telecom provides quality MR content to operators across Europe, the Americas, and Asia, gaining a foothold in the global market.

Global presence has also been vigorously pursued by South Korean operators. They have entered into partnerships with operators abroad to develop and distribute premium content. SK Telecom has been providing quality MR content to operators across Europe, the Americas, and Asia, gaining a foothold in the global market. LG U+ has signed contracts with operators like China Telecom to provide 5G content products and solutions, resulting in a total export volume of USD10 million. In addition, LG U+ has promoted the establishment of the Global XR Content Telecom Alliance.

2.2.3.2 5GtoB Application in South Korea

South Korea has been actively exploring 5G applications in factories, ports, healthcare, transportation, urban public security, and other sectors, and has run a multitude of pilot programs. To date, 5G has been deployed in sectors including industrial Internet, healthcare, smart transportation, urban public security, and emergency response, with groundbreaking applications such as "5G+ AI" machine vision aided quality inspection, digital teleconsultation, pathology and surgery teaching, emergency rescue with remotely-controlled robots or UAVs, COVID-19 prevention robots, and on-premises delivery with 5G-powered autonomous driving.

1. 5G+ Industrial Internet

   As early as in December 2018, 5G services were provided for enterprises by the three telecom operators. Giving full play to the high bandwidth and low latency nature of 5G, enterprises were able to inspect products for defects through the HD images transmitted over the 5G networks, and conduct remote control of robots and equipment, among other things. Both the government and the industry have been proactive in promoting smart factories, which has led to a wider application of 5G in industrial manufacturing. They have also strongly pushed for the convergence of 5G with edge computing, AI, and other technologies, aiming to build next-generation smart factories. According to estimates, as many as 1,050 "5G+AI" smart factories will have been constructed by small- and medium-sized manufacturers by 2026.

   By the end of 2020, machine vision, logistics robots, among other 5G applications had been developed and verified in six factories; automation systems (including sewing robots and gripper robots) had been developed and verified in four sewing factories.

   South Korea will continue its efforts in developing the core 5G technologies enabling smart factories and demonstrating the applications in food production, with hopes to boost productivity and lower the defect rate in domestic food production. To accelerate the construction of smart factories, South Korea will implement an app store that facilitates the development, dissemination, and application of AI-enabled manufacturing solutions.

   South Korean operators attach great importance to 5G applications tailored for industrial enterprises. To incentivize SMEs to build 5G smart factories, SK

Telecom has come to agreement with the Korea Smart Factory Association (KOSF) to promote Metatron Grandview among SMEs. Metatron Grandview is a 5G-based big data analytics solution, which can reduce production costs by up to 15% and extend machinery service life by over 20%. KT has partnered with Hyundai Heavy Industries and U.S.-based machine vision developer Cognex to tap into the 5G smart factory market. Under this partnership, they have been co-developing collaborative robots (cobots), machine vision solutions, and 5G smart factory industrial robots that are expected to relieve workers from challenging and potentially dangerous tasks. LG U+ has built smart factories that integrate 5G MEC and AI technologies in LG's 30 subsidiaries across multiple sectors and 70 factories, including power plants and iron and steel mills.

(1) 5G product defect detection

Myunghwa Industrial, an automobile parts manufacturer, has made use of SK Telecom's 5G connections to implement remote product inspection. With cameras installed near conveyor belts, ultra-HD images can be captured of automobile parts. The images are then sent over the 5G connections to an AI cloud platform for defect scanning.

(2) 5G smart sewing factories

5G has been used in sewing factories for establishing low-latency communication between the machines and the server. In conjunction with MEC and AI technologies, the production process can be monitored, and faulty machines and defective products can be identified in real time. 5G-enabled automated guided vehicles (AGVs) and autonomous robots have also been deployed in the factories to carry items around.

(3) 5G smart cobots

Parkwon Co., Ltd., an automobile parts manufacturer in South Korea, has deployed a private 5G network in one of its factories. Smart cobots have assumed the responsibility of loading, packaging, and other repetitive tasks, which has led to a dramatic decrease in the scrap rate and defect rate, and an enormous boost in productivity. Now 313 cases of automobile parts are produced per hour, up by 39% from the previous value of 225.

2. 5G+ Smart Healthcare

The South Korean government has been pushing hard for the establishment of a "5G+AI" emergency medical system, which is expected to be a fast-response, collaborative, and intelligent system covering the entire emergency handling process. To this end, public-private collaboration between ministries, hospitals, and ICT enterprises was encouraged, and 5G and AI technologies were used to overcome the technological, temporal, and spatial limitations of the existing emergency medical system. With these technologies, biological information and HD videos for diagnosis and treatment can be transmitted quickly to ensure timely medical assistance. By the end of 2020, a 5G-enabled emergency cloud platform had been deployed in Seoul for the transmission of multi-dimensional AI data. With 31 ambulances in Mapo, Seodaemun, Gangdong, and Songpa districts capable of collecting and transmitting data over this platform, a total of 5700+ pieces of image data, 7,500+ pieces of audio data, and 6,600+ pieces of biological data have been collected. In addition, AI-enabled first aid services have been developed for four categories of acute diseases (cardiovascular diseases, cerebrovascular diseases, severe injuries, and myocardial infarction) that are of high morbidity rates and require timely treatment.

(1) 5G smart hospitals

Samsung Medical Center (SMC) has transformed into a 5G smart hospital with the help of KT Telecom. A private 5G network was deployed at SMC, upgrading the operation rooms and proton treatment rooms with delivery robots, digital diagnostic pathology, proton therapy data acquisition, surgery teaching, and patient care. Tests on the new services have been conducted, and delivery robots have proven capable of disposing of contaminated materials and other medical waste, and carrying surgical supplies. The use of delivery robots in operation rooms has reduced the occurrence of secondary infections caused by exposure to medical waste, and freed hospital staff from having to perform waste disposal. The hospital has also set up an AI-based in-patient care system, which enables in-patients to check in on their own health status through voice commands.

(2) 5G pandemic prevention robots

To combat the COVID-19 pandemic, SK Telecom has developed a 5G and AI powered robot and used it at the company's headquarters in Seoul. The robot moved around on its own, collecting data such as visitors' body temperatures and uploading data to the server, reminding people of wearing masks and maintaining a safe distance, asking crowds to disperse, and even carrying out disinfecting work. In 2021, SK Telecom put a 5G disinfecting robot into operation at Yongin Severance Hospital. Thanks to the real-time positioning system of the 5G network, the robot could freely move around within the hospital, monitoring body temperatures and mask compliance. The robot also came with an ultraviolet disinfection system to help rid the hospital of bacteria and pathogens, and was even capable of locating missing patients by leveraging its real-time positioning system and analyzing the density of patients within the hospital (Fig. 2.6).

3. Autonomous Driving

South Korea has built an internationally recognized V2X testbed, a move to encourage enterprises to develop and verify the core 5G technologies behind autonomous driving and remote driving. It has also developed 5G and V2X enabled vehicles and the related equipment, remotely controllable driving cabins, and a cloud-based and software-controlled remote driving platform, making it possible for individuals to verify and experience autonomous driving services for themselves. South Korea is continuing these efforts by building verification environments for V2X services and providing all necessary technical support. Specifically, South Korea is building autonomous driving simulation environments to

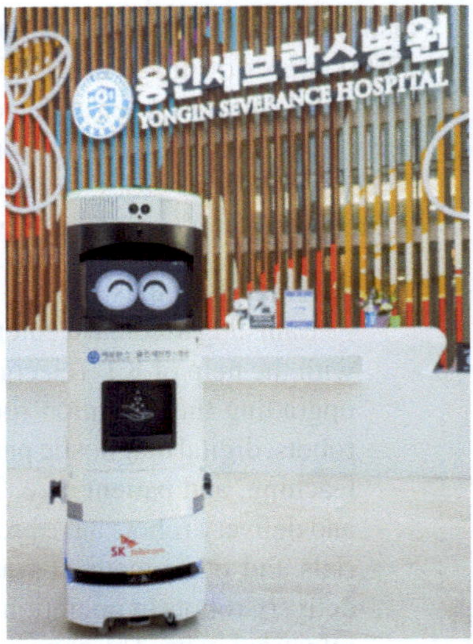

Fig. 2.6 5G pandemic prevention robot. *Source* SK Telecom official website

provide the basic security verification services for the collaborative intelligent transport system (C-ITS) and to provide communications performance verification services in real world driving scenarios. The government has also launched an autonomous driving technology development and innovation project, aiming at commercializing Level 4 (L4) autonomous driving by 2027. The project focuses on creating an autonomous driving ecosystem, applying ICT to road traffic management, and nurturing related services. In addition, the government has been driving the development of the end-to-end technologies enabling the application of autonomous driving to postal logistics, and conducting field verifications of such technologies.

(1) 5G autonomous transport vehicles

KT Telecom has deployed 5G autonomous transport vehicles in its logistics center. These smart vehicles are bolstered by pre-installed indoor maps of the center and real-time feeds of their running status, making an ideal choice for everyday tasks and emergency response. Such vehicles have shortened the distance workers travel by a whopping 47%, reducing workloads while boosting efficiency. KT Telecom expects to use such vehicles and the control systems in small-sized logistics areas in hospitals, libraries, and many other locations in the near future.

(2) 5G autonomous train control tests

The Korea Railroad Research Institute (KRRI) has completed the testing of 5G autonomous train control services on dedicated test rails. The test covered sharing of route, parking mode, and running speed information from trains, as well as real-time response to abnormalities. Compared with the existing Global System for Mobile Communications-Railway (GSM-R), 5G reduced transmission latency between trains while improving transmission capacity and reliability twenty-fold.

(3) The C-ITS test

The C-ITS test, run by SK Telecom and the Seoul metropolitan government, started in early 2019 and concluded in June, 2021. The completion of the test marked the C-ITS beginning of commercialization. During the testing phase, 1,735 5G sensors were installed across 151 kilometers of major roads in the city, for the purpose of monitoring traffic data. Based on the data collected by the sensors, an average of 43 million pieces of pedestrian information were sent to vehicles per day. In addition, a 5G advanced driver assistance system (ADAS) was installed on 1,600 buses and 100 taxis, for which SK Telecom was able to provide traffic information based on the high-precision map it had created.

(4) 5G self-driving delivery robots

SK Telecom has teamed up with Woowa Brothers, a food delivery technology provider, to showcase its 5G MEC-based robot delivery service. Fueled by the 5G edge cloud technology, which features ultra-low latency communication, the robots were able to offer services of higher quality and reliability, including rerouting based on real-time road conditions and responding quickly to unanticipated incidences. The two companies plan to upgrade Woowa Brothers' outdoor robot-based food delivery service launched in a residential-commercial complex in Suwon, Gyeonggi Province with SK Telecom's 5G edge cloud (Fig. 2.7).

4. 5G Smart City

Smart city is one of South Korea's five core 5G+ services. The government has been focusing on two key areas for 5G smart city: enhancing real-time security

Fig. 2.7 5G self-driving delivery robot. *Source* SK Telecom official website

management of major facilities via 5G, digital twin, and other information technologies, and building 5G infrastructure and developing IoT devices for smart ports. In 2020, a number of 5G, AI, IoT, and 3D modeling-enabled applications were demonstrated in Kyungnam and Kwangju, such as fire/smoke diffusion prediction and real-time security management and monitoring. During the COVID-19 pandemic, an air flow analysis service was also launched in a medical center.

2.3 5G Development in the United States

2.3.1 5G Commercial Adoption in the United States

5G networks were put into commercial use early in the United States, featuring high population coverage but low user penetration rate. By June 2021, 5G networks had covered more than 80% of the population, while the user penetration rate was below 10%. In the early phase, with no mid-band spectrum available, telecom operators could mostly deploy mmWave 5G networks. Such high-band networks provide only limited coverage and the deployment is costly. From the end of 2019, low-band spectra were leveraged to build 5G networks for wider coverage, and spectrum sharing technologies were widely adopted to provide 5G services nationwide. In March 2021, the Federal Communications Commission (FCC) completed the auction of C-band (3.7–3.98 GHz) spectra. Afterwards, major telecom operators started to plan for C-band network deployment, which will be the focus in their future 5G network development.

Verizon Wireless, also known as Verizon, started to provide fixed wireless access in four cities based on proprietary standards on October 1, 2018. Later in April 2019, Verizon launched mobile services compliant with 3GPP 5G standards on mmWave 28 GHz and low-band 850 MHz spectra. By the end of December 2020, 14,000 mmWave 5G base stations had been in operation in over 60 cities, and this figure increased to 30,000 in 2021. To date, low-band 5G networks have covered more than 2700 cities and towns and about 230 million people in the United States. Verizon plans to build C-band 5G networks, with the goal of covering 175 million people by the end of 2023, and 250 million people by and after 2024.

American Telephone and Telegraph, or AT&T, launched commercial 5G services in 12 cities in the United States at the end of 2018. The 5G networks run on internationally-recognized NSA 5G standards and were mainly oriented to industry customers in the early phase. By July 2021, mmWave 5G networks had been deployed in 38 cities, mostly in stadiums, arenas, shopping centers, and university campuses; whereas low-band 5G networks had covered 14,000 towns and over 250 million people. In the second half of 2021, AT&T started to deploy C-band 5G networks to enhance its 5G network coverage, aiming to cover 70 to 75 million people by the end of 2022 and more than 200 million people by the end of 2023.

T-Mobile launched 5G commercial services on 28 GHz and 39 GHz frequency bands at the end of June, 2019, and provided nationwide 5G coverage from the end of 2019 by leveraging its existing 600 MHz spectrum resources. To accelerate 5G innovation and deployment, T-Mobile acquired Sprint, which deployed 5G networks on 2.5 GHz, and became a telecom operator in possession of low-, mid-, and high-band 5G networks. To date, its low-band 5G networks and 2.5 GHz mid-band 5G networks have covered 287 million and 125 million people, respectively. With the C-band available for commercial use only in 2023, T-Mobile will continue to expand its 2.5 GHz mid-band network coverage, hoping to cover 250 million people by the end of 2022.

2.3.2 5G Promotion Policies in the United States

The U.S. government regards the development of the 5G industry as a national priority and has laid the foundation for a world-leading 5G industry by means of releasing strategic plans, promoting 5G technology R&D, providing key 5G spectra, and strengthening 5G cyber security.

2.3.2.1 Promotion Policies for 5G Deployment

In October 2018, the FCC released the 5G FAST Plan, a comprehensive strategy for promoting the construction of 5G networks and strengthening America's technical advantages in 5G.

The Plan comprises the following key parts: 1. Make more spectra available to the market, specifically by auctioning mmWave spectra, assigning mid-band spectra, and improving the use of low-band spectra, and create new opportunities for next-generation Wi-Fi in license-free frequency bands; 2. Update infrastructure policies to clear the barriers in 5G development at all levels, with the focus on policies related to small cells, which are widely used for 5G networks. The new policies are expected to accelerate the vetting of small cells and shorten the process for small cell site approval; 3. Amend regulations, including abolishing network neutrality policies, accelerating the approval for connecting new network devices to existing poles in order to expedite the deployment of 5G backhaul networks with reduced costs, promoting investment in next-generation networks and services, lifting regulations on enterprise data service tariffs to encourage investment in optical fiber networks, and prohibiting the purchase of devices and services from companies that pose a national security threat to the American networks and the network supply chain to ensure supply chain integrity.

Apparently, deploying 5G networks in densely populated urban areas can bring huge returns; whereas cost-wise it is hard to motivate telecom operators to deploy 5G networks in rural areas. To prevent widening the digital divide in the United States in the 5G era and bring economic opportunities to the rural areas, the FCC established the 5G Fund for Rural America, which will allocate USD9 billion of funds in two phases over the next 10 years to introduce 5G wireless broadband services to the rural areas. In the first phase, USD8 billion will be provided for 5G deployment in the rural areas in 10 years. This phase is mainly targeted at areas where telecom operators do not have sufficient economic motivation to deploy 5G networks by themselves. In the second phase, at least USD1 billion in addition to any unused funds from the first phase will be used for 5G deployment to support precision agriculture projects.

2.3.2.2 Promotion Policies for 5G Spectra

In October 2018, President Trump signed the *Presidential Memorandum on Developing a Sustainable Spectrum Strategy for America's Future*, proposing that sufficient spectrum resources and effective spectrum management are critical to maximizing the economic driving effect of 5G networks and maintaining national security. Following the release of this document, the FCC announced a plan to release more low-, mid-, and high-band spectra for 5G networks, with high-band mmWave spectra to be auctioned first. Low-band spectra were meant for expanding network coverage, and high-band spectra for increasing network capacity. In the initial phase of 5G development, mmWave frequency bands were used for two reasons. On the one hand, sub-6G mid-band spectra in the United States had all been used for services such as radio, TV, satellite, and radar. On the other hand, given the low optical fiber coverage in the United States, telecom operators opted to replace optical fibers with high-rate and large-capacity 5G mmWave frequency bands to solve the problem of last-mile coverage. Since 2019, the United States has auctioned 28 GHz, 24 GHz, 37 GHz, 39 GHz, and 47 GHz mmWave frequency bands, releasing nearly 5 GHz of 5G spectrum resources to the market.

The mmWave frequency bands, due to limitations of short-range coverage and vulnerability to obstructions, can hardly provide large-scale continuous coverage, which often results in unstable network performance and poor user experience. As such, telecom operators usually deploy mmWave 5G networks only in specific urban areas for hotspot coverage, and deploy low-band 5G networks for wide-scale coverage. However, low-band 5G spectrum resources are limited, and consequently the high bandwidth feature of 5G networks cannot be given into full play.

The United States has early realized the shortcoming of deploying mmWave 5G networks and has proactively taken measures to deal with it. In April 2019, the Department of Defense (DoD) of the United States released *The 5G Ecosystem: Risks & Opportunities for DoD*, which proposed that sharing sub-6 GHz spectra should be prioritized in the 5G spectrum planning to compensate for the insufficient coverage of high-band spectra. At the end of 2019, the Committee on Commerce, Science, and Transportation passed the bill to put 280 MHz (3.7–3.98 GHz) of spectrum resources in the C-band (3.7–4.2 GHz) for 5G systems on public auction. In February 2020, the FCC decided to incentivize satellite companies to migrate out of these spectra with a fund of nearly USD10 billion so that these spectra could be auctioned for 5G network construction.

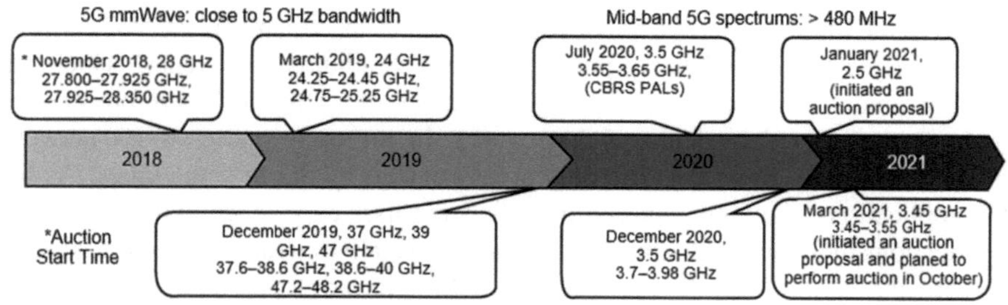

Fig. 2.8 5G spectrum auctions in the United States. *Source* FCC

Since 2020, the United States has paced up auctions of mid-band 5G spectra. Priority access licenses (PALs) for 70 MHz (3.55–3.65 GHz) for the Citizen Broadband Radio Service (CBRS) were auctioned in August 2020; the C-band (3.7–3.98 GHz) spectra were auctioned in March 2021; 100 MHz (3.45–3.55 GHz) of spectrum resources (originally used for military purposes) were put into commercial use in December 2021. Fig. 2.8 illustrates the 5G spectrum auctions in the United States.

2.3.3 5G Application in the United States

2.3.3.1 5GtoC Application in the United States

To address the problem of insufficient optical fiber coverage, U.S.-based telecom operators have been deploying 5G fixed wireless access (FWA) to replace optical fibers for last-mile coverage. This has helped provide better Internet services for households and enterprises while reducing the pipe deployment and maintenance costs. With the high rate feature of mmWave 5G networks, telecom operators were also able to provide enhanced mobile wireless access for consumers in hotspot areas, such as airports, stadiums, arenas, shopping centers, and university campuses. Since October 2018, Verizon has provided mmWave 5G Internet services for households, and by July 2021 the services had been expanded to over 40 cities. T-Mobile started to vigorously explore the 5G FWA market in 2021, aiming at expanding its 5G Internet services to 7 to 8 million households in five years.

Another trend is that cooperation was established between telecom operators and specialized companies, such as content and game companies, to provide VR/AR, HD video, and cloud gaming services. For example, AT&T cooperated with Inception to provide immersive AR book-reading experience for children on Bookful, a 3D and AR reading app, bringing stories into life. It also cooperated with Facebook Reality Labs to provide collaborative video calls and AR experience in Facebook apps, including Instagram and Messenger. AT&T and Verizon both cooperated with Google's cloud gaming platform, Stadia, to supercharge seamless gaming experience using optical fibers and 5G networks.

Given the limited coverage provided by high-band spectra, mmWave 5G networks are leveraged to cover densely populated areas. Stadiums and arenas, which are popular among Americans, feature high volumes of traffic and therefore are the main areas selected to deploy 5G networks. In the early phase of 5G network construction, they were even the only areas covered by 5G in some cities. As for 5G applications, they are centered on enhancing user experience, with the typical applications including high-speed mobile access, HD live streaming of sports events/performances, and VR/AR. For example, T-Mobile developed an app, MLB AR. With 5G-enabled cameras installed on players' hats and catcher's masks, this app provides real-time and true-to-life immersive AR experiences from the perspective of the players, for the viewers regardless of being in the stadium or at home. Verizon has deployed 5G in more than 60 arenas and stadiums. Using 5G and MEC provided by Amazon Web Service (AWS), connectivity in these places is improved and thus user experience is enhanced. Its ShotTracker system uses data from sensors mounted in the stadiums and on the players' bodies to create an indoor GPS, which captures the players' location and movement speed within the stadiums and delivers statistics in real time. Verizon launched an AR mobile game, allowing viewers present at the NFL Super Bowl to throw a football into the pickup box of a pickup truck parked in the center of a field, all in a virtual environment. Another innovation is the AR-enabled app, StatsZone, co-developed by AT&T, Chicago Bulls, and Nexus Studios, an XR team. The app provides visual statistics of the players and some customized content for the viewers, creating an enhanced viewing experience. In addition, AT&T, the official 5G wireless network partner of NBA, provided viewers with unique multi-dimensional views of concerts with its 5G Courtside Concert Series.

2.3.3.2 5GtoB Application in the United States

The United States attaches great importance to building its advantages in 5G technologies and ensuring the security, reliability, and availability of 5G networks. To date, no government policies have yet been released with regard to the application of 5G. 5GtoB applications are still under exploration and technical verification, with tests covering applications in multiple sectors including industrial Internet, healthcare, V2X, and smart city. The mmWave 5G networks deployed in the early phase provide a solid network foundation and test environment for developing 5GtoB applications. The industry players, in close collaboration with each other and with the help of innovation-driven organizations such as innovation centers and incubators, are exploring digital technologies such as edge computing and advanced manufacturing, in order to build a sound ecosystem for 5GtoB applications. There are already 5GtoB applications implemented, though on a small scale, such as 4K video aided security monitoring in factories, VR/AR-based employee training and positioning services, and "5G+VR/AR" remote diagnosis and emergency rescue services.

1. 5G+ Industrial Internet

   The concept of industrial Internet was first proposed by General Electric Company (GE). Industrial Internet in the United States focuses on cross-industry universality and interoperability in its architecture, with data analysis at its core. It is an architecture driven by service value for constant E2E optimization, from devices to service information systems. Riding on the high bandwidth and low latency features of mmWave 5G networks, the 5G+industrial Internet applications are mainly aimed at enhancing the manufacturing processes, factory operations, and facility utilization. Meanwhile, explorations are being made to integrate with the advanced digital technologies (such as edge computing, AI, and big data) and manufacturing technologies for new devices (such as robots and wearables) in the applications.

   Telecom operators, equipment vendors, and industrial enterprises have joined forces to develop and test 5G industrial applications, aiming to find ways to strengthening industrial manufacturing with 5G. AT&T has deployed an IoT video intelligence solution in its Manufacturing times Digital (MxD). Powered by mmWave 5G and MEC, the solution can help manufacturers monitor production, implement device O&M, and track inventories through its video intelligence capabilities. MxD is a manufacturing innovation center founded by the DoD, with the mission of promoting digitalization of the manufacturing industry. MxD owns an innovation space spanning 93,000 square meters in Chicago, where manufacturers can experience the digital solutions they may need for optimizing operations, improving security, and lowering costs.

   (1) 5G smart semiconductor factory

      AT&T and Samsung co-built the first manufacturing-focused 5G innovation zone in the United States at Samsung Austin Semiconductor. The new research space is designed to explore a wide range of 5G industrial applications, to showcase that the benefits 5G can bring to manufacturing and envision the future of smart factories. The main demonstration cases include emergency rescue (using 5G and sensors to help first-aid responders better locate employees and speed up response in an emergency situation), automated material handling and industrial IoT (using 5G, 4K live video, and IoT sensors to better support the factory automation process), and MR for training (using 5G and MR technologies to provide immersive training experience for employees).

   (2) Ericsson's U.S.-based 5G smart factory

      This factory is Ericsson's first factory recognized by the World Economic Forum (WEF) for its large-scale adoption of the Fourth Industrial Revolution (4IR) technology, and the benefits of deploying this technology are already apparent. Since early 2020, the 5G smart factory team has developed 25 use cases, which were to be deployed on a large scale in 12 months. Compared with conventional factories of the same scale, the 5G smart factory has achieved an increase of 2.2 times in terms of output efficiency per employee and reduced manual material processing by 65%.

(3) GM's dedicated 5G ultra wideband assembly plant

Factory ZERO, General Motors' all-electric vehicle assembly plant, is the first automotive plant in the United States to install a dedicated 5G ultra wideband network. Verizon's 5G networks are used to automate material delivery by providing fast and reliable connections for machines and devices in the plant, such as robots, sensors, and AGVs.

2. 5G Smart Healthcare

(1) 5G healthcare center

The Emory Healthcare Innovation Hub (EHIH) is the first 5G healthcare innovation hub in the United States, which is deployed with Verizon's 5G ultra wideband network. The innovation hub works with multiple partners to develop 5G healthcare solutions. These include connected ambulances, remote therapy, and next-generation medical imaging. In the hub, EHIH tested 5G AR and VR applications for medical training, telemedicine and remote patient monitoring, as well as instant diagnosis and imaging systems covering from ambulances to emergency rooms.

(2) 5G private network medical research institute

The Lawrence J. Ellison Institute for Transformative Medicine of the University of Southern California has deployed mmWave 5G private networks. Edge computing and IoT technologies are used to obtain and analyze data in real time with data privacy and security ensured, providing support for doctors to make decisions on the spot. Powered by the advanced technologies, doctors can receive the support data quickly and make decisions in near real time, achieving efficiency improvement.

3. Autonomous Driving

(1) 5G-enabled remote control of unmanned vehicles

Halo took the initiative to launch the remotely-operated car service for commercial use. The service, running in Las Vegas, operates on the T-Mobile's 5G network. Halo is a semi-autonomous vehicle, which is remotely-controlled by a professional pilot through a remote control system running on the T-Mobile's 5G network. The pilot drives the vehicle to the pick-up spot through remote control, drops the passengers, and then drives the vehicle to the next pick-up spot, without having to stop the vehicle between the rides. Halo aims to solve the problem of last-mile transportation to public bus stations, through this innovative on-demand vehicle sharing model. In the long term, Halo aims to expand this service to the entire city and connect it to the public transportation system by collaborating with the local authorities. This will accelerate the adoption of electric vehicles, and in turn will help address the issues of traffic congestion and carbon emission.

(2) 5G+ MEC-based C-V2X autonomous driving test

5G and MEC are the cornerstones of autonomous vehicles. Renovo, Savari, and LG Electronics have used the 5G MEC platform co-developed by Verizon and AWS to carry out tests on autonomous vehicles. Renovo's vehicle data platform indexes and filters the ADAS vehicle data in near real time. With the 5G MEC platform, Renovo can verify new network-ADAS safety functions, for example, alerting all nearby vehicles in real time of dangerous situations to engage immediate driver response. Savari is conducting cellular vehicle-to-everything (C-V2X) tests to explore how the high bandwidth and low latency features of 5G and MEC can be used to power applications that provide warning information to drivers and pedestrians in near real time. LG Electronics is piloting its next-generation C-V2X platform, which uses 5G MEC to transmit data in near real time and improve driving safety through information securely transmitted between vehicles, devices, and the infrastructure.

2.4 5G Development in European Countries

2.4.1 5G Commercial Adoption in European Countries

Since 2019 when the first 5G commercial service was launched, 5G has been put into commercial use in most European countries. By the end of June 2021, 81 operators in 30 European countries had launched commercial 5G services. National mobile network operators in the UK, Germany, France, Italy, and many more countries have deployed 5G networks. At the early stage of commercial adoption, a majority of European operators deployed networks using mid-band spectrum. Since the second half of 2020, some European operators have expedited 5G network coverage by using low-bands and dynamic spectrum sharing (DSS) technologies, thereby improving network availability. For example, by leveraging DSS technologies, Deutsche Telekom has deployed 5G networks covering 80% of Germany's population. High-band 5G networks are advantageous in meeting the ever-increasing demand for high-speed data. Operators in countries such as Spain and Finland have started to deploy networks using mmWave to provide coverage for hotspot areas, enabling 5G applications in specific scenarios. According to the Global System for Mobile Communications Association (GSMA), by the end of June 2021, 5G networks covered 28.51% of the population in Europe and 51.48% of the population in Western Europe.

Operators in the UK, including British Telecom (hereinafter BT), Vodafone, Three UK, and Telefónica UK Limited (O2), all launched commercial 5G non-standalone (NSA) networks in 2019. Since the commercial launch, the four operators have been continuously expanding the coverage of 5G networks. By the end of March

2021, BT had built 5G networks over 160 towns, Vodafone over 100 towns, Three UK over more than 190 towns, and O2 over more than 150 towns. BT plans to deploy 5G core networks (5GCs) to roll out standalone (SA) networks in 2022 and to leverage functions such as ultra-reliable low-latency communication (URLLC) and network slicing to enable key applications such as autonomous driving and large-scale sensor networks in 2023. Vodafone deployed 5G SA networks in Coventry University in 2020. Three UK has deployed 1,250 5G base stations geared towards consumer-centric applications, with hopes for them to also enable enterprise applications. O2 deployed 5G base stations in densely populated areas, such as railway stations, shopping malls, and stadiums.

Vodafone Germany and Deutsche Telekom launched their 5G NSA commercial networks in 2019, and O2 in 2020. The coverage rate is growing rapidly due to the scaled use of DSS technologies. Deutsche Telekom had installed 50,000 5G antennas by mid-June 2021, covering 80% of Germany's population, and the number increased to 60,000 by the end of 2021, uplifting the population coverage to 90%. Deutsche Telekom started 5G SA tests at the beginning of 2021. O2 (Telefónica Germany) embraced scaled 5G network deployment in 2021. By the end of June 2021, 5G networks with about 20,000 5G antennas had been deployed in 80 cities. Vodafone Germany had reached 25 million people with its 5G networks by the end of May 2021. In April 2021, Vodafone launched 5G SA networks in 170 cities and municipalities. Also, it has shut down its 3G networks nationwide and will refarm 3G spectrum for 4G/5G service improvement.

In France, Altice France (which owns SFR), Bouygues Telecom, Orange France (hereinafter Orange), and Free Mobile (hereinafter FREE) all launched 5G NSA commercial networks at the end of 2020. By the end of June 2021, the French National Frequency Agency (ANFR) had authorized[2] 26,084 5G base stations, of which 15,343 were technically operational. By the end of May 2021, the four aforementioned operators had 15,392 5G base stations in commercial use.

2.4.2 5G Promotion Policies in European Countries

Building a single digital market and improving competitiveness in the digital economy have been the European Union (EU)'s top priorities. 5G is considered a foundation for the development of the digital economy and a key element in building a single digital market. Since 2012, the EU has been promoting 5G network deployment, commercial adoption, and industry application innovation by way of formulating a 5G development roadmap, coordinating research between member states, setting up government-initiated research projects, and more.

[2] Authorized base station: A base station authorized by ANFR for implementation. Technically operational base station: A base station capable of emitting radio waves, which may not yet be commercially available. Commercial base station: A base station that is accessible to mobile users.

2.4.2.1 Promotion Policies for 5G Development

The EU released the *5G Action Plan* in 2016 to boost 5G network investment, drive the construction of innovative ecosystems, and improve Europe's 5G competitiveness as a whole. The Plan provides guidance for the EU member states on laying out 5G development strategies and roadmaps. It aims to coordinate network deployment, spectrum allocation, cross-border service continuity, and standard formulation among member states to create a critical mass for 5G innovation in the EU's single market. The EU's 5G development goal is to achieve early-stage 5G network rollout by 2018, large-scale commercial adoption by the end of 2020, and seamless coverage for all urban areas and major ground transportation routes by 2025.

By June 2021, more than 10 European countries, including Germany, France, Spain, Denmark, and the UK, had released comprehensive 5G strategies or roadmaps. Other countries such as Hungary and Portugal, though not yet released comprehensive 5G strategies, had developed spectrum auction plans and fully or partially completed 5G spectrum auctions.

In March 2017, the UK released the *Next Generation Mobile Technologies: A 5G strategy for the UK*, which sets out the UK's ambitions to be a global leader in the development of 5G mobile networks and services, in hopes that the potential of 5G technologies can be leveraged to create a world-leading digital economy that works for everyone. The strategy outlines the key steps that the government will take to pursue the ambitions. The steps relate to business model, trial deployment, regulatory policy, local deployment, network coverage, network security, spectrum strategy, standard formulation, and intellectual property rights.

In July 2018, France released its 5G development roadmap, outlining its aim to be one of the world's first countries to adopt 5G in the industrial field, with milestones of allocating new 5G spectrum and rolling out commercial 5G services in at least one major city by 2020, and providing 5G coverage for the main roads in France by 2025.

In July 2017, Germany released the *5G Strategy for Germany* to promote the development of Germany to become a leading market for 5G networks and applications. The strategy aims to achieve full 5G connectivity and promote both adoption in vertical industries and innovation, and it highlights digital transformation and economic development as Germany's primary goal, alongside having high-performance 5G services by 2025.

In December 2018, the EU officially released the European Electronic Communications Code (EECC), which laid the underlying framework for the current regulation model. The EECC set the goal of ensuring the availability of pioneer 5G spectrum in the EU by the end of 2020 and laying out a lightweight authorization framework for small-area wireless access points (SAWAPs). The EU required its member states to incorporate the EECC into their national law by December 21, 2020, in order to better promote the development of 5G networks in EU countries.

On March 9, 2021, the European Commission emphasized in its publication *2030 Digital Compass: the European Way for the Digital Decade* that 5G is a key element of secure, high-performance, and sustainable digital infrastructure. The

Digital Compass sets out the targets that by 2030, all European households will be covered by Gigabit networks, with all populated areas in Europe covered by 5G, and at least 10,000 climate-neutral and highly secure edge cloud nodes will be deployed.

2.4.2.2 Promotion Policies for 5G Adoption in Vertical Industries

The EU regards 5G as a key enabler of industry transformation. Since 2016, the EU has released a series of policies and projects to engage pan-European stakeholders in the trial and adoption of 5G in industries. These policies and projects cover respective areas of focus, ranging from the study of 5G technology's applicability in key industries, to deployment of large-scale platforms for testing, onsite verification in multiple vertical industries, making breakthroughs in 5G hardware innovation and core technologies, and building a competitive industry ecosystem. We can say that the EU has forged a clear path to applying 5G into vertical industries.

In 2017, the EU started the 5G Infrastructure Public-Private Partnership (5G PPP) phase 2 projects. In this phase, the aims were to make breakthroughs in key 5G technologies, introduce 5G technologies into vertical industries, and develop solutions for 5G adaptation to industries and standardize these solutions. Since July 2018, the 5G PPP phase 3 projects have been launched one after another. 5G infrastructure test and verification projects were among the first ones. These projects aimed to establish a pan-European verification platform, an E2E test platform, a 5G demonstration system, and more. Later on, large-scale tests and pilots were launched for connected and automated mobility (CAM) along three of the EU's 5G cross-border corridors, building infrastructure for V2X and autonomous driving. In 2019, 5GtoB application projects were launched to explore the applicability of 5G technologies in various vertical industries, including smart manufacturing, healthcare, energy, automobile, aviation, railway, logistics, food and agriculture, media and entertainment, public safety, smart city, and tourism. In September 2020, industrialization projects were launched. These projects aimed to promote the popularization of 5G networks and applications by making breakthroughs in 5G core technologies and innovating 5G hardware as well as to expand V2X tests to another three cross-border corridors in Europe and improve the 5G V2X ecosystem.

The EU has invested over EUR400 million in the 5G PPP projects, driving private investment of over EUR1 billion [4]. All these moves, including strategy layout and project funding, reflected the EU's endeavor to establish a world-leading position in 5G industry applications. European countries have also launched policies and projects and set up various funds to support the deployment and application of 5G in industries.

The UK government earmarked GBP200 million through the National Productivity Investment Fund (NPIF) for the implementation of the 5G Testbeds and Trails Programme (5GTT), which includes testbed projects for tourism, rural applications, V2X, telemedicine, industrial production, and other fields. The Industry 5GTT programme (I5GTT) was established in 2019 under 5GTT to further explore the application of 5G in vertical industries, and to develop, demonstrate, and showcase 5G digital solutions for the UK's strategic industrial fields.

Germany has unique advantages in 5G application in vertical industries, especially in the industrial field. The German government's main efforts to promote 5G industry application include allocating 5G private network spectrum, providing financial support, and setting up pilot projects. These efforts are all targeted at building a 5G technology, R&D, and application ecosystem to make Industry 4.0 a reality.

With such support, European operators, industry enterprises, and research institutes teamed up in groups to explore the application of 5G and carried out a large number of tests on applications in autonomous driving, the industrial field, healthcare, and other sectors. Furthermore, for industries with enormous potential for commercialization, 5G application innovation clusters were formed in countries or cities to showcase innovative applications. These clusters also function as a platform for sharing technologies, data, and interfaces, with the ultimate goal of fostering new applications that are easily replicable. Examples include the exploration of 5G adoption in private networks and industries in Berlin, Germany, future factories in Belgium, Ireland, and Norway, smart energy in Italy, and smart city in Ireland and Norway.

2.4.3 5G Application in European Countries

2.4.3.1 5GtoC Application in European Countries

5G FWA is an important supplement to optical fiber broadband in Europe, and this has become the driving force for operators to develop 5G. At the early stage of 5G network construction, Sunrise, a Swiss-based operator, used 5G as an alternative to asymmetric digital subscriber line (ADSL) and very-high-bit-rate digital subscriber loop (VDSL) in areas that were not covered by optical fiber networks. These areas generally have lower priority in optical fiber deployment but could yield greater return from upgrading the existing mobile networks while keeping radiation within limit. Telefónica conducted tests and verifications on replacing wired broadband with high-band 5G in Germany and Spain. Operators in many European countries, including the UK, Italy, and Norway, have launched 5G FWA services for households and enterprises. O2 has unified service tariffs for multiple access technologies in Germany, and launched all-in-one access services. It has also used 5G FWA as a substitute mobile solution for fixed network access so that high-speed mobile access is made available in places that are short in fixed broadband infrastructure.

European operators have also explored 5G technologies in rolling out differentiated content services such as cloud gaming and VR/AR. Operators such as Vodafone, Koninklijke PTT Nederland (KPN), and Telecom Italia have tied up with specialized companies to launch 5G cloud gaming for consumers. O2 Germany has successfully tested VR games on 5G SA networks, during which the game players immersed in interactive 3D game environments wearing VR glasses. Mobile VR games require high on network performance, including extremely short latency, fast data rates, and high reliability. O2 will launch mobile VR games nationwide after 5G SA networks are fully deployed.

For European operators, an important ToC service is ultra-HD live streaming, which provides consumers with an immersive experience when watching performances and games. Featuring low latency and large capacity, 5G is particularly well-suited for deployment in various stadiums. The current applications such as onsite scene capture, real-time HD video transmission, and VR/AR are changing the live reporting and watching experience of sports, culture, and other activities, paving the way for the transition from traditional broadcasting to more 5G-based applications. For example, Vodafone aired the Milan e-Sport final and multiple concerts live over 5G networks. COSMOTE TV live streamed the Greek Cup Final using 5G networks. Two camera robots were remotely and expertly controlled to ensure that not a single moment of the sporting event was missed, and HD audio and video streams were transmitted to the TV studios in real time over COSMOTE 5G. European operators will continue to expand 5G coverage for sports and cultural venues, and provide live streams of key sporting events, such as football, basketball, ice hockey matches, and skiing.

2.4.3.2 5GtoB Application in European Countries

Among the vertical industry application tests, media and entertainment, autonomous driving/road transportation, and Industry 4.0 are most frequently tested.

1. 5G+ Industrial Internet

   5G industrial application is one of the most important vertical industry applications in Europe. The current 5G applications in the industrial field, however, are mostly for production support, and have gradually expanded to R&D design, production and manufacturing, warehousing and logistics, enterprise management, and more. So far, multiple application scenarios have seen small-scale implementation.

   (1) 5G application in oil refineries

   Vodafone Spain and Capgemini have jointly developed two new 5G use cases for one of Cepsa's refineries in Andalusia, Spain, which demonstrated the suitability of 5G technologies in industrial environments. One use case enables professionals to identify transportation pipes through Capgemini's AR application on Vodafone's 5G network. Professionals are able to access pipe information in real time, identify pipes with the dedicated application, and even get support from experts through video streaming when necessary. The other use case enables professionals to monitor the status of rotating devices through sensors connected to the 5G network. By now, Cepsa has installed more than 300,000 sensors in its facilities, generating more than 170,000 signals a day.

(2) Mercedes-Benz Factory 56 (a 5G smart factory)

In June 2019, Telefónica Germany and Ericsson started collaboration to build a 5G network for Mercedes-Benz Factory 56, which is located in Sindelfingen, Germany. The factory has a production area of more than 20,000 square meters and is equipped with indoor 5G antennas and a central 5G hub. The 5G network is meant to enhance data connections between various networked machines and track products on the assembly line in real time, which will lead to optimization of the existing production process as a whole. In September 2020, Factory 56 became operational. The new digital infrastructure, incorporating a high-performance wireless local area network (WLAN) and a 5G network, has laid the foundation for an all-around digitalization of the factory. Many Industry 4.0 applications were deployed in Factory 56, such as smart devices and big data applications. Compared with the S-class assembly line, the new production line improves efficiency by 25%. 5G use cases implemented in the factory include automated quality control that allows cars to be tested on the production line instead of being tested after production, AGV, and 5G networked screwdrivers. The underlying motive for implementing the 5G private network at Factory 56 is to optimize the existing production process. With 5G-enabled applications, all procedures can be optimized and become more robust, and adjustments can be made quickly based on the market requirements. In addition, 5G facilitates smart connections between production systems and machines to ensure the efficiency and precision required for production. Another advantage of using the 5G private network is that sensitive production data does not have to be exposed to any third party.

2. 5G Smart Healthcare

Many European countries are embracing 5G in the medical field. In 2019, medical institutions in Spain, the UK, Italy, and other countries, in cooperation with telecom operators, successfully conducted 5G-based remote surgery tests. As 5G technologies become more mature and more popular, the range of medical applications is widening, including pre-hospital emergency rescue, in-hospital diagnosis and treatment, nursing, robotic surgery support, medical training, health monitoring outside hospital facilities, and remote guidance.

(1) 5G hospital in Portugal

NOS, a telecom operator based in Portugal, has partnered with Grupo Luz Saúde to deploy a 5G network covering the core areas of a hospital in Lisbon, including operating rooms and training centers. The 5G network will be used for the education and training of specialized staff and students. By leveraging VR and AR applications, new scenarios and virtual environments can be created for training, diagnosis, and treatment. Besides, the 5G network will make the hospital's operations and technical means more agile, enhance cost effectiveness, and improve the services.

(2) 5G emergency healthcare

Orange España, Ericsson, and Cisco have partnered with the local police in Sabadell, near Barcelona, to demonstrate the strengths of 5G technology in emergency health care provision. In the trial, a camera was mounted on a police officer's headband, enabling the police officer to attend to a patient who had a sudden epilepsy seizure in the street. The police officer connected directly with the Parc Taulí Hospital, where doctors gave the officer real-time advice on how to stabilize the patient before an ambulance arrived.

(3) 5G remote nursing

The UK's West Midlands 5G (WM5G) has been working with Tekihealth, a telemedicine company, to carry out a trial on 5G remote diagnosis and monitoring for care center residents in Coventry. In the trial, general practitioners used 5G-powered diagnostic tools alongside high-resolution images, videos, thermometers, otoscope readings (an otoscope is a medical device used to look into the ears), portable electrocardiograms, and lung capacity data, to provide treatments during the COVID-19 pandemic.

3. Autonomous Driving

CAM is considered as Europe's flagship 5G application in vertical industries. Its goal is to deploy 5G along the transport paths in Europe and create a complete V2X ecosystem centered on vehicles, from road safety assurance or digital rail operations to high-value commercial services for road users and train passengers. The EU member states signed, in 2017, a Letter of Intent with the view to strengthen cross-border cooperation for large-scale testing, including live tests of 5G for CAM. Generally speaking, 5G autonomous driving in Europe is still in the pilot phase.

(1) 5G autonomous shuttle

Ericsson and Telia carried out a trial on 5G-based unmanned electric shuttles in downtown Stockholm. On a 1.6 km route with 5G deployed by Ericsson and Telia, the unmanned shuttles in the trial traveled between the National Museum of Science and Technology, the Maritime Museum, the Nordic Museum, and the Vasa Museum. The control center sent instructions over the 5G network to the 5G-powered shuttles, which reacted to the instructions in real time.

(2) 5G-connected bus

In Milan, Italy, the local government, public transport companies, telecom operators, equipment vendors, and universities jointly carried out a 5G-connected bus project. The project was meant to test assisted driving empowered by various hybrid cloud technologies and 5G technologies. With the 5G-connected bus solution, buses, with sensors mounted, can connect to the road infrastructure such as traffic lights through an application system, thus improving the flow of traffic and laying a foundation for autonomous driving. Data storage and connection devices would be installed on street lamps, traffic lights, and bus stops. Drivers could be informed about the traffic light status and the recommended driving speed to hit green lights, and could also receive information about the road traffic, obstructions (if any) on the road, and the number of passengers waiting at bus stops through an application system.

(3) 5G cross-border service continuity test for connected vehicles
Ericsson and Volvo succeeded in their first test of connected vehicles crossing over 5G mobile networks in two countries. The test was part of the EU-funded Fifth Generation Cross-Border Control (5GCroCo) project. The success of the test proves that the current infrastructure can ensure cross-border continuity of 5G network services. Maps are updated in real time thanks to 5G connectivity, and this will facilitate autonomous driving and provide environment information beyond the range of vehicle sensors. Updating the maps with information collected by the sensors enables the connected vehicles to choose a driving lane more wisely. However, maps that are just refreshed may already be outdated. In this case, vehicles can send real-time updates to the mobile edge cloud for vehicles coming up to access.

2.5 5G Development in Japan

2.5.1 5G Commercial Adoption in Japan

In March 2020, NTT DOCOMO, KDDI, and SoftBank launched 5G commercial services in Japan. Due to the COVID-19 pandemic, network verification was delayed and Rakuten Mobile postponed rollout of 5G commercial services from June 2020 to September 2020. To speed up 5G network deployment in rural areas of Japan, KDDI and SoftBank agreed to share base station assets.

By the end of 2020, there had been less than 10,000 5G base stations in Japan. Since 2021, operators have accelerated network deployment and they were able to deploy 15,500 new base stations in six months. By the end of June 2021, a total of 24,400 5G base stations had been deployed in Japan. Fig. 2.9 illustrates the number of 5G base stations in Japan.

Four Japan-based network operators have obtained both 5G mid-band spectrum and mmWave spectrum. In the early stage of network construction, the mid-band spectrum (3.7 GHz and 4.5 GHz) is mainly used to deploy networks. Low-band spectrum is used for wide coverage. For example, in 2021, KDDI launched 700 MHz 5G services as a supplement to existing 5G services, thereby enlarging 5G network coverage and improving indoor and outdoor mobile services. Rakuten Mobile is also planning to use the newly granted 1.7 GHz frequency band to deploy 5G in small and medium-sized cities. mmWave (28 GHz) is mainly used to provide the capacity layer in hotspot areas. In September 2020, NTT DOCOMO launched mmWave services at

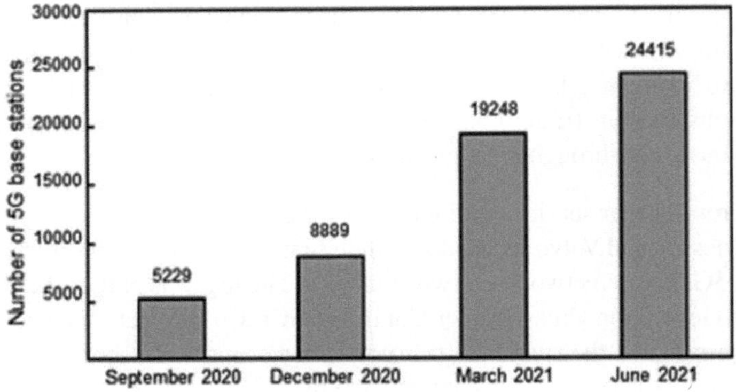

Fig. 2.9 Number of 5G base stations in Japan. *Source* The Japanese Ministry of Internal Affairs and Communications (MIC)

a maximum downlink rate of 4.1 Gbit/s and a maximum upload rate of 480 Mbit/s. SoftBank launched 5G mmWave services in densely populated urban areas, reducing the total cost by up to 35% and providing higher data capacity for FWA and enterprise access.

2.5.2 5G Promotion Policies in Japan

The Japanese government has formulated a clear roadmap for 5G development and is promoting technical tests, spectrum allocation, and commercial deployment. In September 2014, Japan established the Fifth Generation Mobile Communications Promotion Forum to strengthen cooperation between industries, academia, and the government on 5G basic research, technology development, and standard formulation, and to further promote international cooperation. In 2016, the Japanese government released the *Japan's Radio Policy to Realize 5G in 2020* to guide the promotion of 5G in industries. The Policy set the goal of carrying out 5G RAN, core network, and 5G application tests from the fiscal year 2017, allocating 5G spectrum in 2019, and putting 5G into commercial use during the Tokyo 2020 Olympic Games. Government policies focused on strengthening the R&D of key technologies, 5G policy environment, collaboration between industries, education institutions, and the government, and active participation in the formulation of international standards.

At the end of June 2020, the MIC of Japan released the *Beyond 5G Promotion Strategy*: a document which sets the goals of accelerating 5G commercial deployment and rapidly promoting large-scale early-stage 5G deployment and application expansion in industrial and public fields, establishing 5G use cases with international and national influence within the next five years, and generating new value added of JPY44 trillion by 2030. To achieve these strategic goals, Japan plans to expand 5G network coverage, and promote 5G investment and accelerate 5G network construction by offering tax subsidies, introducing favorable tax policies, etc. By 2023, 210,000 5G base stations will be deployed to achieve 5G coverage in all cities.

2.5.3 5G Application in Japan

Guided by the vision of "Building an Intelligent Society 5.0", the Japanese government is actively promoting the mutual promotion and convergence of 5G and AI, IoT, and robotics, alongside large-scale early-stage deployment of 5G and application expansion in key fields. According to the *Information and Communications White Paper* released by the MIC of Japan on July 9, 2019, Japan will adopt 5G in major scenarios, including healthcare, remote education, UAV transportation, autonomous driving, agricultural and industrial production, and disaster relief by leveraging 5G features, such as ultra-high speed, multi-point access, and low latency.

From 2018 to 2019, the Japanese government supported more than 40 comprehensive 5G application test projects, involving entertainment services, disaster protection, tourism, medical care, agriculture, and transportation. From 2020 onwards, the Japanese government focused on supporting 5G application in the following fields: industry, agriculture, healthcare, autonomous driving, and smart city. In 2021, the MIC of Japan invested JPY21.95 billion to build advanced communication infrastructure for remote office, remote education, and telemedicine applications, including helping 5G network construction in areas with unfavorable geographical conditions and supporting local enterprises in building local 5G systems.

2.5.3.1 5GtoC Application in Japan

Japanese operators mainly provide consumers with 5G home broadband and high-speed mobile access services, as well as content and entertainment services such as VR/AR, games, and HD videos. Similar to operators in other countries and regions, Japanese operators are also actively carrying out concept verification for 5G-based watching of sporting events.

In 2019, NTT DOCOMO launched a 5G pre-commercial service, and provided special smartphones compatible with 5G networks for attendees at the Rugby World Cup to help them watch the game from multiple angles. The attendees enjoyed an engaging, and completely immersive watching experience. KDDI integrated 5G and other advanced technologies into the main stadium and fan communication systems of the Kyoto Purple Sanga FC to improve the spectator experience. Rakuten Mobile cooperated with Vissel Kobe to use AR to display statistics and real-time tracking data, and to provide low-latency and multi-angle video services on a mmWave 5G network at Noevir Stadium Kobe.

Japanese operators were attaching, and still attach, great importance to the development and application of HD video applications. NTT DOCOMO has developed a 5G-based 8K VR live streaming system. KDDI has completed 4K video transmission tests using 5G UAVs, and is exploring the application of UAVs in public safety and surveillance services, agricultural monitoring, and disaster response.

XR is another important consumer-oriented application in Japan. KDDI was one of the initiators of the Global XR Content Telco Alliance, which was established in September 2020. The Alliance cooperates extensively with partners to propel the application of VR in multiple scenarios. They launched au XR Door, a smartphone application which enables users to enjoy XR by opening the door that appears on smartphone screens without having to wear VR glasses. Users are provided with a 360° immersive VR experience, such as hotel selection and reservation experience and virtual shopping with 8K videos. In 2021, KDDI used the 5G network covering the top of Mount Fuji to provide visitors with a real-time virtual tour of the mountain top. In the 5G era, KDDI established au VISION STUDIO, which is oriented towards next-generation media and entertainment content and application. It provides users with a unique experience by using technologies such as 5G, XR, and MEC. To date, their virtual human "coh" has been developed using an HD 3D model.

2.5.3.2 5GtoB Application in Japan

1. 5G+ Industrial Internet

   (1) 5G steel factory

   JFE Steel Corporation (JFE for short) has deployed a 5G network in a steel factory. The 5G network is used to transmit real-time 4K HD videos for analysis, facilitating inspection of the production process and the quality of products. This promotes stable production line operations, and realizes digital and intelligent transformation. The production environment requires high on network performance, including real-time performance and stability. With 5G technologies, a range of data collected by various sensors is transmitted to the central control room in real time. This way, devices can be remotely controlled, optimizing the manufacturing process. JFE plans to carry out further application trials based on 5G networks, such as supporting free changes in factory layouts and collaboration between devices and workers.

   (2) 5G+ AI machine tool chip removal solution

   DMG Mori Seiki (a machine tool manufacturer) and KDDI has jointly carried out research and field test activities, during which HD images were taken by cameras inside the machine tool and transmitted to a data analysis platform in real time over a 5G network. The "AI Chip Removal" system evaluated the chip position and amount, and generated the optimal cleaning path. Featuring high speed and large capacity, the 5G network can speed up the automatic collection and transmission of a large data set of images of machine tools, helping to verify the effectiveness of the AI solution.

(3) Fujitsu's 5G smart factory

In April 2021, Fujitsu started operations of a 5G private network at a factory, to implement smart manufacturing applications and verify the suitability of 5G technologies and possible applications on the private network. The factory's 5G private network consisted of a 4.7 GHz SA network and a 28 GHz NSA network, and the applications were as follows:

a. Work training and remote support: Mixed reality (MR) equipment provided field work training and remote support. 3D product models were created in the factory's edge computing environment, which together with instructions, were projected to MR equipment. This way, experts and developers remotely guided and assisted workers in the field. In addition, a large amount of data could be drawn on MR equipment in real time, improving the efficiency of remote work guidance and support.
b. Real-time work confirmation: The 28 GHz network transmitted images of products and work procedures taken by a large number of 4 K cameras installed in the factory to an edge computing environment at high speed. AI analyzed the images in real time and provided workers with instant feedback on whether correct actions were being performed during assembly. Aided by the edge computing environment and the manufacturing execution system (MES), AI distinguished between the operation personnel's hands, part boxes, and parts from the assembly operation images captured by multiple HD cameras, and determined, based on a program, whether the correct part was being taken out of the specified part box and installed in the correct position on the circuit board. The result was fed back to the operation personnel in real time through a display screen and audio instructions, helping the operation personnel perform correct work and improving the efficiency of inspection tasks and quality control.
c. Automated transportation: Fujitsu has introduced a 4.7 GHz network, which can cover the factory, to automate position measurement and path control through real-time communication with AGVs traveling inside the factory during the transportation of parts within facilities. Images were taken by HD cameras inside and outside the factory as well as unmanned vehicles and transmitted to the edge computing environment with low latency, and AI analysis was performed to recognize the positions of unmanned vehicles and control their paths with high precision in multiple dimensions. This reduced transportation costs by automating transportation inside and between buildings and the loading and unloading of parts and products.

2. 5G Smart Healthcare

(1) 5G remote examination

At the beginning of 2019, NTT DOCOMO carried out tests on 5G remote medical examinations. During a test on a remote medical examination service established using 5G to connect Wakayama Prefecture and

Wakayama Medical University, over a distance of about 30 km, a 5G network transmitted HD images of patients to the university in real time. This enabled real time communication and sharing of images taken by 4K close-up cameras, HD echocardiographic videos, and MRI images using an HD videoconferencing system.

(2) 5G first aid

Multiple Japanese hospitals have carried out 5G first aid trials. In these trials, HD images of patients were transmitted to ambulances and hospitals in real time over 5G while the ambulances were on the way, so that doctors could make remote diagnoses and provide expert guidance on how to handle the case. In addition, electronic cases were imported into a system to help doctors quickly obtain patients' historical medical records.

3. Autonomous Driving

(1) Empirical tests on 5G self-driving taxi

Multiple Japanese companies have cooperated to conduct empirical tests on 5G-powered self-driving taxis in Tokyo. In 2014, five enterprises from different industries, i.e. KDDI, Tier Four, Mobility Technologies, Sompo Japan, and Aisan Technology, announced their cooperation to boost the commercialization of self-driving taxis. The plan was to start drive tests in 2020, to establish operation service models in 2021, and to achieve commercialization after 2022. By the end of 2020, these enterprises had conducted two demonstration tests on 5G-powered self-driving taxis in Tokyo. Multiple taxis (JPN TAXI) equipped with an autonomous driving system were running on public roads. A 5G-based remote monitoring system monitored the status of these taxis in real time. Multiple stops were chosen along the best route between the departure and destination and participants could specify their pick-up and drop-off locations. Test cases included autonomous driving with driver seat unattended and autonomous driving with driver seat attended. According to the test results, these enterprises will further propel the commercialization of self-driving taxis and this will impact society by tackling the shortage of public transportation and improving the mobility of populations, and more.

(2) 5G assisted driving

DENSO Corporation and KDDI have started joint research on the application of 5G in autonomous driving to achieve safe and secure mobility, free from traffic accidents and congestion. DENSO Corporation and KDDI built a 5G environment in a test course at Global R&D Tokyo, Haneda. The two companies verified driver assistance technology in self-driving vehicles using vehicle-mounted HD cameras and roadside sensors. In the verification project, information collected by vehicle-mounted HD cameras and roadside sensors was transmitted to the control center over low latency connections, which were achieved through edge computing technology, to build a system for distributing the ever-changing road conditions to self-driving vehicles in real time to verify driver assistance technology.

(3) 5G remote driving

NTT DOCOMO and Sony jointly tested 5G remote driving. During the test, they succeeded in remotely controlling Sony's Sociable Cart (SC-1) entertainment vehicle carrying passengers in Guam from a base in Tokyo, over 2,500 km away, using NTT DOCOMO's 5G network. Videos of the vehicle's surroundings were captured by Sony's image sensors, which were mounted on the front, rear, and sides of the vehicle, and transmitted in real time to a Sony office in Tokyo, where the vehicle was remotely driven while the operation personnel watched a monitor, by leveraging 5G's low latency, large capacity, and ultra-high-speed connections.

In the autonomous driving era, remote control is becoming increasingly important, and cross-border vehicle operation is expected to facilitate global mobility, benefiting populations across different time zones.

(4) 5G-based C-V2X

Subaru cooperated with SoftBank to verify onsite safe merging assistance using 5G NSA and C-V2X communication networks. This verification included two use cases.

In the first use case, they verified safe merging assistance for a self-driving vehicle assumed to merge with traffic on a highway from a ramp when there was enough space to merge. A 5G network transmitted information about vehicles on the main lane to the MEC server near the base station. The MEC server then used the information received to predict the possibility of collision between the self-driving vehicle and a running vehicle on the main lane. If a collision was predicted, the MEC server sent a warning and deceleration information to the self-driving vehicle. The self-driving vehicle then used the information received and information about its surroundings obtained through its own vehicle-sensors to calculate its own appropriate velocity for lane merging to avoid collision. In this test case, the 5G network and MEC server were utilized to offer low latency and high reliability that were necessary to help the self-driving vehicle to successfully merge onto the main lane.

In the second use case, they verified safe merging assistance for a self-driving vehicle assumed to merge with traffic on a highway from a ramp when there were traffic jams. The self-driving vehicle on the merging lane sent a deceleration request for merging to vehicles on the main lane using C-V2X. The vehicle that received the request calculated the optimal merging position for the self-driving vehicle and sent an entry permission in right time. In this use case, C-V2X enabled the vehicle-to-vehicle communication that was necessary for the self-driving vehicle to successfully merge onto the highway when there is not enough time or space.

References

1. CAICT. China Mobile Phone Market Report, June 2021
2. What stopped China's 5G upgrade surge from going global: Canalys' review and insight of H1 2021 global 5G market, 2021
3. IDC. Worldwide Mobile Phone Forecast Update, 2021–2025, 2021
4. EU kicks off final phase of 5G PPP research trials [EB], 2020

Chapter 3
Challenges, Phases, and Trends of 5GtoB Large-Scale Replication

3.1 Key Challenges of 5GtoB Large-Scale Replication

Currently, the digital economy is developing at an unprecedented rate, both in terms of scope and capabilities. By creating opportunities for technological and industrial revolution, digital transformation has become a national strategic focus for developing new competitive advantages. From a supply and demand perspective, digital transformation is expected to improve manufacturing and supply, so that enterprises are better able to use information and communications technology (ICT) to meet diverse consumer requirements. When taking data elements and implementation into account, digital transformation encompasses the entire data management process, including data generation, transmission, analysis, and transaction. Throughout the process, a wide range of ICT technologies is required, specifically sensing technologies (involving sensors and devices), connection technologies (involving mobile communication networks, wired broadband, satellite communication, IoT, and cloud computing), intelligent technologies (including big data and artificial intelligence (AI)), and trust-related technologies (including cyber security, information security, and blockchain). There has been accelerated innovation in technologies such as 5G, big data, cloud computing, AI, and blockchain during recent years. Individual technological breakthroughs are facilitating technological synergies and paving the way for collective evolution. These technologies are becoming the core engine of economic development in addition to being a fundamental driving force. Next-generation ICT technologies are gradually being used in various areas of economic and social development, and therefore are becoming a key force in reorganizing global resources, reshaping the global economic structure, and spurring changes in the global competitive landscape.

5G has taken its place at the core of next-generation ICT technologies due to its strong penetration and compelling power. It accelerates the convergence and iteration of various technologies, and benefits both the consumer and production fronts,

P. Sun, *A Guidebook for 5GtoB and 6G Vision for Deep Convergence*,
Management for Professionals, https://doi.org/10.1007/978-981-99-4024-0_3

promoting chain evolution and generating the multiplier effect during industry transformation. 5G needs to be able to cope with societal expectations which are significantly different from those expected of previous generations of mobile communication technologies. 1G, 2G, 3G, and 4G networks mainly deal with connections between people whose main expectations on networks are a continuous and consistent communication experience. Therefore, these networks share similar characteristics and the devices running on them take on similar forms. 5G, however, focuses more on connections between devices, which come in different forms, computing capabilities, and computing purposes, and therefore the network capabilities required on 5G vary greatly from those required on the previous generations of network. 5G is expected to drive the development of industries, and enable massive connectivity. 5G is powerful enough to support diversified applications, of which enhanced Mobile Broadband (eMBB), Massive Machine-Type Communications (mMTC), and ultra-reliable low-latency communication (URLLC) are the three typical application scenarios. And after years of research, exploration, and pilot demonstration, the industry has come to understand more deeply that 5G can empower industries, being aware that it is important to carefully analyze the common requirements, the basic services, and the main scenarios, and to work out comprehensive solutions, develop the core capabilities, and monetize on them.

Industries have been focused on finding the best way to build capabilities matching the requirements, apply 5G technologies in vertical industries, and realize large-scale 5GtoB application. This is also a challenge the ICT industry is eager to resolve.

5G has seen growing adoption across various industries, and the key sectors and typical application scenarios have stood out. However, there is still a long way to go to achieve full-fledged application: many challenges need to be addressed in terms of network construction, application convergence, supply chain management, and ecosystem buildup.

3.1.1 Challenges with 5G Network Construction

5G networks are expected to provide different capabilities based on industry requirements. For example, uplink 4K/8K video transmission is necessary in industrial machine vision and live broadcasting. The upload rates of 4K and 8K videos need to reach around 50 Mbit/s and 150–200 Mbit/s, respectively, for each of the four to six channels. The required uplink bandwidth is thus much higher than the downlink bandwidth. Therefore, special uplink and downlink slot configuration must be engineered and the corresponding technologies and network solutions must be studied. For industry enterprises, customizing private networks is costly, and charging by traffic may be unaffordable. Moreover, it is not feasible to use public infrastructure because it is more costly and some functions are redundant. Operators have to bear high network construction and operation costs, and yet the profit model is not clear. Such high costs were common during the early stages of 5G commercial use. At present, most demonstration projects receive financial subsidies and earn revenue through intense publicity, which mitigated the investment risks to some extent. However, if there is not enough market or a clear profit model, a positive business loop will not be formed for 5G, and consequently, it will be impossible to drive investment in this area.

3.1.2 *Insufficient Convergence of 5G Technologies into Industry Applications*

5GtoB convergence is still in the initial stage, in that 5G has not found ways to carry the core industry services. The production devices used in industries apply diversified protocols and interfaces, most of which are defined by vendors outside China. As a result, the transformation required for converging 5G technologies into industry applications is rather costly and takes a long time. And the convergence of 5G technologies with other technologies (such as industrial control technologies) is quite challenging. Till now, 5G technologies are mainly used for production supporting services and information management services. Most of the core production control services are still carried on traditional networks such as industrial Ethernet and field buses. As a result, a myriad of transport networks co-exist, making management more complex. There is thus an urgent need to carry out technological innovation and test verification in order to achieve a deeper convergence of 5G technologies into industry services.

5GtoB convergence calls for transformation of multiple links on the industry chain, for which further exploration is required. 5G technologies are expected to promote the development of intelligent devices. However, China's global market share in high-end devices (such as digitally controlled machines) is still small, and China enjoys little advantages in intelligent devices such as next-generation sensors, automated production lines, and industrial robots. 5G technologies will drive the cloudification of processing and computing functions, with which the traditional onsite treatment of programmable logic controller (PLC) devices can be carried out over the cloud. This area, however, is still under exploration. Furthermore, 5G security performance is still not convincing enough in industrial use cases, and this also causes insufficient adoption of 5G in industrial production, which is another challenge for 5G to replicate across industries.

3.1.3 *Insufficient Industry Supply Capabilities*

As 5G converges into industries, new components, particularly industry-specific 5G modules and chips, become part of the industry chain. However, the corresponding supply capabilities are still weak because of the required R&D investment is now rather high due to various reasons. As a result, such modules and chips are still highly priced and their mass production is hard to achieve. Another new component, which is equally important, is customized 5G virtual private networks. For the time being, the cost of deploying customized networks is high and O&M is difficult. In summary, the empowering capabilities for the 5GtoB industry need to be improved.

In addition, the 5G converged applications have posed higher requirements on 5G technologies. The requirements include enhancement on the conventional performance indicators, such as uplink bandwidth, delay, and reliability, as well as assurance of the new performance indicators, like delay variation, network time serving, and positioning. 5G, with its current level of technical standards and commercial devices, is not able to meet these requirements, which in turn has slowed down its convergence into industries. Therefore, extensive research has to be done into the 5G technologies, including 5G enhancement, 5G time serving, 5G positioning, 5G time sensitive networking (TSN), and 5G local area network (LAN), and the related devices, in order to facilitate deeper convergence of 5G into industries.

3.1.4 Lack of Standards for 5GtoB Applications

The public 5G SA networks combined with edge computing can meet the requirements of most industrial use. The industry enterprises, however, have concerns over trusting the public infrastructure with all of their production and development. The concerns focus on whether they have full control of the production and operation data, whether they can access network upgrade and maintenance services in time, and obtain the network costs over the entire lifecycle with precision, as well as whether they can be assured of a stable network performance. In addition, capital flow is a discouraging factor. According to a research on large-scale manufacturers in China, most enterprises hope to recoup their investment within three years. By contrast, the business models for 5GtoB applications are still under exploration, and it is likely to take a long time for enterprises to realize economic benefits and get a complete return on investment. Another thing to note is that most of the players in the Chinese market are small- and medium-sized enterprises (SMEs), which are faced with challenges including weak foundation for informatization and digitalization, high costs of investment and financing, and high pressure on cash flow, and are therefore unlikely to become the main 5G investors in the short run.

The 5G Applications Industry Array (5GAIA) organized the formulation of the *General Technical Requirements for 5G Industry Virtual Private Networks*. This document defines the overall architecture, service capabilities, key devices, and key technologies of 5G industry virtual private networks. In response to the requirements for lower cost and joint O&M, the array is also driving the development of a series of network device standards, such as the customized user plane function (UPF) and service capability platforms. In addition, research has been carried out on wireless Service Level Agreement (SLA) assurance and the key 5G LAN technologies to meet the requirements for network performance assurance and integration with existing networks. Standards tailored for industries, including electric power, steel, and mining, are also in progress, with the goal of standardizing the industry requirements on 5G networks, the network architecture for 5G convergence into industries, and the key network capabilities to provide. Early completion of these standards calls for coordination between ministries, industries, and sectors, which is not efficient yet. Given the variety of use cases 5G must be fitted into, the lack of standards aggravates the process of reaching consensus across industries on 5G devices, modules, chipsets, security, and related tests and accreditation. All this has made the large-scale promotion of 5GtoB applications a challenging task.

3.1.5 5G Industry Ecosystems Need Strengthening

The presence of multiple new information technologies results in unpredictable opportunity costs. Take industrial networks as an example. The factories are typically built with various radio access technologies (RATs), making technology selection difficult. Also, some technologies are quite similar to 5G in terms of technical features and application scenarios, such as TSN, industrial passive optical network (PON), industrial software-defined network (SDN), and Wi-Fi 6. Enterprises worry that investing in 5G will lead to them losing the potential benefits of using other technologies. This also makes the enterprises hesitant about the adoption of 5G.

In contrast to the rapidly evolving ICT technologies, some of the information equipment in vertical industries can no longer meet the requirements of digital transformation, but deploying 5G networks requires replacing the assets on the current networks, resulting in sunk costs. If the enterprises choose to continue using communications devices such as Ethernet, field buses, and Wi-Fi, driven by short-term benefits, the investment in and deployment of 5G in vertical industries will be postponed. What is also worth mentioning is that the industry enterprises have already achieved enormous returns, established solid supply chains and partnerships, and have come to depend more on the established collaborations. This is also a constraint to the cultivation of 5G industry ecosystems.

Effort is still required to work out the models for the players in the 5G industry to collaborate. Telecom operators are actively embracing the opportunities 5G has brought to the ToB markets, but are confronted with a tricky situation with uncertainties, where equipment manufacturers, Internet enterprises, solution providers, and industrial application makers compete and collaborate with each other. The industry landscape is quite complex, with players trying their own ways to seize the strategic heights. There are a multitude of industry platforms, which are independent of each other, rendering high costs from repeated development. The industry is also short of a mature end-to-end (E2E) solution, covering from the network to security, modules/devices, platforms, and software and hardware. Such a solution is expected to integrate the information technology (IT), communication technology (CT), and operation technology (OT), connect supply and demand faster, promote the transformation of 5G technologies into achievements, and eventually lead to the establishment of a 5GtoB industry ecosystem featuring deep convergence.

3.2 Phases and Trends of 5GtoB Large-Scale Replication

3.2.1 *Foundations*

The commercial use of 4G in China happened three to four years later than in the first batch of countries. That was time when the industry had achieved certain level of maturity, with numerous consumer oriented applications already deployed. China was able to learn from these examples and launch large-scale commercial use of 4G applications without having to spend much effort on the exploration. Unlike in the 4G era, China cannot bypass the exploration phase in the 5G era, since 5G is mainly oriented to industry scenarios, with 70% to 80% of the applications involving vehicle-to-everything (V2X) and industrial Internet. China is one of the first countries in the world to use 5G commercially, without precedents, lessons, and experiences to learn from when it comes to technology, industry development, and application. This is particularly true for converged applications targeted at the industry and the real economy as a whole. Therefore, extra caution must be used to grasp the features of 5G, follow the rules regarding network construction, mobile communication technologies evolution, standards formulation, and market development, in order to realize large-scale application of 5G in industries.

1. 5G Infrastructure Construction Outpaces 5G Application

As is common with the previous generations of communication technologies, the construction of 5G infrastructure outpaces its application. High-quality 5G infrastructure is the foundation for the innovation of 5G applications, and the key to the success of these applications. If we look back at 3G and 4G, the killer applications even received doubts at the early stages of their commercial use. Innovation in mobile communication applications requires a solid network and a vibrant market. Network coverage and user penetration generally take two to three years or even longer before reaching the satisfactory level. For example, killer apps such as microblogs in the 3G era and short video apps in the 4G era all appeared two to three years after the network rollout. The success of such killer apps opened up the market space and in turn attracted more resources (capital, talent, R&D, etc.) to innovation. Therefore, the principle of "building roads before vehicles" must be upheld to accelerate the innovation of 5GtoB applications as the 5G technologies continue to mature and the network construction expands.

2. 5G International Standards Are Introduced in Phases

High rate, low latency, and massive connectivity are the defining features of 5G. The International Telecommunication Union (ITU) has defined three usage scenarios that correspond to the three defining features, respectively: eMBB, URLLC, and mMTC. In the eMBB scenario, 5G provides supreme experience for mobile Internet users, supporting consumer-oriented applications, such as ultra-high-definition (UHD) video, virtual reality (VR), and augmented reality (AR), and traffic-intensive and high-rate industrial applications, such as machine vision aided detection and real-time production monitoring. In the URLLC scenario, 5G enables industrial applications that require higher on latency and reliability, such as industrial control, telemedicine, and autonomous driving. In the mMTC scenario, 5G is designed to power applications that focus on sensing and data collection, such as smart city, safe city, smart home, and environmental monitoring.

LTE Release 8, issued in 2009, was the first set of specifications on 4G. It established the main framework and technical solution for 4G. The subsequent releases such as Release 9 and Release 10 enhanced the 4G performance. As for 5G, 3GPP issued the first full set of 5G standards, Release 15, in June 2018, which defined multiple basic functions, with a focus on eMBB services. Since then, the 5G standards have kept upgrading. Release 15 defined the 5G network architecture featuring unified air interfaces and flexible configurations. It focuses on eMBB applications and supports some low-latency and high-reliability scenarios, laying an important foundation for 5G applications. Release 15 specifications are able to meet the requirements of more than half of the 5G application scenarios (basically all ToC scenarios and most ToB scenarios). Release 16 was released in July 2020, upgrading the 5G experience from being "usable" to being "easy to use". With a focus on URLLC, Release 16 meets the requirements of V2X and industrial Internet applications. As an enhancement to Release 15, Release 16 supports a full range of low-latency and high-reliability scenarios and enhances the TSN services, which has given rise to applications requiring low latency and high reliability, such as industrial Internet. Release 16 also defines the specifications for positioning of meter-level precision. Release 17, issued in June 2022, puts emphasis on mMTC and supports medium- and high-speed massive connections. By then, a full set of standards for 5G capabilities was completed.

3. Standardization Is a Prerequisite for 5G Industrialization, and Industrialization Is the Basis for 5G Convergence Application

The industrialization of 5G standards is implemented in phases. Currently, 5G products are mostly 3GPP Release 15 compliant, supporting an E2E latency of 30 ms with 99.99% reliability. This performance can meet the requirements of digital scenarios in many industries, except those requiring lower latency, such as intranet control of production lines in factories, robot control, and motion control. Release 16, supporting an E2E latency of 20 ms, lays a foundation for supporting an E2E latency of 4 ms or even 1 ms. Industry experts believe that it will take three phases spanning at least three years to achieve industry maturity after the finalization of Release 16. The first phase is product development. Driven by technologies, product development is expected to take 1–1.5 years. Release 16-compliant technologies and products will be launched soon. The second phase is industry incubation. Industry-oriented technology verification and adaptation are the main focus. Convergence with industrial control technologies is studied to meet specialized requirements. This phase, driven by the ecosystem, takes 1.5–2 years. The third phase is large-scale expansion. This phase, driven by the market, takes 1–2 years. It aims to eliminate the industry barriers, reduce costs, and implement large-scale application. When the third phase comes,

5GtoB applications are likely to experience a transition from quantitative change to qualitative change. The same phases will repeat for Release 17.

In summary, each set of 5G standards has its particular focus in terms of network performance and functions. 5G technology and industry development is a progressive process. Phased standardization leads to phased industrialization, which in turn results in phased 5G converged application.

4. Function Selection and Development Pace of Subsequent 5G Releases Depend on Market Conditions

Based on experiences with the previous mobile communication standards, both technology roadmaps and market requirements can be critical factors into function selection of technical specifications. Generally, 3GPP issues a release every 1.5–2 years, and 1–1.5 years later, most network devices and chips are put into commercial use. But after Release 15 was issued, products are designed with functions that were originally planned in Release 16 and Release 17 based on the market and customer requirements, which does not necessarily follow the pace of standard release. Standard releases can be categorized into major releases (like Release 15) and minor releases (like Release 16 and Release 17). Release 18 is expected to be a major release that opens the 5.5G era.

5. 5GtoB Development Needs to Keep Pace with Each Industry's Digital Transformation

Currently, 5GtoB applications in China are mostly deployed across the secondary and tertiary sectors (mainly the service industry), where the level of digitalization was relatively high. According to the McKinsey Global Institute (MGI) Industry Digitization Index, the tertiary sector (which includes media, entertainment, public utilities, and healthcare) has taken the lead in digitalization, and the secondary sector (which includes high-end manufacturing, oil and gas, and ore smelting) follows with only a narrow gap, while the primary sector is quite a distance away. This can be reflected by the candidate projects of the 4th "Bloom Cup" 5G Application Competition held in 2021, 42% (down from 54% in 2020) projects were from the tertiary sector, and those from the secondary sector accounted for about 30% (up from 28% in 2020). This is a result of the relatively advanced level of digitalization in most industries in the tertiary sector and the key industries in the secondary sector. In the meantime, as digitalization was pushed forward, the number of projects from the other industries also showed an increasing trend.

3.2.2 Phases and Key Factors

3.2.2.1 Industry Application Development Rules

According to the preceding section, the development of 5G applications cannot be achieved overnight. Gradual introduction from technologies, standards, to industries

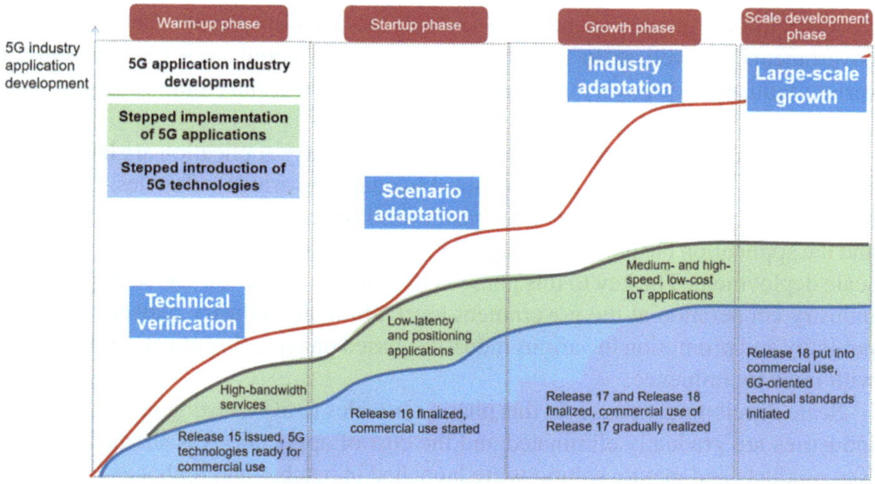

Fig. 3.1 Analysis of 5G industry application development phases (*Source* China Academy of Information and Communications Technology (CAICT))

is a must. 5G technical standards are not mature enough for commercial use at any one time. Instead, they are re-introduced and improved with each release. The functions and performance quality of each iteration of the 5G standards are limited. These iterations together form a phased 5G application development path. We have researched 5G applications in leading tertiary industries (including healthcare, entertainment, city management, education, transportation, and emergency security) and secondary industries (including steel, industrial manufacturing, and metallurgy and mining). Considering the development cycles of 5G and other new technologies, the level of digitalization, and the development of 5G applications in each industry, our research has been able to divide 5G industry application development into four phases: the warm-up phase, the startup phase, the growth phase, and the scale development phase. Figure 3.1 illustrates these phases in more detail.

Warm-up phase: During this phase, 5G standard formulation and R&D are in progress. Requirement analysis and scenario-specific technical discussions are held in industries. The key to this phase is to complete the commercialization of 5G technical standards as soon as possible and realize preliminary cooperation between the 5G industry and other industries in order to lay a foundation for future development. Take Release 15 as an example. After the release was issued, the 5G industry, driven by technologies, started to carry out preliminary cooperation with other industries and perform R&D of Release 15-compliant products.

Startup phase: In the second phase, industry leaders start in-depth cooperation with 5G. In particular, they jointly explore 5G application scenarios and product requirements, perform large-scale scenario adaptation, and start small-scale pilot projects. The 5G converged application industry chain starts to take shape, and the upstream and downstream parts of the industry chain begin to cooperate with each other. The key to this phase is to establish industry cooperation platforms (such as competitions, alliances, and cross-industry associations) under the guidance of governments, as well as to promote small-scale 5G application pilot projects in various industries, discover the real industry requirements, and eliminate requirement uncertainties.

Growth phase: Solutions and products for 5G industry application are launched in small batches and continuously optimized to adapt to various industries. Once they have fully adapted to the industries, customized industry requirements can be met, and the application business models gradually become clear enough to achieve small-scale deployment. The key to this phase is to promote and accelerate in-depth cross-industry cooperation at the government level, eliminate industry barriers, and start publicity and promotion in various industries based on typical application showcases with industry influence.

Scale development phase: In this phase, obstacles to 5G convergence into various industries are gradually eliminated and the cost of applications is sharply reduced. Key products and mature solutions are launched in batches and replicated on a large scale. Their application scope is extended from leading enterprises to SMEs, significantly enabling key industries to reap the benefits. 5G has become a critical part in the digital transformation of leading industries in terms of enablement and profitability. The key to this phase is to give full play to the market, develop replicable and low-cost products and solutions based on the digital level of various industries and enterprises, and achieve fast and high-quality delivery to accelerate the popularization of 5G industry applications and extend the application scope.

3.2.2.2 Current Phase and Key Factor Analysis

In general, there have been notable improvements in the breadth and depth of 5G application practices and technological innovations in China. However, the application standards, business models, and industry ecosystems are not mature enough. 5G applications are still being piloted mainly in top enterprises and have not been applied on a large scale. Some leading industries have entered the growth phase while some potential industries are still in the startup stage. To determine which of these industries are leading ones and which need to be cultivated, they need to be evaluated based on the industry digital transformation, industry development driven by 5G technologies, and converged ecosystem.

In the healthcare field, digitalization of the healthcare service system is the basis of 5G applications. At present, the hospital information system covers 100% of first-class hospitals and 80% of second-class or lower hospitals in China, indicating that wide-range IT-based healthcare will be realized soon. 5G is adopted in this field to implement interconnection and interworking between different medical information systems, such as 5G remote consultation, 5G remote ultrasonic inspection, and 5G emergency treatment. In the future, 5G will help hospitals improve the intelligence level and maximize the value of medical big data. 5G+ AI real-time auxiliary diagnosis and 5G remote surgery, for example, will be able to further accelerate the digital process of healthcare. However, the depth of 5G applications depends on the digital transformation process of the healthcare industry, and this requires changes in the "Internet + healthcare" mode and the standardization of inter-hospital and cross-device data.

Content informatization and network-based communication channels in the media industry are the basis for 5G applications. Currently, the media industry is actively embracing 5G with content and networks already digitalized. Therefore, 5G is mainly being used to replace wired or satellite transmission and bridge the "last mile" of content transmission. For example, 5G backpacks and 5G outside broadcasting (OB) vans help reduce the costs of major events, sports events, and culture and entertainment domains for the media industry. To realize the full use of 5G in the production, ingesting, editing, and broadcasting phases of the media industry, the collaborative and converged development of multiple technologies is needed. IP-based and cloud-based content production, management, and release, which are both critical to full-time information sharing, require the support of new technologies such as cloud computing and artificial intelligence. The application of these technologies in the media industry is still in the initial stage. In addition, the popularization of 5G+ VR/AR, 5G+ 4K/8K, and 5G+ 360° video watching cannot be realized without the rapid development of ultra-HD video and VR/AR industries. The upgrade in industry chains related to production and broadcasting devices, terminal products, and display panels must be accelerated to achieve promotion and innovation of 5G applications in the media industry.

In the industrial field, digital transformation is the basis for 5G converged application development. The penetration rate of 5G converged applications is relatively high in secondary industries which have a good device digital foundation and enterprise informatization level, such as high-end manufacturing, oil and gas, basic product manufacturing, and ore smelting. Meanwhile, 5G is the driving force of industry digitalization. Industry enterprises are encouraged to accelerate digital transformation by seeing the effects of 5G applications on labor saving, production efficiency improvements, and lean management. For industry 2.0 enterprises, only peripheral and auxiliary 5G applications are used due to low digital levels, for example 5G+ machine vision for quality inspection and 5G+ automated guided vehicle (AGV) for logistics. For industry 3.0 and later enterprises, 5G applications are adopted to realize fully-connected enterprise devices, platforms, and production factors, and intelligent production and operation. 5G is expected to penetrate into core phases, for example, 5G+ remote control in the production and processing phase and 5G+ predictive maintenance in the operation and management phase. Converged industrial enterprise services are developed in a point-to-plane manner from peripheral phases to core phases. 5G is a key technology that supports network-based industry intelligence. Although it takes a long time to integrate 5G with the existing industrial infrastructure, 5G converged applications can fully reflect the effects of digital transformation, accelerating the visibility, transparency, prediction, and adaptability of industrial intelligence and promoting intelligent enterprise upgrade.

Large-scale 5GtoB development is affected by many key factors (as illustrated in Fig. 3.2), which can be classified into demand-side factors, supply-side factors, and factors related to the development environment. From a demand-side perspective, the digital level of the industry, the acceptance of new technologies, the clarity of 5G application scenario requirements, the visibility of application effectiveness, and the activeness of core enterprises are all key factors. From a supply-side perspective, the key factors include the relative advantages of 5G technologies, the support level of the 5G industry, and the cost matching degree of industry applications. In addition, factors related to the development environment involve the business models, the policy environment, the promotion channels for 5G industry applications, and the standardization of converged applications.

Based on the analysis of these factors in key fields such as manufacturing, healthcare, energy, and culture and tourism, the color palette for large-scale application development can be represented in Fig. 3.3. A deeper color represents a higher factor level.

The figure above illustrates that industrial manufacturing, electric power, and healthcare are pioneering industries; culture and tourism are industries with great potential; and education and agriculture are industries to be cultivated. V2X, as one of the fields in the transportation industry, is full of potential since it is currently being used predominantly in relatively closed areas such as ports.

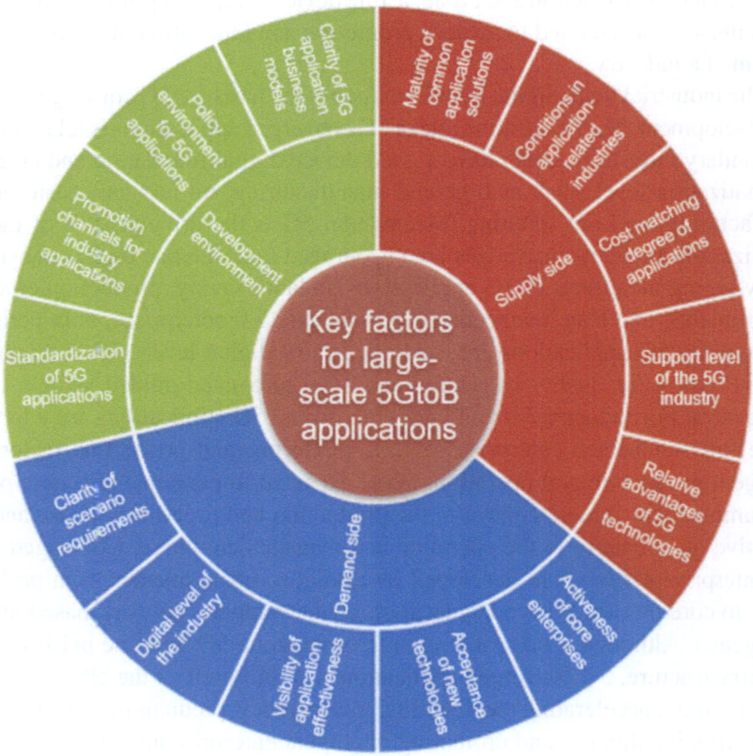

Fig. 3.2 Key factors for large-scale 5GtoB development (*Source* CAICT)

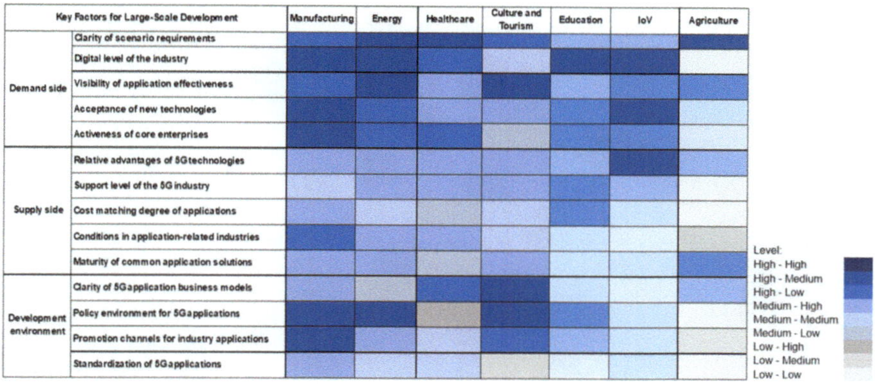

Fig. 3.3 Color palette for large-scale application development in key fields (*Source* CAICT)

Based on the digitalization level and 5G application maturity of various industries, the overall development of 5G applications in these industries, and the analysis of applications in key fields, industries may fall into one of four types: pioneering, having potential, to-be-explored, and to-be-cultivated. Figure 3.4 illustrates industries with different 5G application levels. Pioneering industries are highly digitalized and have clear requirements for 5G. 5G has made some achievements in the digital transformation of these industries, guiding the development of other industries with large-scale replication and promotion. Industries with potential have rather low digitalization levels, but industry enterprises are willing to invest in 5G applications. There is potential for the development of industry converged applications. Industries that ought to be explored have relatively high digitalization levels and a solid digital foundation. However, industry requirements for 5G are not clear and need to be explored further. Developing 5G converged applications in these industries is challenging. Industries which should be cultivated have low digitalization levels and the requirements for 5G in these industries are also unclear.

According to the overall development rules of 5G applications, pioneering industries in China, such as industrial manufacturing, electric power, and healthcare, have entered the growth phase. 5G application products and solutions are continuously adapting to these industries and engaged in business exploration. In potential industries, such as culture, tourism, and transportation, industry requirements and application scenarios are specified for product and solution development and scenario adaptation. Currently, most industries are in the startup phase. Industries to be cultivated and explored, such as education, agriculture, and water conservation, are gradually developing towards the startup phase through active technology verification. According to the phases of key industries in the large-scale development of 5G industry applications (illustrated in Fig. 3.5), it can be concluded that the convergence of 5G and industries is a gradual process which encompasses pilot demonstration, large-scale promotion, and finally large-scale application. It is important to understand the complexity and difficulty of this process.

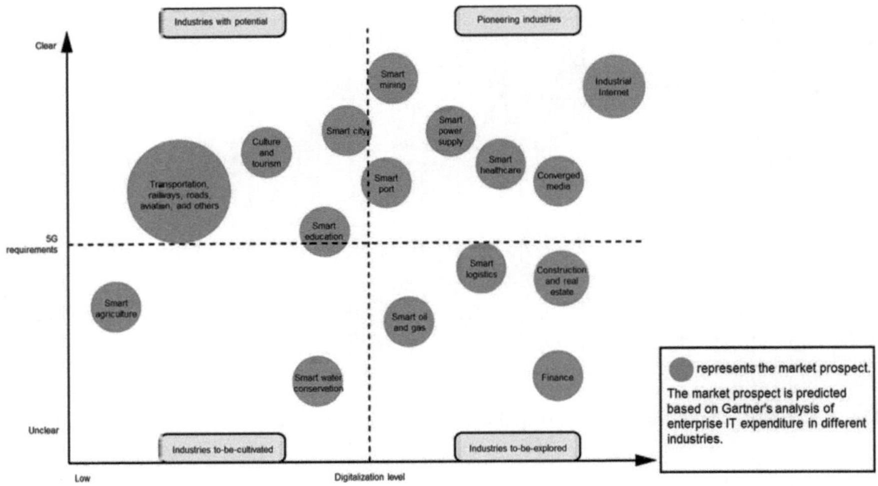

Fig. 3.4 A quadrant chart for 5G application development in key industries (*Source* CAICT)

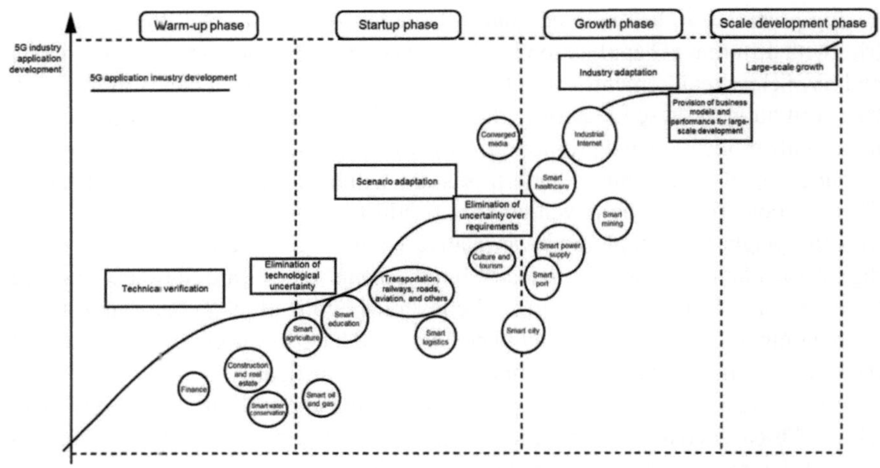

Fig. 3.5 Phases of key industries in the large-scale development of 5G industry applications (*Source* CAICT)

3.2.3 *Trends*

5G is a powerful driving force for information communication development. According to candidate projects in the 4th "Bloom Cup" competition (as listed in Table 3.1), over 40% of projects involve positioning, big data, edge computing, cloud computing, virtual private network (network slicing), and AI. Technologies related to positioning, virtual private network (network slicing), uplink enhancement, and 5G LAN are also gaining significantly increased attention in many projects. Continuous improvement in key technical capabilities has made 5G solutions more comprehensive and brought smoother convergence of 5G into industries. Combining 5G with technologies and industries such as AI, IoT, big data, cloud computing, and HD video will promote breakthroughs in the development of products such as autonomous driving, smart robots, and VR/AR, and also accelerate innovation in application scenarios such as smart factory, smart city, smart transportation, and smart healthcare. However, these technologies and industries are not mature enough, and the corresponding industry applications are still developing and evolving.

Therefore, 5G and other next-generation information technologies form the future trend for industry application by mutual promotion. Connectivity technologies, represented by 5G, work with cloud, intelligence, and computing, triggering a multiplier effect on the development of industry applications and the creation of more diverse applications and services. With the maturity and integration of 5G and other next-generation information technologies, coordinated development is a must for technologies and industries. The future form must be 5G+*X*toB and systematic innovation must be realized to enable digital transformation across industries.

Connectivity technologies, and in particular 5G, have become one of the key factors enabling industry transformation and upgrade in the future. The major goal of 5G is to provide intelligent connectivity which boasts ubiquitous gigabit, a deterministic experience, and hyper automation. In the future 5G+*X*toB form, *X* can represent one of the next-generation information technologies such as cloud, computing/storage, or intelligence. Cloud and computing/storage will become the foundation of the digital world as they provide powerful computing support. AI, as a new engine, will provide true intelligence for enterprises by integrating AI algorithms, models, and intelligence requirements. This will help enterprises reduce costs and improve quality and efficiency. These technologies are intended to work together as well. Specifically, connectivity and computing coordinate and associate with each other

Table 3.1 Analysis of key technologies related to projects in the 4th "Bloom Cup" competition 2021

Technology	2018 (usage/ranking)	2019 (usage/ranking)	2020 (usage/ranking)	2021 (usage/ranking)
Positioning	N/A	N/A	N/A	58%/1↑
Big data	18%/3	44%/2↑	52%/2=	52%/2=
Edge computing	20%/1	38%/4↓	43%/3↑	52%/2↑
Cloud computing	20%/1	38%/3↓	40%/4↓	51%/4=
Virtual private network (network slicing)	N/A	N/A	19%/5↑	47%/5=
AI	13%/4	55%/1↑	55%/1=	46%/6↓
Uplink enhancement	N/A	N/A	N/A	38%/7
5G LAN	N/A	N/A	N/A	12%/8

Source CAICT

through intelligence; intelligent connectivity is used to transmit data for computing; and computing provides support for intelligent connectivity. The convergence of 5G with key technologies such as cloud, intelligence, and big data, provides connection assurance for industry applications and makes the connections among people, among things, and among people and things easier and faster. It also accelerates the reach of ICT technologies from the consumption field to the production field and from the virtual economy to the real economy, ushers in a new Internet of Everything (IoE) era, and develops a new digital economy.

3.2.4 Significance and Values

As the Chinese saying goes, the last leg of a journey marks the halfway point. Now is the time for 5G. 5G is not just a communication technology. It is the infrastructure of the new, prosperous digital economy, and it enables the digital transformation of various industries. According to CAICT, by 2025, 5G will drive network construction investment of about CNY1.2 trillion, information consumption of CNY8 trillion, and economic growth of CNY293 million.

5G application in industries has a multiplier effect and presents huge developmental potential. 5GtoB large-scale replication not only promotes digital industrialization, but also greatly improves the level of industry digitalization in China. Digital industrialization refers to the production and use of information, and it involves technology innovation in the production and supply of information products and services. It is associated with information industry departments and new models in information technology services. 5G industry applications empower industries, reshape industry development models, and create new value. With the evolution and industrialization of 5G standards, the technological spillover effect of 5G will be far greater than that of previous generation mobile communication technologies. More application scenarios and business models will emerge. By connecting resources across the entire industry and value chain, data flows will drive information flows and promote the restructuring of capital, material, talent, and technology flows, which in turn will drive the transformation of business models and organizational forms, reshape industry development models, and inject vitality into digital industrialization. Industry digitalization is the application of information technologies in traditional industry departments, and it is reflected in the increased output, improved quality, and enhanced efficiency that soon follows. The increased output contributes significantly to the total digital economy. 5G applications differ from previous generation mobile communication technologies in that they are distributed based on the 80–20 rule and will be used mainly in vertical industries. It is estimated that every one unit of investment in the 5G industry in China will generate six units of economic output. This significant spillover effect leads to more robust drivers and promotes reforms of quality and efficiency in industry development. Facing informatization requirements in all sectors of society, 5G is not only a new set of opportunities for the mobile communications industry, but also a new field waiting to be explored. Promoting the large-scale development of 5G industry applications will have a profound impact on all industries and ultimately build a new digital economy.

To achieve this ambitious goal, the key to 5G industry application development lies in innovation, transformation, and ecosystem construction. Further work is required to facilitate digital transformation in vertical industries through technological innovation and active investment, and this in turn will accelerate convergence through an open ecosystem.

Chapter 4
Principles of Scaled 5GtoB Promotion

Since 5G was put into commercial use in 2019, 5GtoB has been deployed on industrial networks more slowly than ToC. According to statistics, wireless networks accounted for 4% of industrial Ethernet in 2016. Five years later, by 2021, the proportion of wireless networks in the entire industrial field had only increased by three percentage points. This is because 5GtoB is oriented towards enterprises and is different from ToC which is oriented towards the mass market. According to economic theories, enterprises are economic organizations which aim to make profits, often centering on "economic benefits". This means that heavy assets like network infrastructure are constructed and reconstructed at a slow pace, hindering the 5G application progress. As society steps into the era of economic and social digital transformation, new 5GtoB breakthroughs are expected to speed up scaled development.

4.1 Strategies and Approaches to Scaled 5GtoB Promotion

4.1.1 Scaled Promotion Strategies

For the purposes of scaled 5GtoB promotion, it is necessary to strike a balance between investment constraints and expected benefits, specify appropriate 5GtoB service development principles and strategies, and formulate paths and paces for scaled promotion. On the demand side of ToB, each industry has its own characteristics and industrial customers have varying requirements. 5GtoB should not be promoted at the same pace across all industries. Instead, the key issues that affect the scaled promotion of 5GtoB should be considered, and promotion should begin in a few industries, and then expand to other industries. On the supply side of ToB, all information and communications parties should embrace converged applications based on 5G+ *X*toB technologies, and deepen ecosystem cooperation across the phases of the real economy, virtual economy, and information technologies and

P. Sun, *A Guidebook for 5GtoB and 6G Vision for Deep Convergence*,
Management for Professionals, https://doi.org/10.1007/978-981-99-4024-0_4

services. This better supports exploration of the vast potential of 5G+ industry digital upgrade.

4.1.1.1 Industry Selection

Industries are selected for 5GtoB based on three factors: first, that industry requirements can be met and industry pain points can be tackled; second, that the future market space is broad; and third, that 5G network capabilities can be fully leveraged. Machine vision can raise the accuracy of quality inspection to 99%, and remote control can reduce onsite operation personnel by 80%, thus tackling some key industry pain points. The entire market scale of machine vision has reached CNY6.5 billion and that of automated guided vehicle (AGV)-related industries has reached CNY6.1 billion. This means broad market prospects. Featuring high bandwidth and low latency, 5G helps to achieve machine vision and remote control.

4.1.1.2 Promotion Pace

During scaled 5GtoB promotion, it is important to gradually promote 5GtoB in industries by applying the "simple-to-complex and peripheral-to-core" principle. For example, in the first phase, promotion should focus towards industrial customers with a high digital level and very clear requirements. Applications that require a few changes to existing systems take the lead because customers are willing to reconstruct their systems. In the second phase, promotion should be oriented towards industrial customers with a high digital level but less clear transformation requirements. In some scenarios, convergence with existing application systems requires changes to the existing systems, and as a result, promotion is relatively slow and time is required for verification. In the last phase, promotion should be targeted at industrial customers with a low digital level and low marginal demands. Convergence with existing systems is likely to result in high reconstruction costs. In the application of 5G+ *X*toB, terminals are deployed on a large scale. However, industry terminals lack industry characteristics, and their prices are high. In addition, from the technical perspective, R&D is targeted at enabling industry terminals to have both general and industry characteristics. Therefore, industry terminals tend to be diversified, resulting in obvious differentiation. All of this means that industry terminals are highly customized.

4.1.1.3 Ecosystem Cooperation

A new round of technological and industrial revolution is driving digital development in various fields of the real economy. This is being dubbed the Fourth Industrial Revolution. Currently, the application of 5G+ *X*toB is ushering in a critical phase of scaled development. The sustainable and healthy application of 5G cannot be achieved without the joint efforts of all parties in the industry, and this is why it is important to develop a favorable ecosystem and a cooperative atmosphere. The first task is to consolidate the foundation for developing applications. The necessary steps for realizing this include promoting 5G standalone (SA) network construction by adhering to the "moderately advanced development" principle and considering the requirements of different application scenarios; continuing to propel the R&D and industrialization of 5G advanced technologies and standards; strengthening key system breakthroughs; embracing the deployment of key technologies such as 5G network slicing and edge computing; making full use of 5G, new capabilities, and new features; accelerating filling industry chain weaknesses such as 5G chips, modules, and terminals; and enhancing basic support capabilities for industries. The second task is to deepen the innovation of converged applications. The key steps for realizing this include continuing to enable 5G to power Internet of Everything (IoE) and integrate 5G in various industries; strengthening collaborative 5G innovation with vertical industries to explore 5G applications and build a sound ecosystem for healthy 5G development; cultivating new models and new forms of services such as 5G+ smart grid, 5G+ industrial Internet, 5G+ vehicle-to-everything (V2X), and 5G+ smart agriculture; promoting information consumption upgrading; carrying out demonstration and showcase projects in fields such as smart transaction, smart healthcare, smart culture and tourism, and smart city; promoting the implementation of 5G applications; and cultivating the pattern of scaled replication. The third task is to maintain a fair and open cooperation environment. It is important to carry forward technology orientation, friendly negotiation, and mutual benefit and win–win strategies; further deepen multi-level exchanges and cooperation between the government, industry organizations, and enterprises in terms of international standards, technology R&D, network construction, and application development; and create an open, fair, and transparent market environment. Take the industrial Internet as an example. In addition to promoting the convergence and upgrading of the real industry, the industrial Internet, as an industry and application ecosystem achieved through all-round and deep convergence of next-generation information technologies and industrial systems, has become a new driving force for national economic growth. Driven by the rapid development of the industrial Internet, China's industrial manufacturing will usher in next-generation digital infrastructure construction. This will be accompanied by the interwinding of innovative technologies such as 5G, IoT, and blockchain, and deep convergence with OT. Industrial manufacturing is a new high ground for the development of 5G. It is also the key field in which the new advantages of digital economy development can be fully leveraged.

4.1.2 Approaches to Scaled Promotion

5G is key infrastructure for comprehensive economic and social digital transformation. It has become the core driving force in the development, transformation, and upgrade of China's digital economy. Converging 5G with technologies such as big data, artificial intelligence (AI), IoT, cloud computing, and edge computing unleashes the potential of 5G+ *X*toB, brings about significant improvements, and enlarges the range of capabilities. This fully exerts the value of digital applications in improving industry efficiency. However, implementing a 5G-centered capability system in the industry comes with a number of challenges, such as low level of product standardization, shortage of implementation experience and deep application in core domains, weak cross-domain technology convergence, and incomplete products and ecosystem. Therefore, it is necessary to take a standardized and scientific approach to 5GtoB expansion in order to promote the collaboration of all industry parties and give full play to the role of 5G in the industry.

Based on the 5G technology system's features, commercialization characteristics, and the challenges of industry implementation, as well as the previous 5GtoB service experience, a "3 + 5 + 5" 5GtoB promotion approach (see Fig. 4.1) is being proposed. "3" represents the three core elements of 5GtoB service development, including incubating high-quality showcases, accelerating 5G capability productization, and scaled replication. The first "5" represents the five key steps that are necessary to build mature 5G products based on the 5G technology system and industry service requirements. These steps include determining the target domains, incubating showcases, closed-loop optimization, developing commercial products, and scaled replication. The second "5" represents the five key steps that are necessary for service delivery. These include business opportunity exploration, 5G product system establishment, solution delivery, operations and billing, and O&M support. These all contribute to the operations of mature products.

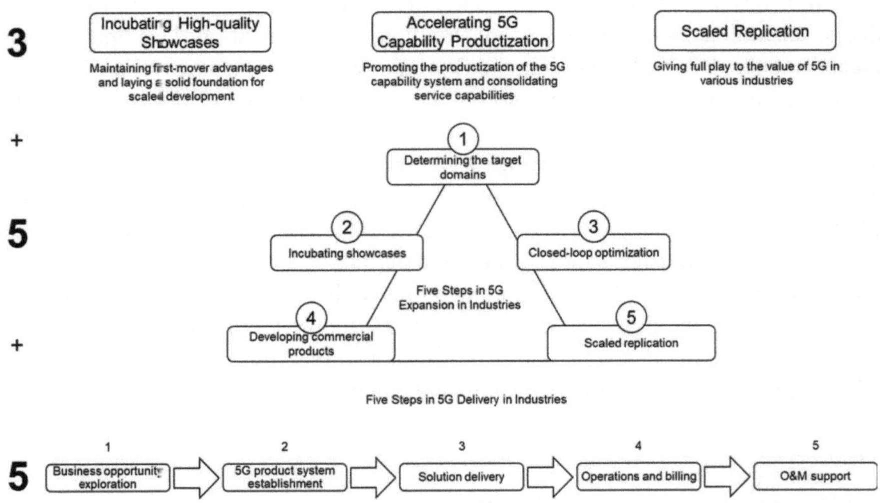

Fig. 4.1 The approaches to 5GtoB promotion

4.1.2.1 Three Core Elements

1. Incubating High-Quality Showcases

Building showcases is a process of developing and designing 5G services and products during which service providers leverage 5G technologies and capabilities to offer solutions tailored to industry service requirements. Key steps to incubate high-quality showcases include the understanding and conversion of service requirements, empowerment using 5G technical capabilities, and designing solutions. The understanding of service requirements is crucial to realizing the value of solution implementation. Requirement conversion is a structured and systematic service analysis process which is used to gain a clear insight into services, and based on these, the required technical capabilities are identified. Empowerment using 5G technical capabilities is a process during which services and technologies are combined in order to deduce whether 5G's inherent capabilities meet service requirements. The effects of empowerment using technical capabilities are based on the breakdown of service requirements and whether the optimal capability is enabled. Solution design is the building of 5G showcases. High-quality showcases can be built and a number of leading demonstration projects can be completed to enhance the understanding of the industry, build product capabilities, innovate valuable business models, and streamline internal processes. Each of these contributes to commercialization. At the same time, a refined and quantitative management system can be built to provide unified standards.

2. Accelerating 5G Capability Productization

Productization of customized projects is the key to giving full play to the core advantages of the powerful 5G capability system. After the achievements in building 5G showcases, it is important to promote the productization of the "private network + platform + application + terminal" 5G capability system. This lays a solid foundation for scaled replication. The productization process starts from the incubation of commercialization. To advance mature market-oriented products, product upgrades are mandatory. Furthermore, it is imperative to specify commercial strategies, such as the product system, pricing system, marketing system, channel system, competition strategies, and the sales and service systems. Moreover, market-oriented products call for systematic collaboration between service providers. This highlights the necessity for clear cooperation between the front end, middle end, and back end, and the importance of clear responsibility matrices for product production, supply, and service. These are the fundamentals of star products which have strong competitive edges.

3. Scaled Replication

5G capability productization is a "zero-to-one" breakthrough, while scaled expansion means one-to-N large-scale replication. The crucial task is to build a standardized operations system. On the basis of this, a healthy expansion ecosystem can be achieved through solution standardization, process standardization, and team standardization, thereby facilitating business development and lowering costs for service providers, and improving business development efficiency and service quality. Furthermore, it is essential to continue to explore targeted industry segments, win influential projects, sign contracts with influential customers, carry out influential events, and work with key partners. All of this will contribute to breakthroughs in deep and broad 5G adoption in key fields, and better support scaled replication and monetization.

4.1.2.2 Five Steps in 5G Expansion in Industries

1. Determining the Target Domains

Selecting the optimal target domains requires cooperation with ecosystem partners, capability readiness, and the understanding of different industries. Determining the target domains is essentially the analysis and selection of target markets and targeted market segments. It is difficult to build product capabilities to suit all domains. Instead, it is vital to focus on several target domains. Target market selection is based on external factors, such as the value space, competitive environment, and service requirement complexity in the industry market. It is also based on internal factors, such as capability and ecosystem readiness, and product maturity. It is widely recognized that a high-quality market has a large value space and a loosely competitive environment.

2. Incubating Showcases

It is necessary to develop standards for 5G industry application showcase projects, establish good partnerships with customers, and develop customized products based on frontline operations. The incubation of showcases is a process of concentrating resources. That is, high-quality product development resources and efforts are concentrated for flagship projects to develop showcases suitable for promotion and replication in accordance with solution standardization, process standardization, and team standardization. The breakdown of service requirements and offering of the optimal capabilities are the core of product incubation. In order to deliver high-quality showcases, it is necessary for industry experts to conduct frontline practice and surveying, and to cooperate with technical experts when designing solutions.

3. Closed-Loop Optimization

Closed-loop optimization based on repeated scenario-specific solution verification and cooperation with ecosystem partners both facilitate replication of the experience obtained from showcases in commercialization on a large scale. Repeated verification, practice, and optimization are essential for offering competitive products. This means that to make products commercially available, closed-loop optimization needs to involve multiple rounds of efforts. These efforts include deep cooperation with one or more partners to realize the systematic deployment and operations of incubated baseline products, collection of operation feedback, such as fault prompts and requirement iterations, and product optimization and iterations.

4. Developing Commercial Products

Solution standardization, process standardization, and team standardization lay a solid foundation for scaled replication. Solution standardization promotes standardization tailored to application scenarios in industry segments by formulating standards for technical requirements, service requirements, and business requirements. Process standardization specifies construction and delivery standards for the pre-sale, in-sale, and post-sale phases. Team standardization, which happens as a result of collaboration between front-end, mid-end, and back-end teams, will contribute to product construction.

5. Scaled Replication

As products mature and vertical industries are explored, scaled replication and promotion will be realized. To achieve this, it is vital to consider factors such as business models and promotion methods. The business model best-suited to industry-specific service products is different from that which is suitable for the mass market and usually involves complex service elements. In most industry scenarios, a "consultation + solution + implementation + operations + service" converged solution is needed, imposing higher requirements on the comprehensive capabilities of service providers. With regard to promotion methods, quick market entry can be realized by carrying out public events such as hosting industry forums and releasing industry product white papers. Such events advertise products and build a brand reputation.

4.1.2.3 Five Steps in 5G Delivery in Industries

1. Business Opportunity Exploration

It is vital to have a deep understanding of 5GtoB product-based solutions and service requirements, and to proactively explore, track, and expand commercial opportunities in vertical industries. Opportunity exploration is a series of systematic marketing activities, including opportunity insight, opportunity operations, opportunity tracking, and the provision of feedback. Opportunity insight is a process in which marketing personnel seek out opportunities through various efforts based on a deep understanding of product capabilities. The opportunity operations process aims to filter opportunities using a converged evaluation system and then allocate the high-quality opportunities identified to the best-suited operations team. Opportunity tracking is a process in which the operations team tracks operations and continues carrying out marketing activities. The provision of feedback relates to the assessment of marketing efforts and then optimizing them as necessary in order to improve the efficiency of customer acquisition.

2. 5G Product System Establishment

Commercial opportunities are converted into project opportunities and solution portfolios are formed based on a standardized product system. Existing standardized solutions are optimized and re-engineered to meet customer requirements, meaning that multiple standardized products and solutions are flexibly combined and customized to offer tailored systematic products and services featuring convergence of "cloud + network + platform + application + terminal". It is also important to propose commercial solution-level operations services to better support the construction of a product system.

3. Solution Delivery

Solution delivery requires communicating with customers and involves the following two phases: project planning and project implementation. In the project planning phase, a range of activities are carried out, including understanding requirements, project positioning and planning, prototype planning, understanding the needed functions, and technical selection. In the project implementation phase, following up on a project's development with timely communication helps development personnel to better understand the project and maintain stable progress and development.

4. Operations and Billing

Configuring related operations and billing policies is necessary to propel the shift from solution design to business operations services. The business model is evolving towards operations and billing, and this is the key to driving the transformation from unsustainable services to sustainable services. In most industry scenarios, it is vital for service providers to aim to provide sustainable services, and the breakthrough point for this is in the operations services following product delivery. In this regard, it is necessary to clarify detailed operations service items for customers and formulate billing policies.

5. O&M Support

O&M support includes the deployment, acceptance, rollout, and the configuration of personnel, products, solutions, and standards that are required for O&M assurance, as well as version iterations and maintenance. In the O&M phase, the most important thing is to maintain product stability. Emergency response must be provided if and when an emergency occurs, in order to retain customers. On top of this, it is important to focus on user experience, especially the correlation between user experience and function iterations after version iterations.

4.2 Standardized Capability Building for Scaled 5GtoB Replication

Scaled 5GtoB replication is a challenge that the entire industry is facing. There are two difficulties in overcoming this challenge: scenario-specific capability accumulation and ecosystem construction. Currently, different roles within the industry chain participate in project delivery in typical industry scenarios. In this situation, having different methods and standards results in repeated work, which hinders continuous capability accumulation and business close-loop. How to better overcome the difficulties is a vital factor for scaled replication of 5GtoB.

Inspiration can be drawn from the revolution that took place in the automotive industry more than 100 years ago. At that time, automobile production adopted manual and customized manufacturing, leading to a long production period and high prices. In this context, only a small number of households could afford automobiles. In 1913, Henry Ford introduced standardized production lines, which had typical features (standardized parts, processes, and ports), in the automotive industry. In just six years, the production cycle of automobiles had been shortened to one eighth of the original cycle, and the price had been reduced to less than one tenth of the original price. Since then, automobiles, which had become a symbol of the Second Industrial Revolution, had truly become a staple in a large proportion of households.

At this critical moment in the modern-day industrial revolution, combined with the challenges of realizing scaled replication of 5GtoB and integrating 5G applications into thousands of industries, it is imperative to achieve scaled replication by accumulating and standardizing various capabilities incubated in projects and thus form "standardized modules" with proper granularity which can become replicable and marketable standardized products/commodities.

It is important to note that adopting 5G in industries involves an ecosystem which is significantly more complex than Ford's standardized production line. However, solution standardization, ecosystem standardization, and industry specification standardization are essentially the same as the standardization of parts, processes, and ports.

4.2.1 Solution Standardization

5GtoB involves multiple stakeholders. Cooperative relationships, transaction models, and business models are complex and vary across industries. For this, a one-stop standardized solution oriented towards sales, operations, and services is needed to improve the efficiency of collaboration, innovation, and operations. Solution standardization can be divided into four levels. The first level is related to basic capabilities, such as speed, latency, jitter, and packet loss rate. The second level is related to a general solution which combines basic capabilities such as differentiated services and quality of service (QoS) assurance. The third level is related to standardized industry applications jointly formed by ecosystem partners. The fourth level is related to solutions tailored to industries.

At the same time, 5GtoB introduces three shifts: from public networks to private networks, from the extranet to the intranet, and from office networks to production networks. These three shifts also raise new network requirements. To meet these requirements, 5GtoB requires SLA-oriented all-scenario solutions, which include network planning, delivery, and O&M services. To achieve scaled replication of 5GtoB, a set of SLA-oriented full-lifecycle standardized operation tools needs to be established.

4.2.2 Ecosystem Standardization

Ecosystem standardization has two facets: industry terminals and industry applications. Industry terminals are classified into basic connection terminals, general-purposed terminals, and industry-specific terminals based on application scenarios. 5G terminals are currently short in diversity and their performance varies. Terminal and module prices are still high. To tackle these difficulties, the entire industry chain needs to work together to promote the classification of chips, modules, and terminals, and to embrace terminal standardization, such as establishment of a terminal ecosystem cooperation platform, sharing of software and hardware resources, and joint terminal design, development, and iterative verification processes.

5G is enabling thousands of industries. Prosperous industry applications are necessary for scaled replication of 5GtoB. The standardization of industry applications usually involves the derivation of network SLA based on the modeling of complex service scenarios. This is followed by 5G network interoperability tests and verification on adaptability with network operations enablement platforms and cloud-based infrastructure as a service (IaaS)/platform as a service (PaaS) platforms for industry applications.

Another key step in ecosystem standardization is the filtering of industry terminals and applications that have been tested based on whether their application scenarios can be replicated and whether they are best practice. Those of qualified ecosystem partners can be released on clouds and one-stop subscription can be provided to accelerate business value monetization and to expedite application in the ecosystem.

4.2.3 Industry Specification Standardization

Industry specification standardization is an important basis for the scaled development of 5GtoB. From a scope perspective, industry specifications are classified into enterprise specifications, group specifications, national specifications, and international specifications. From a function and service perspective, industry specifications are classified into device specifications, network specifications, data specifications, and security specifications. Four Chinese government departments, including the National Development and Reform Commission, jointly issued the *5G Application Implementation Plan in the Energy Field* to accelerate the development of 5GtoB in the energy industry. Industry chain partners are also working together to promote specification formulation in the steel, port, and healthcare industries. They have completed the formulation of standards centered on 5G-based remote control of port equipment and 5G unmanned container trucks, and released 5G-based hospital network construction standards, medical equipment communication specifications, and more.

Chapter 5
5G+ Smart Manufacturing

5.1 Overview

The industrial revolutions and accompanying scientific and technological evolution have led to four large-scale manufacturing industry migrations around the globe and the emergence of five industrial zones in Western and Eastern Europe, North America, on Japan's Pacific Coast, and on Asia's eastern coast. The current industrial revolution, the fourth, is empowered by accelerated innovation, rapid iteration, and breakthrough of next-generation information technologies. To take advantage of new technologies, such as 5G, cloud computing, and big data, on traditional manufacturing, major economies around the world have launched digital transformation strategies to enhance the core competitiveness of traditional industries. The US, aiming to form technological alliances and maintain its global leadership in communications and network technologies, data science, blockchain, and human-machine interaction, released the *National Strategy for Critical and Emerging Technologies*. The European Commission proposed the *2030 Digital Compass: the European way for the Digital Decade*, presenting a vision for Europe's digital transformation in the upcoming decade. Japan launched the *6th Science, Technology, and Innovation Basic Plan* to adapt to new scenarios, promote digital transformation, build a resilient economic structure, and take the lead by achieving a super-intelligent Society 5.0.

Manufacturing is one of China's most important industries. China's government attaches great importance to the convergence of the digital economy and the real economy. The digital economy, all industry links and university researchers agree, is of critical importance. Over the past decade, China's manufacturing industry has progressed rapidly, developing increasingly comprehensive capabilities. It has significantly contributed not only to China's economic and social development, but to the world economy as a whole. Statistics from the Ministry of Industry and Information Technology (MIIT) show that China's manufacturing value added has been the highest in the world for 11 consecutive years since 2010. Between 2012 and 2021, China's industrial value added increased from CNY20.9 trillion to CNY31.3 trillion, accounting for 30.8% of China's total economy. Manufacturing

P. Sun, *A Guidebook for 5GtoB and 6G Vision for Deep Convergence*,
Management for Professionals, https://doi.org/10.1007/978-981-99-4024-0_5

value added increased from CNY16.98 trillion to CNY26.6 trillion, with its global share increasing from 22.5% to nearly 30%. Driven by the dual circulation development pattern and supply-side structural reform, the manufacturing industry will play an even greater role in the future of China's economy. Key industries such as photovoltaics (PV), new energy vehicles, home appliances, smartphones, communications equipment, and high-speed rail are already among the world's best and recognized globally as a result.

5.1.1 Driving Force for Manufacturing Industry Development

According to Zhou Ji, President of the Chinese Academy of Engineering, digital and networked smart manufacturing is at the core of the latest industrial revolution. Smart manufacturing is the most challenging area in industrial transformation, and should be the focus of it. Smart manufacturing is the only option for, and the driving force behind, the development of China's manufacturing industry. Four effects come with smart manufacturing. The first effect is convergence between information technologies and manufacturing technologies and between emerging industries and traditional industries. As the share of the service sector increases significantly, the convergence of the manufacturing and service industries can optimize national economic structure and improve production efficiency. The second is what may be called the pilot effect. Smart manufacturing worldwide is experiencing rapid growth. If China takes the lead in the promotion of smart manufacturing and its large-scale application around the world, breaks down key technical barriers, and combines smart manufacturing with its blooming Internet industry, it can seize a competitive edge. The third is the collaboration effect. Widespread application of smart manufacturing in enterprises will transform labor division and collaboration modes. Emerging technologies such as mobile Internet, big data, cloud computing, and Internet of Things (IoT) will penetrate into all phases and fields of traditional manufacturing. Enterprises will gradually put in place an intelligent, collaborative manufacturing network, thereby improving the entire manufacturing industry chain. The last effect is innovation of enterprises' production and manufacturing technologies and modes. Sunsetting obsolete technologies and modes will bring many new requirements and investment opportunities, which in turn will accelerate the growth of the manufacturing industry and the economy as a whole.

5.1.2 Architecture and Features

Booming information technologies are improving the intelligence of the manufacturing industry. Smart manufacturing is a new way of production that features self-sensing, self-learning, self-decision-making, self-execution, and self-adaptation throughout all production links—from design to production, management, and service provisioning. According to the intelligent manufacturing blueprint (2016–2020) released by China's MIIT and Ministry of Finance, smart manufacturing is built on a deep convergence of next-generation information and communications technology (ICT) and advanced manufacturing technologies. The smart manufacturing architecture can be divided into five layers: device, production line, factory, enterprise, and collaboration. The device layer executes production tasks and uploads onsite data. The production line layer preprocesses the onsite data and reports the data to the factory layer. The factory layer further processes the data and reports the production status to the enterprise layer. The enterprise layer uses production management software to analyze and process the data, afterwards delivering a work plan, which is passed downwards layer by layer to achieve effective control and monitoring of devices at the bottom layer. In this way, these four layers form a multi-faceted system capable of realizing efficiency at both macro and micro scales. The top layer, collaboration layer, enables a single enterprise to communicate in real time with other parties in its ecosystem. This helps establish a comprehensive data platform, achieving Internet of Everything (IoE) and optimizing the entire industry chain. Fig. 5.1 illustrates the architecture of smart manufacturing.

Smart manufacturing redefines the manufacturing system by enabling cross-organizational collaboration, efficiency improvement, and business ecosystem reconstruction. Traditionally, manufacturing had a process that looks like a straight line, consisting of consumer requirement analysis, product development and manufacturing based on customer pain points, product rollout, and marketing. This kind of

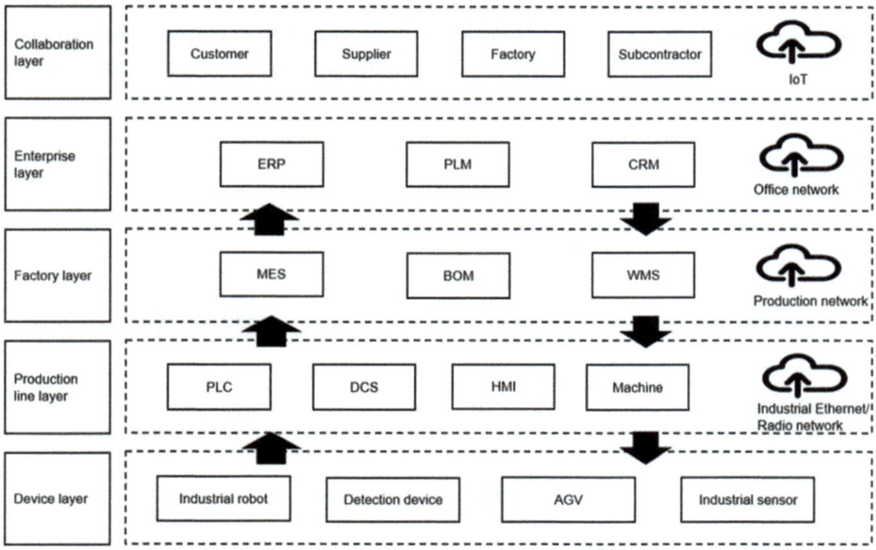

Fig. 5.1 Architecture of smart manufacturing

system, with limited and unidirectional interaction between each two parts as well as slow information feedback, is not very efficient. Smart manufacturing, by contrast, has a ring structure allowing interaction between all parts, transforming the system into a comprehensively integrated unit. Information feedback is in real time, facilitating bidirectional interaction with R&D and production. This makes it easier for R&D and production departments to develop products based on consumer requirements. In addition, automatic procurement reduces production costs and cycles, while flexible production and marketing activities enable real-time information exchange. Production can efficiently respond to market changes, avoiding oversupply and undersupply. Overall, smart manufacturing helps enterprises improve production efficiency and energy utilization, as well as reduce operation costs, product defect rates, and R&D cycles. It will reshape China's competitiveness in the manufacturing sector and also play a key role in environmental protection and sustainable development by decreasing energy and water consumption.

5.1.3 Manufacturing Industry Chain

The manufacturing industry chain consists of three parts as illustrated in Fig. 5.2. They are: upstream, midstream, and downstream. The upstream contains core component suppliers, which supply hardware or software. The midstream has smart manufacturing enterprises, which provide smart manufacturing equipment or solutions. In the downstream are verticals with high demand for automated production equipment, such as heavy industry, 3C electronics, and automobile manufacturing.

Looking closer at the upstream, the hardware layer consists of traditional hardware and emerging hardware. Traditional hardware, for example, lens modules, mostly come from domestic suppliers, while the new types of hardware, such as

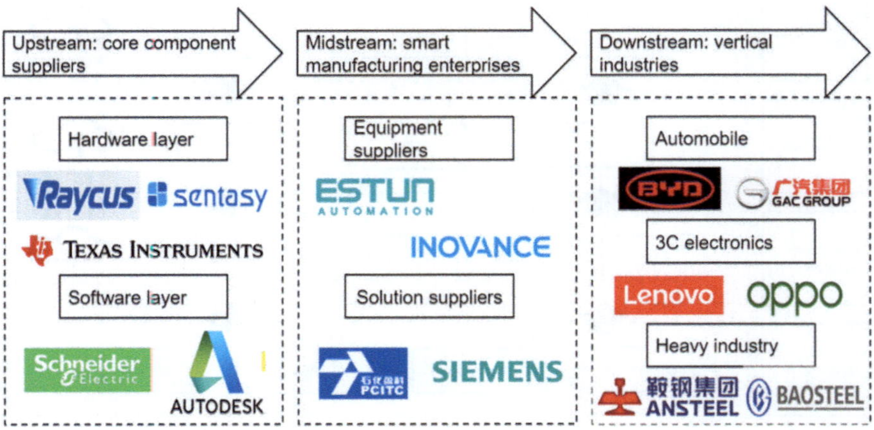

Fig. 5.2 Manufacturing industry chain

smart sensors and high-power lasers, are mostly imported from suppliers abroad. Looking at sensors, they are essential components commonly used by smart manufacturing devices. While China's traditional sensors are balanced in terms of supply and demand, China lags behind in some core technologies of smart sensors. But it is rapidly iterating and innovating, with its share of the smart sensor market increasing with each year. Motivated by the prospect of new industries such as self-driving cars, several dedicated sensor startups have emerged in China, stimulating upgrade and innovation of smart sensor technologies. Future R&D breakthroughs in key technologies will help China hold a higher market share of the smart sensor field. Another new hardware in the manufacturing industry is lasers. China's laser market is experiencing continuous growth and year-on-year shipment increase. Over 90% low-power lasers are supplied domestically, while the rate is only less than 40% for high-power lasers required by smart manufacturing devices. As a result of such limitations, China's smart manufacturing devices are still heavily reliant on imports.

Moving on to the software layer, this is mainly industrial software, most of which is used for R&D and design or production management and control. Industrial software for R&D and design is core to smart manufacturing. It controls data throughout the product lifecycle, shapes the entire manufacturing process, and provides enablement of R&D, management, production, and products. Chinese enterprises are behind in industrial software for R&D and design, occupying only about 20% of the market, with their market share of industrial software for production management and control similarly low. Enterprises with a strong industry foundation and technological background, such as Siemens, Schneider, and General Electric, have the lead in this area. In recent years, however, Chinese vendors have begun to catch up, with enterprises such as Nari Technology, Baosight Software, and HollySys occupying some share of the industrial software for production management and control market. In the future, local supply of industrial software is quite possible.

In the midstream, equipment suppliers face a number of challenges. They are fairly dependent on upstream component suppliers that need to customize production to meet the unique requirements of different industries, and therefore cannot reduce costs through large-scale production. Smart manufacturing equipment suppliers also face high R&D costs: too many industry segments make it impossible to develop a one-size-fits-all device. Instead, there is a need for expert R&D engineers to make industry-specific adjustments. R&D and hardware are the major areas of costs for emerging smart manufacturing equipment, accounting for about 50% and 35% respectively. Smart manufacturing solution suppliers also face significant challenges, because manufacturing enterprises do not perceive a big difference between purchasing smart manufacturing equipment and purchasing smart manufacturing solutions (in this case, factory reconstruction is performed by third parties) during their intelligent transformation. In addition, providing smart manufacturing solutions has no profit advantage over directly selling manufacturing equipment. Solution suppliers have to increase investment to customize solutions. As smart manufacturing expands and achieves greater industry penetration, however, smart manufacturing enterprises can accumulate more experience, leading to higher profit margins for smart manufacturing solutions. At that time, solution provisioning will prevail in the smart manufacturing market.

The penetration rate of smart manufacturing increases when going down the manufacturing industry chain, until it reaches the highest in the downstream. The high demands for smart solutions in the downstream are increasingly driving smart software and hardware development in the midstream and upstream. Downstream industries, such as automobile manufacturing and 3C electronics, are more market-oriented. They can receive timelier consumer feedback and are more proactive in terms of analyzing consumer preferences and requirements in order to upgrade their products. Moreover, these enterprises have higher production efficiency requirements, which stimulate manufacturing technology development and increase the penetration rate of smart manufacturing. Automobiles and 3C electronics have a high demand for automatic production devices. Of the two, automobiles have the higher requirement in terms of the number of devices required for manufacturing and the expenses on them. As a result, in the downstream applications of smart manufacturing, the market share of automatic production devices is higher in automobiles than in 3C electronics with a higher penetration rate. For industries such as metal smelting and material manufacturing, where the upstream process and technologies are updated slowly, smart manufacturing has a lower penetration rate. These industries are labor-intensive and the generally harsh working environment leads to high risks of accident. The introduction of industrial robots, which can work in harsh environments for long periods, can effectively reduce these risks. As smart manufacturing technologies develop and device costs decrease, industrial robots will be widely applied in heavy industry, creating huge opportunities for smart manufacturing.

5.2 Digitalization Trends and Challenges

With the rapid development of next-generation information technologies, digitalization, networking, and intelligence have covered almost all traditional industries, including manufacturing, energy, healthcare, and transportation, and they have gradually developed their respective trends. A series of new models and business forms, such as smart manufacturing, unmanned driving, and smart energy, is enabling various industries and catalyzing the development of new trends. In the digital era, adventurous and innovative technology companies and digital giants are continuously improving their business models. They take advantage of emerging digital technologies, treat data as a differentiated resource, break through the physical boundaries of traditional industries, and as a result are considered to be strong competitors for traditional manufacturing enterprises. In the manufacturing industry, accelerating digital transformation is undoubtedly the only way for traditional manufacturing enterprises to consolidate their advantages and be proactive about development.

Digital transformation of the manufacturing industry cannot be achieved without top-level guidance from the government. The 14th Five-Year Plan has clearly explained industrial digitalization and proposed a deepening of "the digitalized application of R&D and design, manufacturing, operations management, market services, and other links," alongside the cultivation and development of "new models such as personalized customization and flexible manufacturing," and the acceleration of "the digital transformation of industrial parks." Digital transformation of the manufacturing industry can accelerate independent data flows and accomplish the multi-dimensional connection of all elements, industry chains, and value chains in the manufacturing industry. The digital transformation of manufacturing enterprises aims to extend business longevity, and maintain stronger and more sustainable competition in the internal and external environments of continuous transformation and development. To achieve continuous development in the manufacturing industry, it is vital to understand the pain points and breakpoints of traditional manufacturing in digital transformation.

5.2.1 Innovation Subjects Lack Cohesiveness

At present, technological progress in the manufacturing industry in China still primarily depends on technologies introduced from developed countries. Most domestic technological innovations are developed via a process of introducing a technology, imitating it, and then re-innovating it, and as a result, China's manufacturing industry is relatively weaker when it comes to technological innovation. From an innovation chain perspective, China's innovation activities have long been focused on improving the quality and efficiency of processing and assembling, and as such are still at the periphery of the global innovation chain. There are many technological innovations which show quick results despite starting with little investment in China's manufacturing enterprises. However, few original, disruptive technological innovations are seen.

There are four reasons for this. Firstly, the traditional export-oriented original equipment manufacturing (OEM) development mode enables many manufacturing enterprises to earn stable profits without investing in R&D, design, and branding. As a result, enterprises tend to develop insufficient R&D and design capabilities. Secondly, the manufacturing industry's previous innovation policy did not provide sufficient support or guidance for enterprises to invest in working with and developing cutting-edge technologies. As a result, China's manufacturing enterprises' design and equipment manufacturing capabilities did not keep pace with their growth, and they became trapped in a vicious cycle of "technology introducing, technological backwardness, technology re-introducing, and technological backwardness again". Thirdly, insufficient protection of intellectual property rights leads to "easy infringement and difficult rights protection", which affects the incentives for enterprises to innovate. Fourthly, many enterprises are trapped in the dilemma of scattered innovation subjects and insufficient motivation to innovate, even in groups, because of the lack of shared resources.

The R&D innovation in the manufacturing industry can be divided into two parts: high-end equipment R&D and low-end equipment innovation. The pattern of re-innovating through learning that is commonly seen in China's manufacturing industry has been affected by the increase in the number of wholly foreign-owned high-end manufacturing enterprises. The reverse R&D mode restricts the accumulation of basic, original knowledge in China, and the core technologies of high-end equipment manufacturers still lag behind the manufacturing powers. It is difficult to acquire core technical knowledge related to high-end equipment through research after dismantling, and the integration complexity of high-tech knowledge is difficult to acquire in a reverse manner. As such, it is difficult for the high-end equipment manufacturing industry to acquire core technologies through the reverse R&D mode (technology introducing, absorbing, and re-innovating). In low-end equipment manufacturing, the combination of reverse R&D and low-cost labor in China meets the requirements of consumables and low-end industrial goods both in China and in the international market. However, breaking the industry bottleneck of high-end equipment manufacturing requires long-term accumulation of forward R&D and manufacturing knowledge.

Accumulating technical R&D data is very important for technological innovation. The data is not only an indication of accumulated technical R&D experience, but also the basis of breakthroughs in innovation. By using high-quality databases, R&D personnel can quickly develop new product solutions. In multinational enterprises, R&D personnel spend nearly half of their working time on building R&D databases. Meanwhile, most Chinese manufacturing enterprises do not have database support as part of the R&D process. In addition, due to the lack of access to information and resource sharing mechanisms, China's manufacturing enterprises generally lack experience in R&D accumulation and technical resource sharing. Consequently, there is a chronic and problematic tendency to repeat R&D resource construction.

5.2.2 A Diminishing Demographic Dividend Pushes Costs up

The labor force in China decreases year by year. With the continuous increase in the uptake of higher education, employment preferences are changing. As a result, the manufacturing industry is facing labor shortages and in turn higher labor costs. Labor shortages are more severe in heavy industries with harsher working environments. Traditional manufacturing industries are in urgent need of transformation with the help of smart manufacturing technologies in order to alleviate the impact of labor shortages. The wages of employees in China's manufacturing industry have been increasing each year. In 2021, the average wage increased by 6% year-on-year, and enterprises felt the pressure of these rising labor costs. In addition, the average wages of employees in China's manufacturing industry are much higher than those in Southeast Asian countries such as Vietnam, Thailand, and Malaysia. This indicates that the competitive advantage that China had as a result of low labor costs is being eroded. To regain competitive advantage, the manufacturing industry needs to transform from labor-intensive production to technology-intensive production, and smart manufacturing is the only way to usher in a brighter future for the industry. Fortunately, the boom in next-generation information technologies has led to new industrial robots having a significantly shorter cost recovery period, and the average wage gap between industrial robots and manufacturing workers has gradually narrowed. The replacement effect of industrial robots is palpable in many places. In high risk industries, such as metal smelting and mining, enterprises can utilize industrial robots to effectively reduce the accident rate and avoid labor risks. As the cost of an industrial robot continues to decrease, the penetration of industrial robots in the manufacturing field will increase significantly.

5.2.3 Smart Manufacturing Lacks a Reliable Talent Pipeline

Talent is the key factor supporting the development of the manufacturing industry towards mid-range and high-end manufacturing. According to the *Planning Guide for Manufacturing Talent Development*, which was jointly released by China's Ministry of Education, the Ministry of Human Resources and Social Security, and the Ministry of Industry and Information Technology, ten key manufacturing industries will face a huge talent gap in future. By 2025, the talent gaps in both the next-generation information technology industry and the power equipment industry will exceed 9 million; the talent gap in the high-end numerical control machine tool and robot field alone will reach 4.5 million; the talent gap in the new materials field will reach 4 million; the talent gaps in the energy saving field and new energy vehicle field will reach 1.03 million; the talent gaps in the aerospace equipment, agricultural machinery, biomedicine, and high-performance medical instrument fields will each exceed 400,000; the talent gap in the marine engineering equipment and high-tech ship field will reach 266,000; and the talent gap in the advanced rail transit equipment field will reach 106,000.

In terms of enterprise development requirements, nearly 30% of manufacturing enterprises believe that the biggest obstacle to operating smart devices is talent. They worry that the existing basic measures or talent cannot adapt to the new smart manufacturing process. Enterprises need not only machines to replace simple jobs, but also skilled workers, especially workers who can independently operate various smart robots and senior technicians who can repair machines. An increasing number of enterprises are finding it easy to acquire equipment but hard to acquire true talent.

The talent required for manufacturing development and digital transformation can be classified into four categories. The first category is entrepreneurs with a global vision and innovative thinking. The second category is scientific and technological talent with scientific, technological, engineering, and mathematical backgrounds. They are the main driving force of industrial technological innovation. The third category is the craftsmen who have mastered precise manufacturing techniques. The fourth category is the interdisciplinary talent who have broad multidisciplinary knowledge. At present, these four categories of talent are obviously insufficient in China, and this has severely constrained the progress of China's manufacturing industry towards mid-range and high-end manufacturing. From a historical point of view, the availability of talent is dependent on basic factors such as scientific and technological foundations, cultural traditions, and industrial and economic levels. Specifically, the lack of training in scientific methods, innovative ideas, critical thinking, and hands-on skills, leads to a shortage of talent, and a subsequent lack of breakthroughs.

As the manufacturing industry is becoming increasingly smart, smart manufacturing has transformed the nature of professional positions in enterprises. Professional talent cultivation in the manufacturing industry is missing. The role of some traditional production positions may gradually disappear altogether. New positions related to digital modeling, lean specialists, reverse modeling, 3D printing, as well as precision measurement and inspection are becoming more and more important. Currently, there are no university courses corresponding to these new positions. Employees in these positions are predominantly nurtured and trained by enterprises. However, as smart manufacturing becomes more prominent, the current education and training mechanism cannot meet its huge demand for interdisciplinary talent. The professional knowledge required by smart manufacturing is scattered across academic disciplines. As such, cultivating interdisciplinary talent that can adapt to smart manufacturing poses new challenges for higher vocational education. At present, education and training in China is vocational, and this cannot meet the professional, universal, and integrated skill requirements of smart manufacturing.

5.2.4 Unreliable Financing Channels for Enterprises

The development of smart manufacturing and the smart upgrade of enterprises cannot be achieved without funding. Currently, there are still problems such as incomplete government-led industry funding, unreliable enterprise financing channels, insufficient early investment for startups, and insufficient merger and acquisition funding to facilitate resource integration.

Smart manufacturing is a capital-intensive industry. Capital scale and strength is one of the most significant barriers in the smart manufacturing industry. Take smart manufacturing equipment as an example. A complete set of smart manufacturing equipment is developed through multiple phases, such as R&D, design, production, installation, commissioning, and customer acceptance. The project implementation takes a long time in most cases. The production process usually takes at least several months and even longer than a year, while the development of a new product takes even longer than this. In terms of business settlement and the procurement of standard smart manufacturing parts, full-amount cash settlement on delivery is most commonly used. However, customers usually pay in installments, including prepayment, paying for shipping, acceptance payment, and paying for a warranty. The payment collection period is long. As such, enterprises in this industry are faced with great financial pressure, especially in the rapid service development phase.

5.3 Major 5GtoB Large-Scale Replication Scenarios and Typical Cases

5.3.1 HD Video Applications

5.3.1.1 Service Requirements

The traditional electronic information manufacturing industry, meaning the manufacturing of electronic information products such as TVs and phones, faces several imminent challenges. As demographic dividend gradually disappears and a new generation of labor considers traditional manufacturing work to be repetitive, dull, and having low value added, a labor shortage is becoming more and more prominent. Labor shortage and increased labor costs are key factors undermining the competitiveness of China's manufacturing industry on the world stage. The other major factors are higher upstream customer requirements and downstream industry technological innovation, which compel the manufacturing sector to accelerate the pace of its reforms, making flexible factories a must. Automatic production and smart manufacturing are the only choice for China to eliminate labor shortage, enhance competitive edges, improve production efficiency, and achieve its industry upgrade objectives. HD video, combined with the enormous capabilities of 5G, is an important tool toward this goal.

HD video has already been adopted by many manufacturing enterprises as an important tool for R&D, design, production, quality inspection, and security management. In R&D and design, researchers carry out remote experiments using onsite images and data transmitted via 5G while designers leverage the large bandwidth and low latency features of 5G to access immersive virtual environments through various virtual reality (VR) terminals and remotely modify design drawings. In production and manufacturing, remote device operators accurately control onsite industrial devices in real time based on HD videos and production site data. In quality inspection, HD images are identified and analyzed through machine vision and 5G as well as edge and cloud algorithms so that real-time detection, automatic sorting, and quality source tracing of product defects are implemented. In security management, the intelligent monitoring system collects real-time data through 5G, checks video surveillance images, and reports user-defined alarms if necessary, achieving all-round intelligent security monitoring and management at the production site.

Currently, HD video applications are usually deployed using wired access/4G+hard disk storage. This is not a good mode for the following reasons: (1) Fixed backhaul using video optical transceiver, bare optical fiber or x Passive Optical Network (xPON), and modem is expensive and has poor scalability. (2) Power over Ethernet (PoE) has a low usage and high maintenance costs. (3) Current 4G uplink bandwidth cannot support multi-channel HD videos. Unlike this mode, 5G is well-suited to HD video applications. 5G solves difficult security system layouts and construction, high maintenance costs, and poor flexibility. It provides a higher and more stable uplink bandwidth to support HD video, more precise visual identification, and various other applications. In addition, cloud service capabilities such as storage and computing can be used to support massive data storage and AI-based video analysis.

5.3.1.2 Network Solutions

Monitoring is required in various smart manufacturing scenarios, including industrial robot and production line operation as well as employee work. Video surveillance is no longer limited to the traditional functions: surveillance, video recording, and playback. Instead, it is more intelligent, developing towards character, facial, and object recognition, as well as behavior analytics. These new functions set higher requirements for video stream definition and smoothness.

A wireless video surveillance system consists of wireless cameras, a wireless network, and a video surveillance platform. Onsite surveillance videos are uploaded to the cloud through the wireless network. Videos, images, voice, and data can be transmitted bidirectionally in real time. In addition, AI-based behavior analytics algorithms can be used to automatically identify onsite non-compliant behaviors and report alarms in real time, greatly improving operational security and standardization.

5.3.1.3 Industry Ecosystems

Large bandwidth features arrived not long after the development of 5G. As a result, HD video applications are widely used in manufacturing. Telecom operators provide network, spectrum, and cloud resources to achieve E2E connections. Large-scale comprehensive solution suppliers in the manufacturing industry or 5G solution providers such as Huawei provide communications equipment and E2E 5G solutions. In terms of hardware, video terminals mainly include modules, gateways, and HD cameras.

Since industries have different requirements for service/data isolation and network performance, enterprises can build their own wide area or local area 5G private networks.

In terms of platform construction, enterprises such as Haier and SKYWORTH have already built their own 5G industrial Internet platforms through function integration. These enterprises achieve economies of scale and make profits by providing common platforms and customized services to small- and medium-sized enterprises in the manufacturing industry. Due to the specialization and complexity of industrial systems, they mainly provide personalized value-added services in specific scenarios for specific small- and medium-sized enterprises that aim at digital transformation. In the long run, however, their profit sources will expand to cover more general solutions.

Regarding application development for purposes such as quality inspection and security surveillance, AI solution providers such as Hikvision need to provide AI middle-end to application developers. In this way, application developers can accumulate algorithms, knowledge, methods, and experience to achieve precise prediction and quick troubleshooting through human-machine collaboration.

5.3.1.4 Business Models

Model 1: ICT project delivery. In this model, E2E solutions integrating innovative applications and products are delivered to customers, according to their requirements, as a package.

Model 2: Lease instead of construction. In this model, the government or industry customers plan a project. A telecom operator invests in the project, leasing it after completion.

The operator provides E2E network connections and industrial multi-access edge computing (MEC), charging monthly or annual service fees by local traffic steering private line, leased industrial MEC resources and wireless network services, and 5G connections enabled for industrial terminals.

Industry customers only need to specify their requirements, such as coverage areas. The operator is responsible for network survey, design, construction, optimization, and maintenance. The operator also carries out E2E integration and delivery, providing video-related industrial application services, such as 5G cloud-based machine vision and HD video surveillance.

5.3.1.5 Typical Cases

A bathroom and kitchen brand has recently deployed a detection and identification system in its campuses and factories. This system utilizes video-stream-based smart image recognition as well as the latest AI deep learning and big data technologies to analyze the video surveillance data of campuses and factories, facilitating onsite safety supervision by automatically identifying features such as fires, work clothes, and safety helmets. This way, security analysis of equipment, personnel, and environments can improve the problem awareness, status control, proactive warning, and emergency handling abilities of O&M personnel.

In addition, the enterprise uses 5G+ AI for microwave oven quality inspection. After a microwave oven panel is stamped, its appearance needs to be checked visually. This is done by cameras placed on the front, rear, left, right, and top of the production line, with industrial PCs deployed sparsely. This results in high investment and difficult algorithm optimization. With instant image transmission via 5G and cloud-based AI processing, however, an AI model can be trained in real time, improving inspection efficiency, reducing labor waste caused by product rework, and reducing investment in industrial PCs and O&M costs.

5.3.2 AR Applications

5.3.2.1 Service Requirements

With 5G networks, augmented reality (AR) encompasses a host of new use cases in the industrial manufacturing sector, including AR-based remote inspection and O&M and AR-based guidance on complex assembly tasks.

In modern factories, machines are running 24/7 across diverse production processes. Proper operation of devices has never been more important. Production-oriented enterprises attach great importance to stable and efficient device operation and production workshop safety. Improper inspections which miss device damage and security risks cause huge losses of life and property. The usual practice in smart manufacturing is manually recording issues found during inspection and asking experts to be present at the maintenance site. This is neither efficient nor cheap. By contrast, 5G+ AR remote inspection and O&M systems can exchange onsite audios, videos, images, and texts between AR glasses and the platform in real time. This allows experts to interact with onsite engineers, provide guidance, and carry out investigations and device maintenance as if they are working on-site. The system, with its thorough and reliable monitoring as well as response, analysis, and management mechanisms, provides automatic O&M services throughout deployment and daily use. In addition, real-time data can be compared with background data to implement closed-loop management of application services.

With AR-based guidance on complex assembly tasks, trainers can identify system components and provide module assembly guidance to prevent quality issues. This improves training efficiency while reducing training costs.

Bit rate, latency, accurate positioning and orientation, and mobility are key to an optimal AR application experience. Strong edge computing and storage capabilities can give full play to AR advantages. In these respects, 5G is far more advanced than Wi-Fi. 5G MEC enables AR applications to run on virtual machines rather than traditional AR application servers, eliminating the need to purchase and install server hardware. Simplified system deployment and maintenance will accelerate large-scale development of such applications. 5G+ AR applications will be among the first to be replicated on a large scale in the smart manufacturing industry. Some AR applications have already made significant progress in manufacturing enterprises.

5.3.2.2 Network Solutions

AR applications require ultra-large uplink and downlink bandwidth. A single AR terminal requires 10 Mbit/s uplink bandwidth, with latency lower than 20 ms. To meet such requirements, both VR/AR servers and the user plane function (UPF) of the core network need to be deployed closer to users in order to reduce latency.

The MEC migrates computing, storage, and service capabilities to the network edge so that applications, services, and content can be deployed closer to users

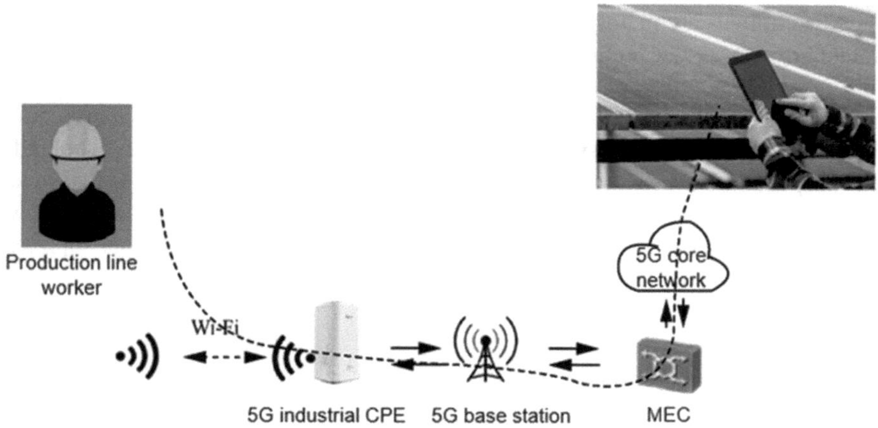

Fig. 5.3 Network solution for AR applications

in a distributed manner. This meets service requirements in 5G scenarios such as enhanced Mobile Broadband (eMBB), ultra-reliable low-latency communication (URLLC), and Massive Machine-Type Communications (mMTC). The MEC can also use network data to implement network context awareness and analysis. This capability is open to third-party service applications, improving network intelligence and promoting deep convergence of networks and services. Fig. 5.3 illustrates the network solution for AR applications.

5.3.2.3 Industry Ecosystems

In the manufacturing industry, networks and spectra required by AR applications are provided by operators, communications equipment and E2E 5G solutions come from 5G solution suppliers, and AR application solutions are integrated by terminal suppliers. To be specific, telecom operators and 5G solution suppliers provide ICT products and services, including networks, clouds, and edge equipment rooms. Terminal suppliers offer basic hardware, including AR glasses, control software, and operating systems.

Telecom operators have two precious resources: 5G spectrum and network. In addition, they have strong ecosystem organization capabilities and a nationwide marketing network covering various industries. They can ensure the key performance counters met for AR services and optimal user experience.

Although computing is currently mainly performed on AR terminals, in the future AR will rely on close device, edge, and cloud collaboration. For 5G solution suppliers, cloud deployment is suitable for massive industrial production data. Edge computing can be used for accurate positioning and orientation, local industrial data, and specific 3D scenario rendering.

AR terminals are mainly used for workshop security protection and monitoring, remote expert service support, and step-by-step guidance for production processes and tasks. They will replace traditional, inefficient electronic spot check systems which display information on handheld devices and only support manual inspection item checking, data input, and data submission to background databases. The traditional spot check process involves a lot of work but, despite that, is unable to provide the real-time data and inspection standards required to locate defects. As AR grows and plays an increasingly vital role in smart manufacturing, AR terminal suppliers will have many excellent options for future development.

5.3.2.4 Business Models

In the 5G era, telecom operators are teaming up with platform enterprises to explore business models. In 5G private network resource provisioning, operators can follow the Basic-Advanced-Flexible (BAF) model to offer a menu of services that enterprises can choose as needed.

In terms of platforms, operators can use their 5G MEC and slicing capabilities to provide enterprises with computing platforms and application-specific network service capabilities. They can charge platform maintenance or app invoking fees for the graphical processing unit (GPU) and AI capabilities provided. In addition, bundle sales of cloud resources can be used to improve service loyalty and increase overall revenue.

As for industry applications, operators can provide an E2E application solution package based on their corporations' common IT service platform. Alternatively, operators can provide services for enterprises' production and management processes through SaaS or app subscription.

Terminals can be deployed directly by industry partners with fees charged per project. Another choice is for operators to act as integrators, delivering products and charging related fees together.

When 5G is applied to production processes, operators and solution suppliers find that single network traffic operation business models cannot meet customer requirements. When beginning 5G commercialization, operators develop products based on technological maturity and market requirements to meet as many customized needs of factories as possible, making the cost-based pricing (cost plus profit) business model an option. With this model, together with ecosystem partners such as AR device suppliers and third-party O&M service providers, they can also provide enterprises with network services. 5G+ AR relies heavily on edge computing services. Edge computing features a short distance from the production site as well as security, reliability, and flexibility. As a result, edge computing has become a major revenue source for operators in AR application scenarios.

The basic business model for AR device suppliers is the sale of devices and auxiliary software and services. There are two product and service provisioning modes. One is to pursue hardware, software, and service sales. The other is to develop customized solutions integrated with other products. AR device suppliers need in-depth knowledge of 5G technologies and challenges facing the traditional manufacturing industry. Industry-oriented customization brings high added values at the cost of a narrowed customer range and decreased number of customers capable of paying high prices.

5.3.2.5 Typical Cases

Haier Group has tried many 5G advanced manufacturing applications, including AR-based remote fault diagnosis and training, in its factories based on the COSMOPlat industrial IoT platform. In AR-based remote fault diagnosis, production line maintenance personnel wear AR glasses, the cameras of which record onsite audio and video. These materials are then transmitted to experts through the 5G network for remote fault diagnosis. The experts perform operations such as AR annotation and frozen-screen annotation and return guidance to maintenance personnel in real time, accelerating onsite device fault rectification.

Haier also has AR-based training on complex assembly tasks. Traditional training for complex assembly, such as front-loading washing machine assembly, usually takes five to seven days. Worse still, a lack of experienced employees to be trainers means there are too many trainees and potentially low training quality. With AR-based training, one or two AR glasses are placed on each production line. The AR glasses connect to the 5G base station through the built-in 5G modules. Because computing power is deployed locally, on the MEC side, AR glasses are lighter. In addition, they can obtain real-time background data such as assembly process models, steps, and annotations from the 5G MEC server in the data center. Trainees can receive step-by-step one-to-one guidance, improving training efficiency.

Preliminary calculations indicate that the new 5G+ AR solution results in significant production efficiency improvements in Haier's factories, helping reduce device assembly time by 25% and maintenance and service costs by 40%.

5.3.3 AGV Applications in Factory Logistics

5.3.3.1 Service Requirements

As warehouse turnover and order numbers increase, the disadvantages of manual forklifts are evident: high costs and a high rate of damaged goods. With the development of industrial informatization, automated guided vehicles (AGVs) in warehouses as well as automated mobile robots (AMRs) have become major players in automatic logistics. AGVs in warehouses not only enable autonomous goods transfer, but also change how goods are picked. AGVs can be integrated with traditional transfer and picking systems to satisfy refined sorting requirements within smaller spaces. AMRs equipped with simultaneous localization and mapping (SLAM) technology can receive data from both lidars and vision sensors and can move automatically without physical directors or markings. AMRs can be deployed flexibly and are extensively used in warehouses and factories. With rising service volume come greater types and numbers of AGV and AMR systems, resulting in soaring traffic volume between devices, control systems, and O&M platforms.

Traditional logistics services in factories face various significant obstacles. Wi-Fi signals are easily blocked by doors, walls, and racks, causing packet loss. Another key issue is frequent handovers between Wi-Fi access points (Aps). The combination of these two problems causes AGVs to pause for 5–10 seconds, and in some cases even interrupts tasks—requiring an employee to come and perform a manual restart. With 5G, such pauses and interruptions, especially when AGVs move between workshop and warehouse, are no longer are a problem. 5G provides stable, reliable connections and avoids the cost of independent Wi-Fi network deployment. It easily provides the bandwidth, reliability, and latency required for AGVs and AMRs. In addition, 5G network slicing technology allows for secure isolation of private and public network services and enables user-defined resource allocation. With 5G, AGVs can operate with 5 mm precision, placing goods exactly where needed.

In factories workshops occupy a large space and are far away from each other. The production process involves many types of heavy materials and finished products, with cross-floor operation quite common. This increases the difficulty and intensity of warehousing operations. Warehousing, logistics, and production systems must be upgraded to integrate more intelligence during factories' digital transformation.

With 5G-capable AGVs, automatic warehousing and logistics systems greatly improve production quality and efficiency. Workshop operation capabilities are upgraded through almost unmanned warehouse transportation, electronic inbound and outbound tickets, visualized warehouse management, and intelligent material entry and exit Unmanned operations decrease labor costs in the following ways:

- Accurate warehouse data is displayed in real time, with financial risks reduced through intelligent warning of losses caused by obsolete inventory.
- Instant sharing of warehousing data and intelligent shipment reminders help achieve shipping and logistics coordination, improving service quality.
- Automatic collection of finished goods information and one-click lot number-based batch tracking enable quick management.
- Data can be exchanged between PCs, tablets, mobile phones, and large screens, with real-time intelligent analysis of warehouse materials enabling better management and decision-making.
- Industrial tourism is possible in smart manufacturing, changing the focus of brand vitality from low costs to advanced products.

5G, featuring superior mobility, eliminate pauses, system congestion, and even breakdowns of entire production and logistics systems caused by packet loss during AGV roaming. Fig. 5.4 illustrates the technical architecture of 5G-assisted smart AGV dispatching.

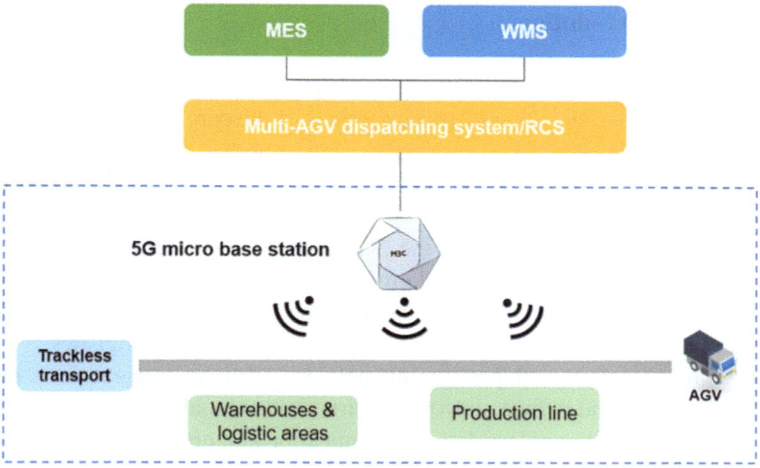

Fig. 5.4 Technical architecture of 5G-assisted smart AGV dispatching

5G has the following advantages in AGV dispatching:

- 5G can alleviate packet loss caused by AGV roaming thanks to its low latency, mobile roaming, and handover features.
- By using a licensed spectrum, 5G eliminates mutual interference caused by unlicensed spectrums in traditional AGV wireless communications.
- Xn interfaces between 5G base stations ensure service continuity during inter-base station AGV handovers.
- 5G works with the MEC to streamline AGV operations.
- With high speeds and enhanced collaboration capabilities, 5G enables AGVs not only to move faster and smoother, but to develop their own self-organizing and collaboration capabilities.
- 5G, with high reliability and ultra-low latency features, enables AGVs to be aware of workers' actions in real time, flexibly responding to and cooperating with them—for example, keeping a safe distance.
- 5G strengthens remote, real-time AGV control. Workers in a monitoring center can instruct remote AGVs to perform operations in certain production environments not suitable for humans, such as those with a high temperature or voltage.
- 5G implements real-time data collection and analysis to assist digital twins, by allowing an enormous number of production devices and key components in factories to interconnect with each other for instant production data collection. This also streamlines production processes and facilitates energy consumption management. Last but not least, production logs generated by 5G-enabled intelligent AGV systems can reflect, in real time, operational statuses such as the current movement track and docking status.

5.3.3.2 Network Solutions

Since AGV services require low latency and high reliability, it is recommended that AGVs, AGV servers, and the user-plane gateway of the core network be deployed at a local data center so that they are closer to users. Another important step, to ensure high reliability, is redundancy backup for the access network, bearer network, and core network.

The network solution format is: the 5G network covers terminals in the entire area and connects to the positioning platform, enabling personnel and material information to be displayed and managed based on basic location information. Fig. 5.5 illustrates the network solution for AGV applications in factory logistics.

The communication and positioning networks can be integrated to improve network efficiency and reduce network construction and maintenance costs. One 5G network fulfilling both wireless data communication and precise positioning service requirements can effectively reduce customers' total investment. Mobile devices equipped with 5G communications units can be positioned based on the 5G wireless positioning technology.

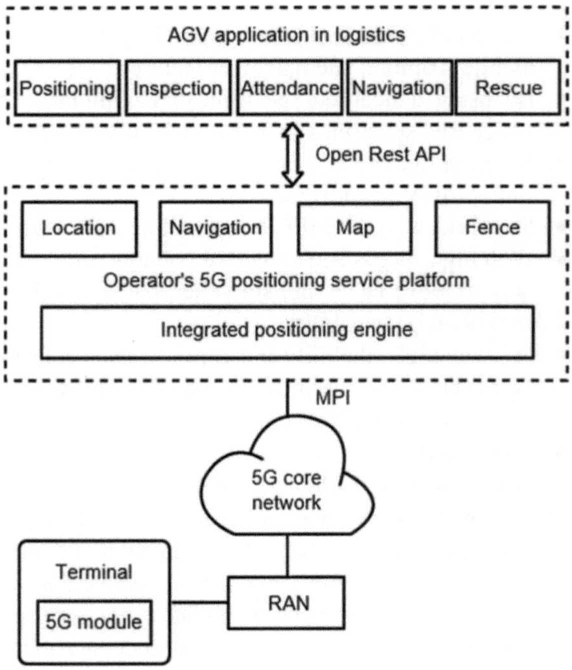

Fig. 5.5 Network solution for AGV application in factory logistics

5.3.3.3 Industry Ecosystems

In the manufacturing industry, networks and spectrums required by AGV applications in factory logistics are provided by operators, communications equipment and E2E 5G solutions come from 5G solution suppliers, and AGV application solutions are integrated by terminal suppliers. To be specific, telecom operators and 5G solution suppliers provide ICT products and services, including networks, clouds, and edge equipment rooms. Terminal suppliers offer basic hardware, including AGVs, control software, and operating systems.

Operators act as network suppliers to provide connection services and collaborate with 5G solution suppliers on private network integration. Precise AGV control in factories is achieved through a combination of 5G, customer-premises equipment (CPE), heterogeneous networks, and precise positioning.

AGVs are made up of the servo system, control system, and gear reducer. Quick response and drive precision are key to the servo system. SLAM and visual composite navigation technologies of the control system are controlled by enterprises outside China. China's AGVs are extremely homogenous, with manufacturers mainly focusing on low-profit vehicle manufacturing. To dominate the market, a supplier must have independent software R&D capabilities. Currently, AGV suppliers can develop their solutions in two major directions. One is to develop integrated, universal solutions through in-depth cooperation with large enterprises or by leveraging shareholders' strength. The other is to focus on specific scenario requirements.

5.3.3.4 Business Models

The demand by AGV applications for 5G features is growing. These features include stable communications during AGV roaming, visual identification during goods loading/unloading, automatic obstacle avoidance, and remote monitoring. Once these demands are satisfied, AGVs will be smarter when performing transport tasks.

There are four profit models available for AGV manufacturers. (1) They can provide reconstruction and upgrade services for existing customers in specific application scenarios. (2) They can take mature, flexible smart 5G+ AGV use cases and apply them to meet the needs of new customers, fully leveraging their advantages in R&D and innovation to increase project competitiveness and the price premium they charge. (3) They can build on the high reliability and low latency features of 5G to develop cloud-based AGV dispatching systems and controllers, cutting AGV manufacturing costs. (4) They can send out professional engineers to train onsite colleagues or customers' maintenance engineers, making AGV maintenance more efficient and less expensive.

For manufacturing enterprises, large-scale use of 5G-connected AGVs will raise overall production efficiency, reduce workloads, improve working environments, and cut production risks. AGVs can reduce employee costs by taking over manual work, ensure staff safety, and avoid work-related accidents and losses caused by negligence or fatigue. In addition, it is safer to use AGVs in dangerous working conditions, where they also provide more accurate and stable performance than employees.

The growth of AGVs in manufacturing is creating higher demands for network capabilities. Operators will provide major capabilities such as virtual private networks and edge computing. Technology suppliers can offer chip modules, core components, network devices, network tests, terminal design, and related services.

5.3.3.5 Typical Cases

Haier Group has implemented a host of 5G advanced manufacturing practices, including 5G+ AGV, in its Ririshun logistics industrial park in Qingdao. Unlike traditional wireless networks, 5G avoids network disconnections and makes AGV systems more stable with its advantages in reliability and mobility management. In addition, 5G and MEC make the AGV system smarter. Using traditional wireless networks, the AGV trunking dispatching system has several flaws, such as incomplete 5G terminal surveillance information, shortage of video and image data, and non-optimal decision-making and planning. By contrast, 5G and MEC adopt AI algorithms such as deep learning, neural networks, and decision trees to implement 5G terminal management and control as well as make dynamic, instant, and optimal plans in unstructured environments. Another benefit of 5G is production line automation. Visual and wireless technologies are used to achieve converged positioning and obstacle determination in workshops and campuses. Location and movement information is then uploaded through low-latency 5G networks to implement automatic obstacle avoidance and precise material delivery.

The benefits of 5G have made Haier's introduction of 5G+ AGV a great success: production network stability improved by 10% and production efficiency by 27%, accelerating the digital, networked, and intelligent transformation of Ririshun logistics.

5.3.4 Policies and Standards

Industrial use is a key focus of 5G applications. A key form of new infrastructure, the industrial Internet promotes enterprises to shift from closed to open innovation, speeding up innovation and development in manufacturing. China's industrial Internet has entered a critical development period. To stimulate its development, relevant departments have launched a series of policies in recent years to strengthen support for and the regional spread of industrial Internet applications. Table 5.1 lists the recently released policies which aim to foster industrial Internet innovation and development.

In 2022, under the guidance of the *"Set Sail" Action Plan for 5G Applications (2021–2023)* released jointly by 10 departments including the MIIT, industry players will develop more 5G+ industrial Internet typical application scenarios. On the one

Table 5.1 Policies related to smart manufacturing

Release Time	Policy Name	Contents
December 2021	14th Five-Year Smart Manufacturing Development Plan	Boost the in-depth integration of next-generation information technologies and advanced manufacturing technologies; focus on improving innovation, supply, support capabilities, and application levels; accelerate the construction of a smart manufacturing development ecosystem; promote the digital transformation and intelligent upgrade of the manufacturing industry; accelerate the scale deployment of new network infrastructure such as industrial Internet, IoT, 5G, and gigabit optical network
January 2021	Industrial Internet Innovation and Development Action Plan (2021–2023)	Promote "5G+ Industrial Internet"; encourage industrial enterprises to build fully-connected 5G factories; promote the penetration of 5G applications from peripheral links to core production links; speed up the spread of typical scenarios
September 2020	Action Plan for Digital Transformation of the Building Materials Industry Towards Smart Manufacturing (2021–2023)	Encourage enterprises to actively explore "5G+ Industrial Internet" and promote the in-depth integration of industrial Internet and the building materials industry
October 2018	National Intelligent Manufacturing Standards System Construction Guide (2018)	Formulate more than 300 smart manufacturing standards by 2019; cover both basic common standards and key technical standards; gradually put in place a relatively complete smart manufacturing standard system

hand, they will enlarge the application scope of 5G+ industrial Internet, gradually advance 5G and gigabit optical network construction, explore typical production line-level and workshop-level application scenarios, and promote a fully-connected 5G industry. On the other, they will introduce more digital technologies, implement digital transformation initiatives in the manufacturing industry, promote the application of industrial Internet platforms in campuses and enterprises, and cultivate a number of system solution suppliers.

5.4 Summary and Prospects

5G, the next-generation mobile communication technology, provides diverse and high-quality communication assurance for smart manufacturing and production systems, promoting the convergence of massive information in each link. As the convergence of 5G and manufacturing grows deeper, 5G will not only optimize the production processes (for example, improved controllability, boosted operation efficiency, and reduced production costs and energy consumption), but also bring into the industry, and popularize revolutionary products, technologies, and models. In the future, intelligent upgrades of the manufacturing industry will be all-embracing and far-reaching. 5G-centered convergence and innovation will become a powerful driving force and strong support for high-quality development of China's manufacturing industry.

Chapter 6
5G+ Smart Ports

6.1 Overview

6.1.1 Major Processes and Operations

6.1.1.1 Definition of Port Industry

As transportation hubs located alongside bodies of water, ports feature water-land intermodal transportation equipment and facilitate the safe transportation and berthing of ships. Ports serve a myriad of different roles. They are the connecting links between land and water transportation, distribution centers for industrial and agricultural products as well as imported and exported materials, places for the loading and unloading of cargo and berthing of ships, and lading and drop-off points for passengers. Naturally, ports play a decisive role in promoting international trade and regional development.

About 90% of global trade is carried on ships, which depend on ports. Ports are therefore a key economic barometer and lifeline of the modern economy. As economic globalization has accelerated, necessitating the emergence of modern logistics, port capabilities have been converted and upgraded, depending on specific historical and regional characteristics. From 1992 on, the United Nations Conference on Trade and Development (UNCTAD) put forward the concept of first-, second-, third- and fourth-generation ports. Table 6.1 lists the key characteristics of each generation. As implied by the inter-generational progression, ports have transformed from inter-regional commodity circulation centers into commercial and trade centers, and will eventually become all-in-one logistics, trade, industry, and finance units, encompassing a diverse range of services, including cargo warehousing, transportation trade information, and cargo distribution. Since the start of the twenty-first century, ports have become nimble and all-encompassing service centers that harness the power of information and network effects [1].

P. Sun, *A Guidebook for 5GtoB and 6G Vision for Deep Convergence*,
Management for Professionals, https://doi.org/10.1007/978-981-99-4024-0_6

Table 6.1 Inter-generational and essential characteristics of port development

Generation	Economic and social background	Major function	Essential characteristics	Port type
First-generation ports (before the eighteenth century)	Initial wave of globalization	Cargo loading, unloading, and warehousing	Water-water (river and sea) transshipment, water-land transshipment	Hinterland ports
Second-generation ports (early eighteenth century to mid-twentieth century)	Early industrial revolution	Serving nearby industries	Maritime transportation of raw materials and products	Large specialized or cargo owner's wharf
Third-generation ports (1950s and 1960s)	Development of modern logistics and information technologies	Integration with modern services (mainly with logistics)	Service industries such as modern service	Regional service-oriented logistics port
Fourth-generation ports (Late twentieth century to present)	Emergence of container transportation networks and competing logistics chains	Liner transportation centered around the major global maritime routes	Port shipping alliances with upstream and downstream business relationships	Non-territorial or chain wharf

6.1.1.2 Processes and Operations

Port operating processes are closely related, highly complex, labor-intensive, and dependent on human–machine interactions. Traditionally, containers were loaded and unloaded through an operating flow, which could include: ships, quay cranes, container trucks, gantry cranes, horizontal transportation equipment, and container yard operating systems. Figure 6.1 illustrates the port operating flow. It can be divided into six major phases: cargo ship entry/departure (shipping containers to ports), cargo loading/unloading via quay cranes (loading/unloading containers and moving them to horizontal transportation areas), transportation via container trucks (moving containers from quay crane areas to container yard areas), optimized container yard management (stacking containers), container truck entry/departure (transporting containers into/away from ports), and land-port intermodal transportation (conveying cargo via ports and other transportation systems). The loading and unloading equipment tends to vary depending on the phase of operations [2].

1. Ship Loading and Unloading Machinery

   Ship loading and unloading machinery generally refers to quay cranes. Quay cranes are perched on container wharves, and tasked with hoisting and lowering containers to and from container ships. Quay cranes can move along fixed tracks to a number of different destinations. Trolleys are equipped with dedicated lifting appliances, for the purpose of loading and unloading containers in different bays

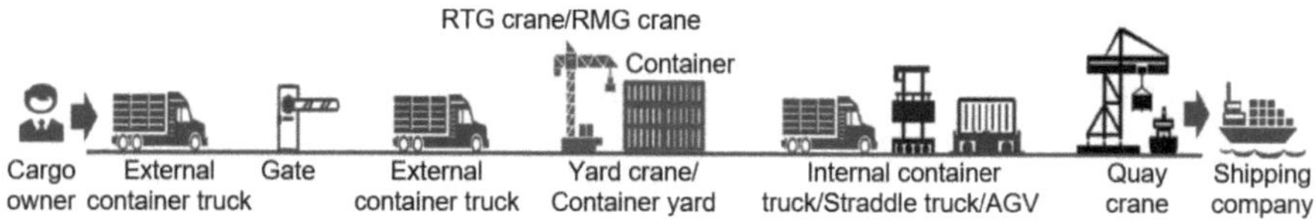

Fig. 6.1 Port operating flow

on ships. The number of quay cranes can vary depending on the port's container throughput and the loading capacity of ships.

2. Horizontal Transportation Machinery

   Horizontal transportation machinery, as the name suggests, is responsible for horizontally transporting containers from railway loading/unloading lines to the container yards or the quayside, from container yards to the quayside, or within container yards. Container trucks are the most common form of horizontal transportation machinery. In highly automated container wharves, automated guided vehicles (AGVs) are used for horizontal transportation, and mature AGV technologies facilitate unmanned container transportation, thereby substantially reducing labor costs. In addition, container chassis trailers, straddle trucks, and reach stackers can also be used in the corresponding loading/unloading systems.

3. Container Yard Machinery

   Container yard machinery handles the containers inside container yards. In China, rail-mounted gantry (RMG) cranes and electric rubber-tired gantry (E-RTG) cranes are mainly used in new dedicated container wharves. RMG cranes are special machines dedicated to the loading, unloading, moving, and stacking of containers at container yards within container wharves and container transfer stations. RMG cranes, which are powered by electricity, have become the preferred choice for new wharves, due to their eco-friendly, energy-efficient, cost-effective, and highly-reliable nature. As lifting equipment limited to container wharves and driven by electricity, E-RTG cranes move vertically and horizontally to turn over or load/unload containers. They feature tires and operate on flat ground within container yards at a wide range of different container wharves, but most often in large container wharves.

   Straddle trucks and reach stackers are used for container yard operations in the corresponding loading/unloading systems as well. Container forklift trucks and empty container handlers are also commonly used in container yards.

4. Train Loading and Unloading Machinery

   Train loading and unloading machinery is used for loading/unloading container trains. RMG cranes and reach stackers serve as the basic and auxiliary machines on railway loading/unloading lines, respectively.

   Ports face many of the same challenges, regardless of where they are located: increasing labor costs, labor-intensive operations, harsh working environments, and insufficient manpower. Automation has become a common strategy for reducing costs and boosting efficiency, and digital technologies like artificial intelligence (AI), big data, the Internet of Things (IoT), 5G, and autonomous driving have facilitated the process. Container wharves apply higher levels of automated equipment to boost productivity and maintain their competitive edge.

6.1.2 Current State

6.1.2.1 Accelerated High-Quality Port Development in China

The Belt and Road Initiative has helped made Chinese ports major players in global industry, and the government's "dual circulation" development strategy has brought new opportunities for high-quality port development. A new emphasis on transportation and the construction of free trade ports has led to a transformation in how ports are used and administered. Rapidly-developing technologies, ranging from the Internet, big data, and cloud computing, to blockchain, harness the massive amount of cargo trade data, allowing for in-depth integration between ports and the larger economy and society, as well as across the board improvements to port services. Most of China's coastal ports have evolved into third-generation ports, and some ports have even made the leap to become fourth-generation ports. Fourth-generation ports are regional or global port service networks in their own right, with the hub port serving as the core and assisted by a number of international wharf telecom operators. These sophisticated networks are a sign of high-quality development in this sector.

According to China's Ministry of Transport (MOT), the cargo throughput of Chinese ports reached 14.55 billion tons in 2020, and the container throughput totaled 260 million twenty-foot equivalent units (TEUs), both ranking first among all countries. In 2020, the volume of freight carried along China's inland waterways reached a staggering 3.815 billion tons. As of the end of 2020, China had more than 120,000 km of navigable inland waterways, the most extensive system of any country in the world.

According to Shanghai International Shipping Institute, China had 15 of the world's top 20 ports in terms of cargo throughput, and the Port of Ningbo Zhoushan maintained its status as the world's busiest port for the 12th straight year, with cargo throughput exceeding 1.17 billion tons in 2020. The Port of Shanghai continued to develop, ranking second despite two years of stagnating throughput. At the Port of Tangshan, the throughputs of containers and bulk cargo, such as coal, iron ore, and grains, continued to grow at a rapid pace. In 2021, the Port of Ningbo Zhoushan reached the annual cargo throughput of 1.224 billion tons, with a year-on-year increase of 4.4%, ranking first in the world for 13 consecutive years. The Shandong Port Group emphasized synergy and resource integration, with the Ports of Qingdao, Rizhao, and Yantai all experiencing steady growth. The Port of Tianjin achieved slight throughput growth while assuming its status as a first-class port. The Beibu Gulf Port benefited from a new western land-sea corridor to reach its potential, with rapidly-growing throughput enabling it to rank 19th.

6.1.2.2 Drivers of Change: Labor Shortages and Cost-Related Pressures

Ports have long been a labor-intensive industry, but a myriad of factors, including career choice changes among younger workers and the rising cost of living in port cities, have made recruitment harder than ever, and labor increasingly expensive.

A notable shortage is in the supply of internal container truck drivers in Chinese ports. According to the China Road Transport Associations (CRTA), most of the internal container truck drivers in Chinese ports are men aged 40 to 50, and there's a 20% shortfall in the number of required drivers. Intensive and oftentimes dangerous working conditions have dissuaded many young people in their twenties from pursuing this career. As the today's middle-aged drivers retire in the years to come, this shortfall will become even more obvious.

Fatigue resulting from intensified work affects the efficiency of logistics transportation. According to the *Big Data-based Analysis Report on China's Highway Freight Transport Industry (2020)*, container truck drivers in China often work irregular hours, and face high levels of stress and poor working environments. The average work week is approximately 49 h, which is higher than the average level in the Chinese labor force.

Nonetheless, the container throughput of Chinese ports continues to grow, resulting in insufficient transportation capacity on many wharves. According to MOT's data, the container throughput of Chinese ports increased every year between 2014 and 2020, reaching 260 million TEUs in 2020, an increase of 1.2% over the previous year. Among them, coastal ports completed 230 million TEUs, up by 1.5%. The Ports of Shanghai, Zhoushan, Shenzhen, Guangzhou, and Qingdao were the top 5 ports in China in terms of container throughput, with different degrees of increases. The annual throughput of the Port of Shanghai was 43.5 million TEUs, with a year-on-year increase of 0.4%, ranking first in China. The annual throughputs of the Ports of Zhoushan and Shenzhen were 28.72 million and 26.55 million TEUs, respectively; their year-on-year growth rates were 4.3% and 3%, ranking second and third. Due to limited manpower and infrastructure, wharves have been hindered by insufficient transportation capacity. *Global Container Terminal Operators Annual Review and Forecast* released by Drewry projects that the handling capabilities of global container ports will grow at an average annual rate of 2.5% over the next five years, but that global demand will grow by an average of 5% over the same period. The average utilization rate of wharf capacity is expected to increase from the current 67% to over 75%, which means that the handling capacity shortage of global container ports will persist well into the future. Inadequate port infrastructure and handling capabilities have become major challenges for the international shipping industry, and therefore reducing costs while boosting efficiency has become a primary goal for ports.

6.1.2.3 Targeted Port Operating Requirements

As enclosed areas that feature an abundance of different items, such as vehicles, cargo, and vessels, and no shortage of people, ports are energy and labor-intensive. The operating status of each wharf needs to be reported on a monthly, quarterly, or annual basis. In addition, mid- and senior-level personnel require accurate and multi-faceted real-time operating data, so that they can make informed and time-sensitive decisions about future operations.

Targeted management of wharf operations, as well as labor and energy allocations, requires that engineers have access to data about personnel, equipment, and energy consumption for the entire port at large, as well as the individual wharves. Data about how resources are consumed should even reach down to the level of individual vessels, quay cranes, and containers, in order to ensure optimal operating efficiency.

6.1.2.4 Multi-Network and Multi-RAT Wireless Communication Systems for Ports

Mainly consists of 2.4 GHz and 5.8 GHz Wi-Fi networks and digital trunking systems within 1000 Mbit/s.

- 2.4 GHz Wi-Fi: Operation data upload networks used by port equipment to upload sensor information, such as operating status, fault, and wind speed. The information is transmitted to the central control room and displayed via 3D animations.
- 5.8 GHz Wi-Fi: Camera data upload networks used for video surveillance related to tallying and operators' cabs in quay cranes.
- 1.4 GHz private networks: Instruction delivery networks, in which the tally clerk receives the tallying videos and delivers instructions to trailers after the assessment.
- 400 MHz private network: Voice trunking communication networks used for voice dispatching in all areas within the port.

With the rapid development and wide use of radio technologies, a number of industries have grown dependent on spectrum resources. Golden frequency band resources have been almost exhausted, and the conflict between spectrum resource supply and demand is now an urgent issue. The 400 MHz narrowband data transmission network uses closed technologies, and has been gradually phased out. 2.4 GHz/5.8 GHz Wi-Fi, due to its technical attributes, uses unlicensed frequency bands, resulting in poor coverage, poor interference resistance, limited services, and high maintenance costs. Carrier-grade reliability standards are difficult to be met. Private networks can meet most wireless data transmission requirements for smart port areas. However, due to bandwidth limitations, the high-bandwidth and low-latency requirements across a wide range of scenarios, such as those involving unmanned container trucks and remote control of mechanical equipment, cannot currently be met at smart ports.

6.2 Digitalization Trends and Challenges

With the application of innovative technologies and deepening of port informatization, automation, and intelligence, the port industry has entered a newly digital era characterized by a rapid pace of change. As ports continue to evolve from single logistics nodes to catalysts of economic development, trade, and financial activities, their operations increasingly affect the regional economy and even hinterland activities. Competition between ports has gradually shifted from how much capacity they can handle to how innovative they are, with the integration between logistics, shipping, and peripheral resources. More attention is being paid to developing collaborative innovation, comprehensive services, and ecosystems. In recent years, many major global ports have made strides toward becoming technology- and knowledge-intensive smart ports while centering on digitalization.

6.2.1 Race to Digitize Among Global Ports

In 2011, the Port of Rotterdam proposed a strategic vision of port development for 2030, from 10 key factors: port distribution and transportation, logistics, investment and financing environment, space, eco-friendliness, port-industry-city convergence, human environment, innovative capabilities, policy environment, and the regional economy, and outlined the technological roadmap for smart ports based on global trends. In recent years, this port has strengthened its digital technology applications and port ecosystem, making substantial progress in the following areas in particular: operating efficiency, distribution and transportation, value chain service innovation, international trade facilitation, port-city convergence, and sustainable development.

Likewise in 2015, the Port of Singapore proposed its Next Generation Port 2030 (NGP 2030) initiative to cope with global competition, utilize next-generation technologies to improve port efficiency and land utilization, and ensure safe operations. The initiative focused on smart port construction and applied advanced technologies to the Tuas Mega Port, including automated wharves, a smart ship traffic management system, and digital port communities. It also emphasized the use of clean energy, ecological protection of port waters, and coordinated development of port and city, for the purposes of sustainability and efficiency. In June 2020, the Maritime and Port Authority of Singapore (MPA) launched the *Maritime Digitalisation Playbook*, which aims to help maritime companies accelerate their digital capabilities.

6.2.2 Steady Progress in Chinese Ports

To promote the construction and development of smart ports and implement the *13th Five-Year Plan for Information-based Transportation Development*, the MOT issued the *Smart Port Demonstration Project Implementation Notice* in 2017, specifying that a first batch of ports would be selected to carry out smart port demonstration projects, which use informatization to implement smart logistics and dangerous cargo safety management. In the same year, the MOT released a list of 13 smart port demonstration projects. Currently, all smart demonstration projects are being carried out in an orderly manner, and have achieved fruitful results. With the application and promotion of 5G technologies, China's smart ports have entered the 5G era. For example, in July 2019, Xiamen Ocean Gate Container Terminal started 5G application construction for smart ports. By the end of 2019, its port area has achieved full 5G network coverage. In the first half of 2020, 5G+ gantry crane remote control has been applied on a large scale in the Port of Meishan, and 5G+ unmanned container truck applications have been successfully tested.

6.2.2.1 Fully Automated Wharf Construction

The Port of Shanghai is the leader in promoting smart ports in China and even the world. From wharf automation to paperless services, it has always been at the forefront of the industry. The Port of Shanghai, based on the "super brain", Intelligent Terminal Operating System (ITOS), controls the full-scenario and full-process operating of automated container wharves, such as loading/unloading, stacking, transshipment, and entrances and exits. The port has upgraded its intelligent operating system for ultra-large automated container wharves, a first in China, to ensure entirely independent intellectual property rights, which boosts operating efficiency. The average efficiency per quay crane increases by more than 10%. The annual throughput of a single wharf improves by more than 50%. The per capita productivity is promoted to 213% of that of traditional wharves. In 2021, Shanghai International Port (Group) Co., Ltd. (SIPG) and Huawei jointly applied the Fifth Generation Fixed Network (F5G) technology to ultra-remote port control scenarios for the first time in global ports. In this way, bridge crane operators in Shanghai offices can ensure efficient operating of automated wharves hundreds of kilometers away.

On May 11, 2017, Asia's first fully automated container wharf went into operation in the Port of Qingdao. At present, the average crane efficiency reaches 36.2 move/h, and the maximum efficiency reaches 47.6 move/h, surpassing those of manually operated wharves. The operating efficiency is over 50% higher than that of similar automated wharves outside China. At the Port of Qingdao, phase-2 project uses the world's first automated hydrogen-powered RMG crane, which features light weight, low energy consumption, environment protection, and safe and efficient operations. It also integrates the world's first 5G+ automation technology to achieve full coverage of wharves, automatic control of quay cranes and RMG cranes, and big data upload of HD videos via 5G networks.

6.2.2.2 Collaborative Port and Land Transportation Services

The Port of Xiamen debuts the first container smart logistics platform at ports in China. The platform collects and shares real-time logistics information from booking to wharf gate entrance times for containers and from inward manifest processing to lading and container return, by centering on electronic Equipment Interchange Receipts (EIRs) and breaking the barrier after integrating information resources from six major service parties, including wharves, shipping companies, shipping agencies, customers, container yards, and logistics companies. When a container trailer passes a wharf gate, images are automatically collected and recognized, cargo damage is verified, and trailer passing is automatically allowed. The entire process is unattended, shortening the passing time from 90 to 26 s with a 20% higher logistics efficiency [3].

The Port of Shanghai continuously explores smart port practices and introduces new technologies to improve service quality. First, a unified service platform is built among ports, shipping companies, cargo owners, agencies, and departments of port of entry, provisioning online customer services and reducing logistics costs and times. Second, a one-stop query website is set up. This integrates data of seven container wharves in the Port of Shanghai, eight wharves of SIPG in branches of the Yangtze River, and two wharves in the branches of inland waterways, and better coordinates logistics resources of ships, ports, cargo, and containers in the Yangtze River Economic Belt. Third, an electronic container truck service platform is established to match container trucks and cargo transportation demand, substantially reducing driving of empty container trucks and cargo retention caused by information asymmetry, balancing wharf operation intensity, and improving port logistics efficiency. In 2020, the blockchain technology was introduced to the smart container booking system of the Port of Shanghai to improve and reconstruct existing functions of the platform, such as information release, online payment, and container yard reservation.

With port services as the core, the Port of Dalian integrates multiple intermodal transportation information resources, builds an inland comprehensive distribution and transportation system, promotes operation collaboration and information sharing between upstream and downstream logistics nodes, and effectively supports railway-water intermodal transportation services of containers. A cross-service smart logistics platform is built to implement a new mode of port-centric and one-stop comprehensive information services.

6.2.2.3 Combination of Multiple New Technologies

Ningbo Meidong Container Terminal Co., Ltd.'s smart container trucks have been officially deployed. The wharf is the rare container wharf in China that applies automatic and large-scale operations via a "remote loading and unloading control+ smart container trucks" combination. With no need to thoroughly reconstruct traditional port facilities, smart container trucks can fully adapt to actual operation scenarios, covering 176 working conditions in main scenarios for loading, unloading, and moving of various types of containers. In addition, the autonomous driving system accurately identifies obstacles and predicts behaviors to implement autonomous avoidance, overtaking, and bypassing. The container trucks can run smoothly in complex operation scenarios of port areas, creating a brand-new mode for smart transformation of traditional ports.

Tianjin Port Group independently designs and develops an AI-based smart horizontal transportation management system to collaborate with key resources, such as Terminal Operating System (TOS), yard cranes, quay cranes, Artificial Intelligence Robots of Transportation (ARTs), automatic lock stations, and automatic charging piles. The ITOS for next-generation ports achieves optimal global dispatch, improving the operating efficiency by more than 20%. Besides, already cutting-edge unmanned container trucks are further enhanced, and self-developed ARTs are deployed on a large scale. This group has taken the lead in breaking through the L4 unmanned driving bottleneck at ports. In addition, the first-ever "5G+ BeiDou" converged commercial solution for global ports has been launched, featuring all-weather and all-purpose operations. This has facilitated the development of private networks for industrial communication and centimeter-level positioning systems, laying the groundwork for the ubiquitous deployment of smart port applications, and construction of virtual digital twin wharves. Moreover, this port also builds the world's first zero-carbon wharf, with facilities and equipment autonomously powered by the wind-photovoltaic green power system.

6.2.3 Trends of Technologies Related to 5G Application Scenarios

As ships become larger in size and throughputs grow at ports, automated container wharves become an inevitable future trend. In recent years, automated container wharves of the Port of Xiamen, Port of Qingdao, and Port of Shanghai have been put into use.

6.2.3.1 All-Sensing

Ports are chock full of infrastructure, which ranges from underwater equipment to land-based equipment. IoT technologies are used to sense and collect information about each operation link, facilities, equipment, and cargo at ports. It is the basic means for port digitalization and the prerequisite for smart ports. During container information collection, tracking and monitoring, and supply chain management, a visual recognition system is used to read container IDs. The radio frequency identification (RFID) technology helps read electrical tags on containers and container trucks and collect container information. Wireless communication networks automatically collect and store it to the central information system.

In ports, hundreds of large freighters, thousands of large containers, and tens of thousands of people move every day. In such a complex environment, cameras function as major remote monitoring equipment to ensure the safety of transportation, production, and cargo. They can be deployed on vehicles to identify container IDs based on AI and automatically tally cargo. Cameras can also be used to recognize license plate numbers, faces, and cargo.

6.2.3.2 Wholly-Interconnected

The loading/unloading process of an automated wharf involves quay cranes, AGVs, and Automated Rail-Mounted Gantries (ARMGs). During ship unloading, a quay crane loads containers to designated AGVs for horizontal transportation to the container yard. Then, the ARMG places the containers. During ship loading, the ARMG loads containers to designated AGVs for horizontal transportation to a quay crane operation area. Then, the quay crane places the containers to ships. At present, over 90% of quay cranes and yard cranes are manually operated at heights. Remote control is required in this scenario. Some yard cranes (rubber-tired gantry (RTG) cranes) of new ports are deployed with optical fibers. However, optical fibers are easy to wear, and transformation and upgrade are costly and time-consuming. A few information-based ports use Wi-Fi or LTE-Unlicensed (LTE-U), featuring poor reliability, latency, and rate [4].

Remote loading/unloading is a key 5G application scenario. It fully utilizes the high bandwidth, low latency, and high reliability of 5G networks to implement remote control of quay cranes and yard cranes and HD video upload.

In addition, traditional container trucks are driven manually. Drivers repeat the same tasks over and over, making them prone to fatigue and affecting transportation efficiency and safety. Unmanned transportation is a key part and cornerstone of smart port construction. As port automation develops, AGVs/Intelligent Guided Vehicles (IGVs) and 5G unmanned container trucks are used, substantially reducing labor costs and achieving 24-h operations. 5G's high bandwidth, low latency, high reliability, and massive connectivity facilitate the operating of AGVs/IGVs/unmanned container trucks and real-time road condition upload via high-precision positioning and vehicle–road synergy. In this way, the running data of the AGVs/IGVs/unmanned container trucks can be transmitted to the background control center in real time. The control center monitors the transportation progress and container trucks' locations, posture, power, and load, and views the real-time sensing and planning information of the vehicles. If a container truck is faulty or heads to a temporary area (not in the regular route), 5G remote takeover can be performed to ensure safe transportation.

6.2.3.3 Converged Platform

Each professional wharf has its own LAN, wireless network, and related service systems. A port group consists of multiple branch companies and wharves, which are interconnected through the campus network of port group and the backbone network. Each professional wharf is deployed with a service cloud that meets its own requirements. With an office network and a cloud data center, the port group provides one-stop services for basic resource services, O&M, and application data, and integrates professional clouds of wharves and public clouds of groups to build a hybrid cloud management platform.

5G applications not only bring 5G connections and terminals, but also upload massive amounts of data to the cloud through 5G networks. The massive amounts of data become data assets and key production elements through big data and AI technologies, and accelerate circulation in each production phase, serving as an important basis for enterprise production, sales, and decision-making.

6.2.3.4 5G Converged Development Policies

China has issued a series of policies to promote the construction and development of smart ports, beginning with the *Outline for the Construction of Nation with Strong Transportation System*, which was issued in 2019. With the widespread deployment of 5G technologies, China's smart ports have entered an all-new 5G era.

In November 2019, nine departments, including the MOT, jointly issued the *Guidelines on Building World-Class Ports*, proposing the construction of a smart port system, bolstered by innovation as well as the R&D and promotion of key technologies related to port equipment, automated container wharf operating system, and remote control technologies. The information infrastructure, driven by new technologies like 5G, BeiDou, and IoT, and autonomous driving demonstrations for container trucks should be built to deepen the ties among ports.

In July 2020, ten departments, including the Ministry of Industry and Information Technology (MIIT) of the People's Republic of China, jointly released the "*Set Sail*" *Action Plan for 5G Applications (2021–2023)*, proposing the development of 5G-assisted positioning products applicable to port container environments, to accelerate deployment of automated wharves and container yards. This policy promotes end-to-end (E2E) digitalization during port construction and maintenance, accelerating the construction of smart port infrastructure, and promoting the application of 5G in scenarios related to unattended inspection, remote tower cranes, AGVs, autonomous driving of container trucks, and smart tallying.

In August 2021, the MOT issued the *Action Plan for New Infrastructure Construction in the Field of Transportation (2021–2025)*, which proposes the smart upgrading of existing container wharves at the Port of Xiamen, Port of Ningbo Zhoushan, and Port of Dalian, the construction of next-generation automated wharves at the Port of Tianjin, Port of Suzhou, and Beibu Gulf Port, transformation to remote automatic controls for bridge cranes and gantry cranes, and accelerated applications such as smart scheduling and remote equipment controls at port stations. The policy facilitates the widespread use of unmanned container trucks and AGVs for horizontal transportation. It promotes the construction of a smart logistics service platform for ports, to bolster smart monitoring and hazardous item warning. In the wake of the commercial deployment of 5G, it advances network coverage, promotes the vehicle-to-everything (V2X) and Internet of Ships (IoS), and enhances the construction of smart interconnected ports based on ubiquitous sensing and port-vehicle synergy.

6.2.4 Major Challenges

As port service volumes grow, accompanied by development considerations, port digitalization has new requirements. Automation, digitalization, and intelligence are widely regarded as key to improving competitiveness, as they make logistics more efficient and cost-effective. The COVID-19 pandemic has a profound impact on port production and operating, while further incentivizing the adoption of smart ports and advanced technologies like 5G, blockchain, edge computing, AI, and computer vision.

With the deepening of smart port construction, service coordination, management, and technical convergence need to be balanced. The lack of information sharing, inconsistent standards, and less than seamless collaboration are challenges that are still yet to be overcome. First, the sharing of port distribution and transportation information is inadequate, especially with regard to the sharing of supply chain and highway/railway information, which affects the production and organizational efficiency at ports. Second, port service models are not flexible enough either. Due to complex port systems and inadequate collaboration and information sharing, few personalized services are available, and the customer service experience could stand to be improved. In addition, the collaborative service capabilities in port logistics chain are currently inadequate. Ports have not implemented genuine online collaboration spanning services, organizations, departments, and systems. Due to the challenges associated with sharing logistics information and creating collaborative services, there is still room for developing both door-to-door and E2E visualized information services.

6.2.4.1 High Costs of Port Automation

Currently, automated ports are the newest ports. If automation is required in existing ports, the initial investment needs to be quite substantial. The investment consists of operating equipment purchasing (or infrastructure construction) and manufacturing. Infrastructure investment is mainly used for equipment positioning. For most enterprises, tens of thousands of magnetic nails are embedded in the ground of wharves, forming an invisible track, to provide location and route signals for AGVs on the ground. This physical signal transmission mode requires high construction costs. It used to be the only feasible positioning method. As 5G technologies mature, precise positioning is now possible via the wireless signals from the Global Positioning System (GPS), considerably reducing construction costs for automated wharves.

Considering how complex operating equipment is, the high manufacturing costs are quite reasonable compared with the traditional one. The cost of manufacturing automated wharf equipment will only continue to fall as technologies improve.

6.2.4.2 Inadequate Sharing of Port Distribution and Transportation Information

Many ports feature comprehensive information service platforms. However, it is difficult to share the supply chain information and railway/highway information due to the lack of unified planning. This undermines operating efficiency.

It is imperative for ports to adjust their closed operating models, and work more closely with upstream and downstream partners in the logistics value chain. Ports need to change their development strategies, by shifting their strategic focus from controlling port resources to meticulously managing port resources, from optimizing internal processes to interacting with external systems, and from increasing customer value to maximizing ecosystem value, as well as accelerate the transformation to systematic cross-border port ecosystems.

6.2.4.3 Inadequate Collaboration of the Port Logistics Chain

Smart transformation projects in ports, such as smart port gates and smart quay crane tallying systems, are driven by technology companies and promoted by grassroots wharves, lacking system-level designs. Most of the projects feature redundant construction and a lack of system cohesiveness, making it difficult to form economies of scale. Widespread online collaboration between services, organizations, departments, and systems is not currently the norm.

Port companies should make full use of ports' central position in the logistics chain to further integrate upstream and downstream resources, and to promote service collaboration. They should also accelerate the deployment of interconnected port information service platforms, harness digital technologies to remove barriers between industries in the logistics chain, promote the efficient operations of logistics, information flow, and capital flow in the port service chain, and enhance the experience of cargo owners and logistics participants. Port companies should also actively promote efficient and collaborative logistics chain services that span industries, departments, and regions to optimize resource allocation.

6.2.4.4 Lack of Mature Smart Port Solutions

Currently, China's smart port construction is promoted via demonstration projects. First, construction requirements vary with ports, which requires differing transformation strategies and solutions. The R&D, promotional policies, and collaborative operations models are not highly-coordinated. Second, new technologies are widely used in smart ports, including IoT, cloud computing, big data, 5G, BeiDou, and geographic information systems (GISs). There are many platforms for deploying these technologies, including comprehensive information service platforms, service collaboration platforms, and intermodal transportation platforms. However, technical standards still need to be unified, and reliability and stability to be tested. Lastly, telecom operators at ports all occupy different roles. Automation requires coordination among them, to balance the needs of all stakeholders.

6.3 Major 5GtoB Large-Scale Replication Scenarios and Typical Cases

6.3.1 Remote Control of Gantry Cranes

6.3.1.1 Service Requirements

When cargo arrives at a port, it is packed into containers for loading/unloading, meaning the container's throughput becomes a major indicator of the scale and benefits of modern ports. Traditional container ports tally containers through gantry cranes, and transmit multiple channels of video signals for gantry cranes via optical fiber networks. In container wharves, RMG and RTG cranes are the most widely used types of gantry cranes. RMG cranes move alongside tracks in the container yard, while RTG cranes flexibly transport containers between container yards. At present, RTG cranes are used more in existing wharves, whereas RMG cranes tend to be used in new wharves.

Traditional gantry crane operators work in cabs that are 30 m above the ground for around 10 to 12 h every day. They look down from an angle of 90 degrees to operate heavy machinery, and are prone to fatigue and safety risks in such environment. Moreover, to ensure round-the-clock operations, each gantry crane has three operators on shift, meaning one wharf usually requires hundreds of gantry crane operators. This high labor cost contradicts the development of modern ports with increasing cargo throughput. As the labor cost soars and automation technologies develop rapidly, more container wharf companies are applying remote control technologies to gantry cranes in an attempt to automate traditional wharf equipment. According to the McKinsey, the manual lading efficiency is 25 move/h to 27 move/h at traditional container wharves, and 30 move/h at wharves with higher efficiency. Automated wharves can boost efficiency by more than 30%.

A gantry crane remote control system provides remote communication, video surveillance, single-crane control, and central control subsystems. They run together to implement remote control and semi-automatic operating of a gantry crane, which is as follows: The TOS automatically sends an instruction to the central control subsystem that then decomposes the instruction. The best suitable gantry crane is automatically selected based on the instruction, and the decomposed instruction is forwarded to the single-crane subsystem of the gantry crane. This subsystem automatically converts the decomposed instruction into execution steps, which are then automatically executed. After the instruction is complete, the single-crane subsystem returns a completion message to the central control subsystem that then sends the message to the TOS to complete the workflow and waits for the next instruction to be allocated [5].

6.3.1.2 Network Solutions

Remote loading/unloading depends on the following key technologies:

1. High data rate: To get a full picture of the site, crane operators use cameras to collect video data. High bandwidth is required for ensuring smooth and real-time uplink transmission of HD videos, as well as downlink video transmission of HD VR glasses or MR glasses worn by the operators.
2. Low latency: To guarantee the real-time delivery of interactive instructions between the central control room and gantry cranes, a low-latency network is desired. In this case, the central control room controls the gantry cranes in real-time through sensors.
3. High reliability: During the loading and unloading, the system automatically and wirelessly delivers a remote control instruction to the wireless terminal screen of each machine. After completing the operation, the operator confirms the instruction on the wireless terminal. In this case, a control signal fault will cause a series of misoperations, leading to accidents.

The network deployment between gantry cranes and the central control room is fast and convenient. Though a wired network guarantees latency and bandwidth, the movement of gantry cranes is limited and fast network deployment is not possible during the digitalization of ports.

Before 5G solutions, wireless signal transmission has always been a major issue that hinders smart ports. 4G and Wi-Fi technologies cannot meet the bandwidth and latency requirements of port production. If optical fibers are deployed on gantry cranes, smart transformation is costly and complex, and optical fibers experience wear and tear every year. The attenuation of signals over optical fibers affects the determination of the crane arm position. New technologies such as 5G and edge computing provide the best connection solution for port automation and intelligence, injecting a new impetus into smart port construction.

Ultra-reliable low-latency communication (URLLC) is a typical 5G application scenario. In June 2020, 3GPP officially launched 5G Release 16, bringing higher key performance, application capabilities, and basic network capabilities. Release 16 enhances the key performance of URLLC, which has a one-way latency of air interface that is less than 1 ms, and a reliability of 99.9999%. The specifications improve network data bearing capabilities, especially the mmWave communication capability, with extended mmWave application scenarios. It also strengthens basic network capabilities in Release 15, significantly improving self-organization, automatic operating, and meter-level positioning.

Remote operating stations flexibly control RTG cranes/gantry cranes with multiple HD cameras in *N*-to-*N* mode. The cameras collect operating status parameters of main equipment and lifting appliances of the RTG cranes/gantry cranes and send them to the central control console at the Port of Waigaoqiao over 5G networks. Remote operators determine the operations and deliver control instructions. Figures 6.2 and 6.3 illustrate the 5G remote control network solutions for ports.

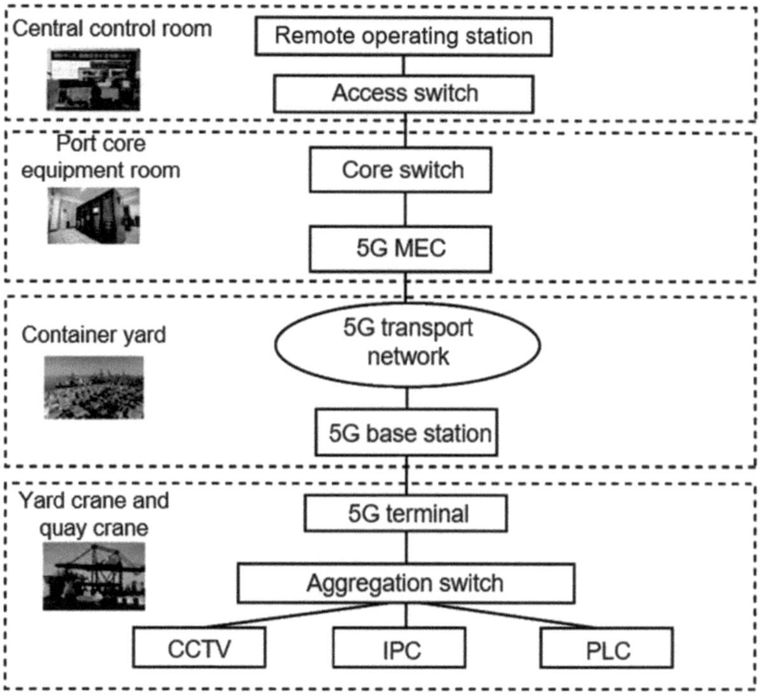

Fig. 6.2 5G remote control network solution 1 for ports

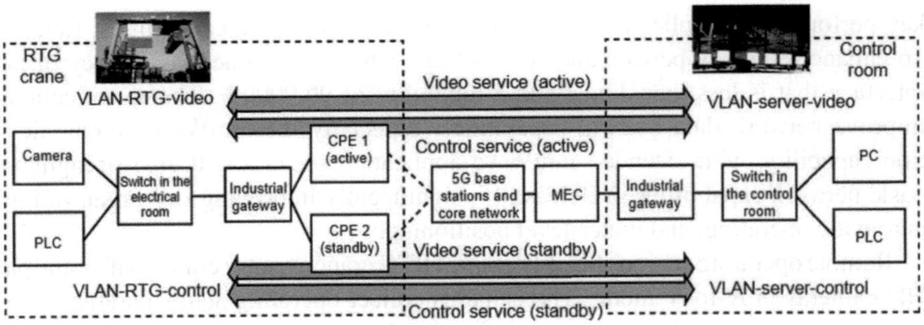

Fig. 6.3 5G remote control network solution 2 for ports

6.3.1.3 Ecosystems and Business Models

Ports connect ecosystem stakeholders and supply chains and require a core competitiveness of sustainable development. Population, sustainable development, new business and trade models, global environmental changes, and new technology development are five impetuses for port modernization. To succeed, ports must employ new technologies for ever-changing needs, realize widespread interconnection, process and utilize a large amount of data more efficiently, establish a comprehensive ecosystem, and make sustainable development a competitive edge.

In gantry crane remote control scenarios, telecom operators provide networks and spectra, 5G solution suppliers come up with 5G E2E solutions, and heavy machinery manufacturers integrate remote control solutions for terminal applications. Telecom operators and 5G solution suppliers offer ICT equipment and services, including networks, clouds, and edge data center. Heavy machinery manufacturers provide basic hardware, including gantry cranes, sensors, cameras, as well as remote control software and operating systems. Three construction modes are used in 5GtoB solutions for ports in China.

Mode 1: Mainly designed for large ports with robust information-based infrastructure, strong demand for digitalization, and port companies who are actively promoting automation. Under this mode, telecom operators work with 5G solution suppliers to integrate 5G communication solutions with applications, while heavy machinery manufacturers jointly transform and integrate the gantry crane remote control system, after which port companies integrate these solutions.

Mode 2: Mainly designed for large ports with strong information-based infrastructure. Under this mode, heavy machinery manufacturers play the leading role in automation and complete the remote control solution for gantry cranes as project integrators. Telecom operators collaborate with 5G solution suppliers to integrate 5G private networks.

Mode 3: Mainly designed for small ports where telecom operators are actively promoting digitalization. Telecom operators, as the system integrators, provide E2E integration services. 5G solution suppliers integrate 5G communication solutions with applications, while heavy machinery manufacturers jointly complete the remote control system transformation solution for gantry cranes.

Based on the industry development, solution maturity, and its understanding of the industry and technology accumulation, an enterprise selects the mode that can make full use of resources and advantages of all parties.

The large-scale deployment of 5G technologies at ports benefits all parties immensely. For supervisory departments, new technologies change the existing port production and operation modes, while making port areas safer. Furthermore, the smart system promotes efficient information exchange and improves management efficiency, enabling better safety supervision.

For ports, the overall labor cost at ports is significantly reduced, realizing revolutionary transformation from labor-intensive to automatic and smart production and operation.

For heavy machinery manufacturers, there is limited new space for ports in China and beyond. Automation brings new development opportunities, accelerating the development of heavy machinery enterprises and related subsystem providers. For example, Shanghai Zhenhua Heavy Industries Co., Ltd. has the biggest global market share of heavy machinery production, such as quay cranes and RMG cranes, at container ports.

5G supports remote control of cranes, bringing new opportunities for network, application, and terminal development. For networks, the construction of base stations is the basis. According to the latest MIIT data, China has built nearly 1.4 million 5G base stations, boasting the world's largest 5G standalone (SA) network. For applications, 5G private networks provide customized high-quality network services for industries while enabling innovation-driven, secure, and reliable 5G services of cloud-network synergy. For terminals, 5G terminals are becoming particularly diversified thanks to emerging 5G industry applications. The industry chain can provide products and services such as chips, modules, core components, and terminal design, which are charged in multiple ways, such as one-off and annual service fees.

6.3.1.4 Industry Standards

In June 2018, the China Ports and Harbors Association (CPHA) officially released the association standards of *Technical Requirements for Quayside Container Crane Remote Control System (T/CPHA 1–2018)* and *Technical Requirements for Remote Control System of Rail Mounted Transtainer & Rubber Tired Gantry Crane (T/CPHA 2–2018).*

The main content of *Technical Requirements for Quayside Container Crane Remote Control System (T/CPHA 1–2018)* is as follows: First, it specifies the composition, technical requirements, and test methods of remote control systems for quay cranes. This standard applies to the design, manufacturing, transformation, and use of these systems. Second, the remote control system consists of the video surveillance, operating and positioning, lifting appliance posture, smart recognition, information management and control, remote operating station, and safety protection systems. Each of these systems involves varying numbers of subsystems and has configuration requirements. For example, the smart recognition system consists of container, container tractor, ship scanning, and hatch subsystems. Apart from the general requirements, this standard specifies technical requirements for each system.

The *Technical Requirements for Remote Control System of Rail Mounted Transtainer & Rubber Tired Gantry Crane (T/CPHA 2–2018)* specifies the composition, technical requirements, and test methods of remote control systems for RMG and RTG cranes. This standard applies to the design, manufacturing, transformation, and use of these systems. The remote control system consists of video surveillance, operating and positioning, safety protection, information management and control, remote operating station, and smart recognition systems. Each of these systems involves varying numbers of subsystems and has configuration requirements. For example, the operating and positioning system contains lifting positioning, trolley positioning, bridge positioning, automatic bridge deviation correction, remote transportation between container yards, container tractor guidance, and automatic container landing subsystems. Apart from the general requirements, this standard specifies technical requirements for each system. For the interface of information management and control system, the standard specifies the content and format requirements of the information exchanged between the wharf production management system and the operation safety control module.

6.3.1.5 Typical Cases

The Port of Ningbo Zhoushan is the largest iron ore and crude oil transshipment base in China. It consists of 19 port areas and about 620 berths, including nearly 160 large berths above 10,000 tons and about 90 large and ultra-large deep-water berths above 50,000 tons. The Port of Zhoushan is an industry pioneer in completing remote control verification of 5G RTG cranes and putting them into production. Through 5G-based RTG crane transformation and verification, 5G can deliver high uplink bandwidth and stable low latency for remote control of multiple RTG cranes. After 5G-based transformation, the gantry crane control solution cuts labor cost by 70% and improves efficiency by 30% [6].

In October 2021, the first 5G remote control excavator of Shandong Port Group officially went into operation on the 10th berth in the east area of Shijiu port area of the Port of Rizhao [7]. The 5G remote control platform uses the excavator posture feedback system to implement 1:1 replication of the excavator's cab on the ship. With the help of environment and excavator posture/status monitoring systems, operators can be fully aware of the real operation environment on the platform and the real-time equipment status, similar to watching a 4D movie. This ensures operator safety while significantly improving the working environment. Besides, operators can switch remote instruction signals to control multiple onsite excavators on the same console, improving operating efficiency. During remote operations, the unique dual-channel control of action instruction signals is used to check misoperations or signal faults as soon as possible. When an exception occurs, interlocking ensures operation safety.

6.3.2 Wireless Video Surveillance

6.3.2.1 Service Requirements

As the global economic integration continues and international markets converge, wharves have shifted into large freight turnover centers with hundreds of large freighters, thousands of large containers, and tens of thousands of people every day. In such a complex environment, the safety of transportation, production, cargo, equipment, and personnel has become paramount for wharf management.

Currently, port cargo inspections depend on the subjective experience of the supervisory personnel. A great deal of manpower is required for monitoring inspection issues, causing a shortage of manpower. Cargo inspection points at ports are scattered and lack information sharing, causing challenges to data exchanges at regional logistics hubs and weak external information services. There are few auxiliary tools for container yard inspection, a lack of advanced technologies and equipment, and low resource intensification. As a result, onsite inspections are in a relatively extensive state and inspections are not sufficiently targeted and operable, which affects the efficiency of container yard inspection to some extent.

Most ports have many vehicles and routine O&M personnel, whose real-time locations are not shared with safety personnel. When personnel and vehicles are close to hazardous areas, no real-time warning is generated, resulting in poor management and control. Considering high humidity and corrosion, large machinery (including quay cranes, RMG cranes, RTG cranes, gantry cranes, and ship unloaders) requires regular inspections to prevent faults. Due to a lack of information-based methods, routine inspections and maintenance are time- and labor-consuming, while blind spots at heights may exist, posing safety risks and greatly increasing the pressure of onsite safe production. There is no unified platform for visualized and refined management of routine inspection information. As a result, the inspection status cannot be quickly and accurately reported, paper records may be lost, and the inspection line status and time cannot be accurately collected.

To boost the all-round management and coordinated handling of emergencies, ports urgently need to improve the integrated management system, upgrade equipment inspection system, enhance the integrated ground and air surveillance of wharves, perform real-time smart analysis, and assist safety and smart operating management.

- Safety protection: cargo recognition management for tripwire crossing detection, removed object detection, real-time facial recognition analysis, and alarm linkage

- Operating management: license plate number recognition, head counting, and real-time crowd density analysis, improving the campus operating efficiency
- Automatic tallying: AI-based recognition of container IDs using crane cameras
- Smart inspection: rapid and smart inspections using unmanned aerial vehicles (UAVs) and robots
- Post-event smart analysis: behavior retrieval and post-event smart analysis of video synopsis, reducing manpower investment and time cost

6.3.2.2 Network Solutions

5G+ video surveillance for port areas has the following requirements:

- Supervisory personnel can remotely view real-time work progress and status of equipment.
- Multi-angle views are available, which helps ensure the safety of equipment operators and their environments. Wireless surveillance broadens operators' horizons and avoids blind spots, improving operation safety.
- Cargo information in container yards is captured for automatic tallying.

The major pain point of video surveillance is the wired connection of cameras because their locations are fixed and are difficult to maintain. Optical fiber deployment in some areas (such as squares) is costly, and cabling in corners is difficult. Technologies such as Wi-Fi have insufficient capacity and poor stability. Current UAV-based inspections are driven by 4G, which have difficulty in collecting real-time data such as images and videos. For example, real-time upload consumes a large amount of traffic, so to ensure real-time performance, the video definition must be reduced. In addition, some ports have built production auxiliary systems, such as the container production information management, site video surveillance, mobile handling machinery monitoring, access control, and entrance and exit control systems. As being developed by different vendors, these systems can only be managed separately. Information-based applications for port areas are basic and the work efficiency is low. 5G enables real-time transmission channels with a wide pipe, provides wireless connections of industrial cameras and visual processors, facilitates automation of container tallying applications, and reduces labor costs.

In the 5G+ video surveillance solution for port areas, HD dome cameras are deployed at port entrances and container yards to capture onsite images, obtain real-time equipment, personnel, and cargo information, and send the information to the central control room through 5G networks for analysis, recognition, and big-data-based processing, implementing operating management, safety protection, and tallying. 5G enables wireless transmission for the visual system unit. 5G virtual private networks implement remote surveillance anytime and anywhere with optimal coverage, while port data can be remotely viewed in the port areas, realizing visualized and multi-dimensional management. Figure 6.4 illustrates the video surveillance network solution for port areas.

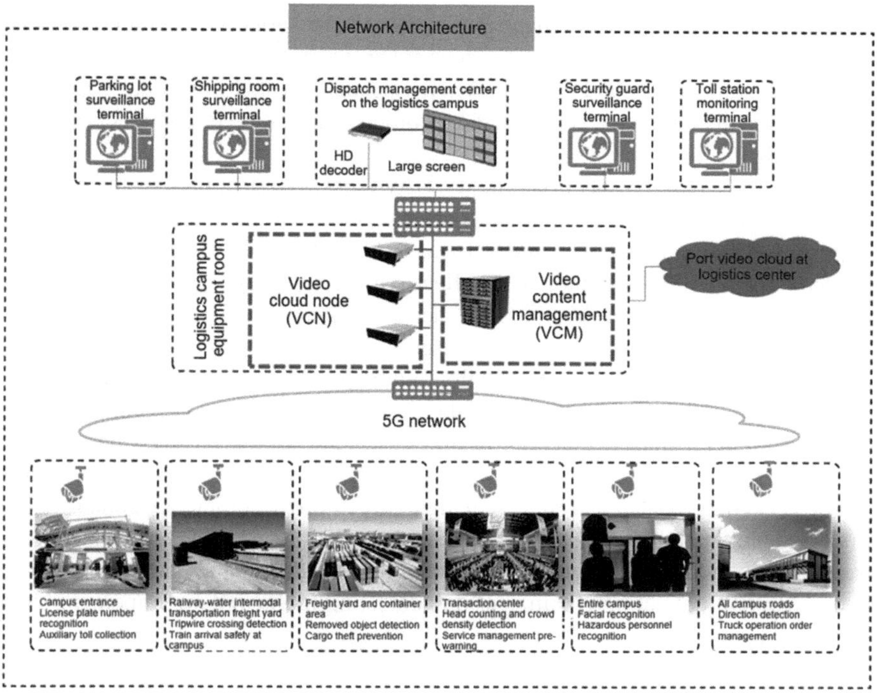

Fig. 6.4 Video surveillance network solution for port areas

For smart gantry crane surveillance applications, the HD cameras installed in the cabs of the port gantry cranes send real-time video streams to the 5G multi-access edge computing (MEC) platform, where the operators' facial expressions and driving status are analyzed. If an abnormality such as fatigue or dozing is detected, a warning is immediately generated. The 5G+ video surveillance solution resolves longstanding issues related to insufficient bandwidth and high latency in 4 K ultra-high-definition (UHD) video surveillance scenarios. It provides clearer video images, which ensure more efficient and responsive intelligent analysis. This reduces accident rates at ports and enhances the overall safety of port operations and personnel.

To facilitate intelligent container or cargo tallying, the HD surveillance cameras installed on bridge cranes collect real-time information related to key operations. This information is analyzed to power a wide range of services, including container number recognition, trailer number recognition, International Organization for Standardization (ISO) code recognition, automatic confirmation of operating status, recording of abnormal operation handling, video recording of warehousing, and container damage detection based on complete container images. It is on this basis that visualized management is provided for container yards. This encompasses container entering/exiting, packing/unpacking and number locking, automatic container number verification, container slot management, expired container management, and container damage repair, with the fees for each operation automatically recorded. In addition, data interfaces are provided to allow data and yard information to be transmitted and viewed when it is needed. Furthermore, daily reports can be automatically generated based on the recorded information of containers entering and leaving the yard.

During UAV-based inspections via 5G, UAVs sense the surrounding environment, plan an optimal inspection path, and perform automatic navigation. 4 K/8 K HD cameras and thermal imaging cameras harness the high rate, low latency, and massive connectivity of 5G to monitor and collect data from bridge cranes, RMGs, ship loading and unloading machinery, and other port equipment. Once data is uploaded, professional-grade image diagnosis is used to detect equipment cracks, rust, and falling parts on a timely basis. Evaluation reports are then automatically generated. 5G UAVs cover 360 degrees to inspect the health of equipment, mitigating the blind spots that hinder manual inspection. Thanks to versatile apps that offer route imaging, automatic inspection, and stable shooting, among other features, ground personnel can inspect equipment at any time, which greatly improves efficiency and precision. UAVs can automatically inspect the 40 detection areas on a bridge crane in approximately 2 h, which is over eight times more efficient than manual inspection, and are also capable of collecting HD images and detecting faults that are imperceptible to the naked eye.

6.3.2.3 Ecosystems and Business Models

In wireless video surveillance scenarios, telecom operators provide networks and spectra, 5G solution suppliers come up with 5G E2E solutions, and visual equipment vendors integrate surveillance solutions for terminal applications. Telecom operators and 5G solution suppliers offer ICT equipment and services, including networks, clouds, and edge data centers. Visual equipment vendors provide basic hardware, including sensors, cameras, as well as control software and operating systems. Two construction modes are used in wireless video surveillance solutions for ports in China.

- Mode 1: Mainly designed for large ports with robust information-based infrastructure, strong demand for digitalization, and port companies who are actively promoting automation. Under this mode, telecom operators work with 5G solution suppliers to integrate 5G communication solutions with applications, and visual equipment vendors work with onsite machinery vendors to deploy wireless video surveillance, after which port companies integrate these solutions.
- Mode 2: Mainly designed for small ports where telecom operators wish to promote digitalization. Telecom operators, as the system integrators, provide E2E integration services. 5G solution suppliers integrate 5G communication solutions with

applications, while visual equipment vendors work with onsite machinery vendors to integrate wireless video surveillance solutions.

Based on the industry development, solution maturity, and its understanding of the industry and technology accumulation, an enterprise selects the mode that can make full use of resources and advantages of all parties.

The large-scale deployment of 5G technologies at ports benefits all parties immensely. For supervision departments and port companies, 24/7 blind-spot-free surveillance can cover the entire port area; real-time monitoring of containers and goods in storage areas helps prevent theft; and loading/unloading areas can be monitored in real time as well. Unmanned port inspection, 24/7 surveillance, HD video, and intelligent analysis have improved safety at smart ports by leaps and bounds.

Visual equipment from visual equipment vendors and subsystem providers cannot currently be deployed on a large scale due to the limitations of communication technologies. Automation allows such applications to be used more widely in facial recognition, behavior recognition, license plate recognition, target classification, and other scenarios. This lays the groundwork for new intelligent security systems, expanding the scope of available services. Gartner has predicted that outdoor surveillance cameras will be the largest market for global 5G IoT solutions by 2022.

6.3.2.4 Industry Standards

In July 2015, China's MOT released the *Technical Requirements for Port Video Surveillance System Networking (JT/T 982–2015)*. This standard determines the technical requirements for the networking architecture, audio/video codec, interface and control protocols, and coding specifications of port video surveillance systems, covering the design, development, O&M of the surveillance systems. It also serves as a benchmark for similar systems.

In December 2020, China UHD Video Industry Alliance released the *General Technical Specification for 5G UHD Surveillance Camera (CUVA 006–2020)*, a standard that details the technical specifications of 5G UHD cameras, such as those related to the resolution, bit rate, frame loss rate, and peak-to-average traffic ratio. The standard lists the technical requirements for 5G UHD cameras in terms of front-end video collection, media compression and encoding, application-layer transmission protocols, and transmission performance. It also provides reliable criteria and methods for testing functions and performance. This standard is ideal for video surveillance scenarios that use 5G UHD cameras, and for other surveillance scenarios that involve 5G transmission.

6.3.2.5 Typical Cases

The Jiangyin port, run by the Fuzhou Port Group, launched the first-ever 5G smart port platform in Fujian province. This smart port employs large-scale 3D surveillance, which runs on an AR-enhanced command platform. It remotely collects real-time panoramic video of the Jiangyin port via AR panoramic cameras. This helps production scheduling personnel deliver real-time commands to operating flows to adjust operations, providing an advanced method for managing port operations. HD surveillance cameras installed in the cabs of bridge cranes rapidly send the collected real-time video images to the MEC platform via 5G, where the operator's facial expressions and driving status are analyzed. If the system detects fatigued or sleeping personnel, an immediate warning is sent out. The 5G video surveillance solution resolves longstanding issues related to insufficient bandwidth and high latency in 4 K UHD video surveillance scenarios. It provides clearer video images, which ensure more efficient and responsive intelligent analysis. This reduces the occurrence of port-related accidents and bolsters the overall safety of operators and port operations [8].

The Rizhao port, run by Shandong Port Group, has deployed a wireless video surveillance system that transmits HD images from hatches to crane cabs, and from cabs to the central surveillance platform, making tallying more secure and efficient. After trucks with the intelligent tallying function are parked at specified inspection areas, the system uses the dome cameras on both sides to compare the trucks and goods against the system data. If the information is correct, the trucks are allowed to pass. The tally clerk only needs to confirm the tallying results. Once the system is completed, tallying of cargo will be safer than ever, with minimal human–machine operations required and no misoperations. The time required to tally each vehicle will be reduced from 3 min to 10 s [9].

6.3.3 *Unmanned Transportation at Ports*

6.3.3.1 Service Requirements

Demand for imports and exports is robust in China. According to MOT, cargo throughput at Chinese ports totaled approximately 12.87 billion tons from January to October 2021, up 7.8% year on year. The country's container throughput stood at 24.52 million TEUs in October 2021. A major challenge for ports is transporting cargo in a cost-effective manner, while significantly boosting throughput.

Another major issue is the shortage of truck drivers. Truck drivers must be qualified and experienced, having obtained driving licenses of class A2 or higher. Labor costs for truck drivers account for over half of total port transportation costs. According to reports, the average labor cost of each truck driver is CNY150,000 to CNY200,000 each year, a number that continues to grow. The total labor cost of truck drivers at Chinese ports is approximately CNY50 billion to CNY100 billion each year. In addition, ports operate 24/7, which means that drivers need to work on rotating shifts. This may result in fatigue driving, posing safety risks. In addition to reducing costs, port digitalization improves efficiency. Under traditional manned driving solutions, E2E scheduling is difficult to perform, and other issues such as chaotic queues, queue jumping, unloaded trucks on some trips, and uncontrollable working time severely hamper transportation efficiency.

Horizontal transportation automation is the core to slashing costs and supercharging efficiency at ports. In recent years, the automation of port operations in China has continued to progress. While vertical transportation at ports has reached a high level of automation due to the use of automatic RMG cranes, horizontal transportation automation remains a major pain point. With existing solutions, reconstructing sites for AGVs remains difficult, and the price of vehicles remains high. Automated straddle trucks cannot be deployed at ports in China due to stacking height-related limitations.

Replacing truck drivers with unmanned driving saves labor costs, while also eliminating a lot of uncertainties. The central control console coordinates the overall operations, loading and unloading processes, running duration and routes, and traffic sequence. Related enterprises have begun to deploy unmanned driving at ports in recent years. A report from Green Pine Capital Partners indicates that ports are the second largest global market for unmanned driving, accounting for 11.3% of the sector, a total of USD 27.9 billion in 2020. Unmanned transportation systems at ports are a key application for 5G technologies. The high speed, low latency, and massive connectivity of 5G facilitate high-precision positioning and vehicle–road synergy, allowing for autonomous truck driving and real-time transmission. In the relatively simple transportation environments at ports, 5G makes autonomous driving easier to deploy.

6.3.3.2 Network Solutions

Ports typically involve two types of transportation: transportation within enclosed areas (transportation of containers and bulk cargo within a port) and transportation within semi-enclosed areas (transportation of cargo between a yard within a port and a yard or warehouse outside the port). Port environments are more complex than trunk roads and logistics warehouses, due to the many container stacks, fixed wharf facilities, mobile tools for hoisting, and goods and passenger transportation methods. The detection and recognition capabilities of unmanned container trucks require higher specifications and strict margins for tolerable errors. Higher precision must be built in to the system design, while factors like high humidity, high salinity, and large temperature differences place high requirements on technologies.

Unmanned transportation solutions at ports are typically specific to the scenario and environment and have three characteristics. First, their intelligence abilities account for all variables within complex environments, including people, vehicles, and objects. Unmanned container trucks come equipped with advanced technologies, such as radar (laser, millimeter-wave, or ultrasonic), HD cameras, and satellite positioning modules. Ultrasonic radar is capable of detecting all types of objects, and is therefore more feasible for truck reversing, making it ideal for port transportation systems. Second, 5G networks and V2X technologies are used to enhance collaborative sensing at the vehicle and road sides. Unmanned container trucks are interconnected with automatic production equipment and systems at ports. In addition, cloud computing and remote monitoring services can be added to allow for real-time optimization, intelligent scheduling, and remote driving, providing multi-dimensional assurance for the secure operating of unmanned container trucks. Third, unmanned container trucks differ greatly from AGVs and IGVs in terms of technologies. For example, magnetic nails must be fixed on the traveling routes of AGVs and IGVs only offer poor positioning accuracy. By contrast, unmanned container trucks overcome the technical weaknesses of AGVs and IGVs, and even run in a more cost-effective way.

Unmanned transportation solutions at ports require real-time functions such as real-time road-side sensing, vehicle–road synergy, vehicle-vehicle synergy, and real-time video streams. A single container truck requires a minimum uplink bandwidth of 20–30 Mbit/s and stable latency under 20 ms. At peak times, ports can put 40 container trucks in operation concurrently, but this requires a total uplink bandwidth of 1200 Mbit/s. 5G technologies provide the necessary network support for such applications. More specifically, 3GPP Release 16-compliant 5G has demonstrated the sufficient performance and stability to meet the wireless data transmission requirements of unmanned container trucks. CPEs can be used to support the dual fed and selective receiving function, further improving signal transmission quality and latency stability. 5G networks are ideally positioned to serve as transmission solutions in support of all applications for unmanned transportation at ports.

6.3.3.3 Ecosystems and Business Models

In unmanned transportation scenarios, telecom operators provide networks and spectra, 5G solution suppliers come up with 5G E2E solutions, and autonomous driving solution suppliers provide autonomous driving solutions by integrating solutions of container truck manufacturers and control systems. Telecom operators and 5G solution suppliers offer ICT equipment and services, including networks, clouds, and edge data centers. Autonomous driving solution suppliers and container truck manufacturers offer street-legal unmanned container trucks, as well as the corresponding control software and operating systems. Two construction modes are used in unmanned transportation solutions for ports in China.

- Mode 1: Mainly designed for large ports with robust information-based infrastructure, strong demand for digitalization, and port companies who are actively promoting automation. Under this mode, telecom operators work with 5G solution suppliers to integrate 5G communication solutions with applications, and autonomous driving solution suppliers work with container truck manufacturers to integrate unmanned driving solutions, after which port companies integrate those solutions.

- Mode 2: Mainly designed for large ports with strong information-based infrastructure where autonomous driving solution suppliers play the leading role in applying automated solutions. Under this mode, autonomous driving solution suppliers facilitate overall project integration to apply a comprehensive unmanned driving solution, while telecom operators serve as network suppliers to provide connected services, and collaborate with 5G solution suppliers on 5G private network integration.

When rolling out an unmanned driving solution, a port should select one of these two modes, which make full use of the resources and advantages of all parties.

The large-scale deployment of 5G technologies at ports is enormously beneficial for all parties. For supervision departments, unmanned container trucks improve operating safety and prevent accidents. Traditional manned driving solutions are unable to perform E2E scheduling for all vehicles, and hindered by issues like chaotic queues, queue jumping, unloaded trucks on some trips, and uncontrollable working times, which severely hampers transportation efficiency. In comparison, unmanned driving solutions are controlled by the central scheduling platform that automatically collects data from each service process at a port, significantly improving scheduling efficiency. Ningbo Daxie Wharf is a good example of this principle. After the port is digitalized, it is estimated that the overall efficiency can be improved by 30–40%, and the time that ships stay at docks can be reduced by 30%. Accordingly, the throughput of the port can grow from 3 million TEUs to 4–5 million TEUs, or even up to 6 million TEUs.

For unmanned driving solution suppliers and container truck manufacturers, ports are typically characterized by enclosed environments and low speeds, making such environments ideal for the commercial deployment of autonomous driving. Chinese ports currently make use of more than 25,000 container trucks, but the majority of ports still depend on manned driving. Although the current penetration rate of autonomous driving container trucks at ports is less than 2%, it is estimated that the penetration rate of level 4 autonomous driving container trucks will exceed 20% by 2025, covering 6,000–7,000 such trucks. The overall market scale of autonomous driving at ports in China is projected to exceed CNY6 billion, accounting for about 30% of the global market. Autonomous driving can be rapidly rolled out at ports with clear business models. Given this, more autonomous driving projects are expected to be put into large-scale commercial use over the next two to three years.

5G offers the communication capabilities that underpin unmanned driving and other service applications. All participants in the value chain stand to benefit, whether it is from one-off charges or annual service fees.

6.3.3.4 Industry Standards

In May 2021, Dongfeng Motor Corporation, COSCO SHIPPING Ports, and China Mobile (Shanghai) ICT Co., Ltd. jointly released *Port Driverless Container Vehicle Performance and Test Methods*, the first standard to specify the performance requirements and test methods for unmanned container trucks at ports. This standard consists of four parts: cooperative driving scenarios and driving behavior requirements, wireless communication and information security requirements, technical requirements for vehicle functions and performance, and vehicle test and test methods. Each part details its application scope and provides normative references, terms and definitions, and specific technical performance requirements. This standard is applicable to the development of the autonomous driving system for unmanned container trucks at smart ports, and has been approved by the China ITS Industry Alliance.

6.3.3.5 Typical Cases

Port of Ningbo Zhoushan, the world's largest port, teamed with the Ningbo branch of China Mobile and Huawei to demonstrate the hybrid operation of 5G smart container trucks and common container trucks, representing a milestone in the port's development. 5G technologies endow smart container trucks with cutting-edge AI, so that they can detect surrounding infrastructure and objects, for example, containers, mechanical equipment, and lighthouses, and make decisions, such as to decelerate, brake, turn, bypass, or park, in response to everyday events and emergencies alike. After containers are placed on these trucks by quay cranes, the trucks will automatically start and transport the containers to the RTG crane operating area, traveling along optimized routes provided by the intelligent scheduling system. Enclosed areas of ports have thereby become efficient zones for horizontal transportation [6].

In May 2020, Xiamen Ocean Gate Container Terminal launched China's first "5G+ unmanned driving solution for smart ports", in which unmanned container trucks from Dongfeng Motor Corporation were unveiled and passed onsite acceptance tests for all 5G smart port service scenarios. The Port of Xiamen has deployed continuous 5G network coverage, with a 5G coverage ratio of 98.89% across the entire port. 5G edge computing and a slice-based private network have been constructed to implement a wide range of 5G applications, such as driving behavior analysis, AGV remote control, autonomous driving, and port equipment remote control, boosting operating efficiency and safety throughout the port [10].

As the largest and most comprehensive port in southern China, the Port of Guangzhou is renowned for its world-class fully automated wharf, that is, the fourth phase of Nansha Port. In May 2021, the port was jointly commissioned. Quay cranes proved capable of accurately picking up containers from ships based on a set of automatic system instructions, and automatically placing those containers on unmanned AGVs or IGVs. The AGVs or IGVs followed automatically charted routes, and transported the containers to the container yard, with the assistance of intelligent algorithms. Then, the RMGs picked up the containers, and placed them in specified positions, also on an automatic basis [11].

6.4 Summary and Prospects

5G technologies have been successfully deployed at ports across a wide range of operations, and are now ready to be promoted on a large scale. The physical space of ports is relatively enclosed, and thus using 5G to construct smart ports is of enormous practical value. The current 5G applications at smart ports have demonstrated that remotely controlled cranes can improve safety and working conditions for crane operators. Intelligent tallying has proved capable of handling a large number of incoming and outgoing goods of all types, and allows for more sophisticated management of items. Smart security and UAV inspection solutions can reduce the strain on overtaxed port security systems, which are dependent on tedious manual inspections. As 5G continues to be deployed on a greater scale, the range of service scenarios will only grow, and more pilot projects will be carried out for the purpose of verifying 5G capabilities in common operations and building flagship 5G applications for vertical industries. 5G smart port solutions will undoubtedly be replicated in other ports, both in and outside China, with the emergence of new technologies and business models, providing a boon for the industry and a much-needed technological upgrade.

References

1. Zhao B, Wang N (2010) Study on the inter-generational essential characteristics and evolution of ports (in Chinese). China Ports, 2010(5):50–51, 44
2. Liu CL (2013) Cargo handling technology at ports (in Chinese). Dalian Maritime University Press, Dalian
3. Improving quality and efficiency and reducing costs to build a port ecosystem (in Chinese) (2018)
4. Gu HH (2010) Conveying machinery and container machinery in ports (in Chinese). China Communications Press, Beijing
5. Liu CM, Zhang CJ, CHEN WB, et al (2021) Development trend of key technologies for domestic coastal automated container terminals (in Chinese). China Ports 2021(1):17–23
6. Huawei Wireless Network. New breakthroughs in 5G smart ports: Port of Ningbo Zhoushan works with China Mobile and Huawei to build a world-class port (in Chinese) (2021)
7. Li SH (2021) The first 5G remote control excavator in Rizhao Port of Shandong Port Group was put into use (in Chinese)
8. State-owned Assets Supervision and Administration Commission of the People's Government of Fujian Province. First 5G smart port platform in Fujian province (in Chinese) (2019)
9. Sui ZW (2021) Intelligent tallying in the tallying company of Rizhao Port of Shandong Port Group (in Chinese)
10. China Mobile Communications Group Co., Ltd. (2020) How unmanned container trucks work? (in Chinese)
11. China United Network Communications Group Co., Ltd. (2021) China Unicom added new driving forces to 5G smart port construction (in Chinese)

Chapter 7
5G+ Smart Mining

7.1 Overview

7.1.1 Current State

Mineral resources are the products of the long-term formation, development, and evolution of Earth's crust. They are formed by the accumulation of natural minerals through unique geological processes under unique geological conditions. Different geological processes nurture different minerals. Mineral resources have the following features: (1) Non-renewability. Mineral resources are limited in reserves and are to a certain extent irreplaceable. (2) Challenging mining conditions. Most mineral resources are formed in unique geological environments. Disturbing these environments can easily cause ecological disorder. (3) Uneven distribution. Often, mineral resources are not rich in areas with high levels of economic development.

Common mineral resources include energy, metal, non-metal, groundwater, and gas minerals. Different kinds of mineral resources have unique properties, uses, and ways of formation. As a result, each requires a mining method unique to itself.

Mineral resources are unevenly distributed around the globe. Reserves of many mineral resources are concentrated in a few countries. The United States, for example, has the world's most abundant coal reserves. By the end of 2020, the uncovered coal reserves of the United States were 248.941 billion tons, accounting for 23.2% of the global total [1]. Mining requires abundant funds and advanced technologies. Having relatively rich reserves but a low level of mine exploration and development, most developing countries tend to encourage foreign-funded enterprises to participate in local mining projects, thereby improving their domestic investment environments for mining. As such, multinational mining enterprises have been developing rapidly, driving mining progress worldwide. They have been exploring mineral resources, expanding their mineral reserves, and extending the service life of resource-based mines, while expanding the business scale and improving the market landscape through mergers and acquisitions. Leveraging their strengths in capital, technology, and management, they have built advantages in the scale, products, and cost of

P. Sun, *A Guidebook for 5GtoB and 6G Vision for Deep Convergence*,
Management for Professionals, https://doi.org/10.1007/978-981-99-4024-0_7

mining. With the recent technological improvements in mineral resource exploration, exploitation, and utilization, many more mineral resources have been discovered and made available. Low-grade mineral resources and symbiotic/associated mineral resources have also been exploited and utilized. Mine exploitation and utilization is gaining more efficiency.

China is an important country for mineral resources and mining. By the end of 2020, 173 types of minerals had been discovered in China, including 13 energy, 59 metal, 95 non-metallic, and 6 groundwater and gas minerals [2]. China has a rather complete category of mineral resources and a mature exploration and development system. Its production and consumption of major mineral products is among the highest in the world. For example, coal production in the Chinese mainland accounts for 50.7% of the world's total and coal consumption for 54.3%, both exceeding the sum of the rest of the world [1]. Given China's large population, however, the per capital reserve of most mineral resources in China is less than half of the global average. In addition, the Chinese mining industry suffers from disorderly and inefficient exploitation as well as overexploitation. There are also some pollution problems and occasional safety accidents. According to the National Mine Safety Administration, there were 123 coal mine accidents in 2020 nationwide, resulting in 228 deaths.

7.1.2 Major Processes and Operations

Generally, the exploitation process of mineral resources includes mining, mineral processing, smelting, and industrial production. Mining methods vary depending on the mineral resource types and the mining areas. There are two common mining methods: open-pit mining and underground mining. Open-pit mining is often used for shallow, thick ore bodies. Underground mining is often used for deep, thin ore bodies. Open-pit mining requires the use of large-sized equipment, but features low capital investment on infrastructures, low production cost, low ore loss and dilution ratios, as well as high production efficiency and safe operating conditions. However, some ore bodies are not suitable for open-pit mining but suitable for underground mining. In some cases, in the same mining area, open-pit mining may be used in early operations, while underground mining may be used in later stages.

Generally, the process of open-pit mining includes: perforation, blasting, mining and loading, transportation, and waste disposal, in sequence. The first step, perforation, involves drilling holes in pre-selected places using the rotary blast hole drills, downhole drills, and related equipment. This step prepares the site for blasting. In blasting, explosives are filled into boreholes. Detonation of the explosives causes ore rocks to collapse. The next step, the central part of open-pit mining, is mining and loading: using single- or multi-bucket excavators, bucket wheel excavators, and related excavating equipment, ore rocks are collected across the entire base rock or from loose blasted muckpiles and loaded onto transportation vehicles. Transportation's primary function is taking the mined ores to a mineral processing plant, crushing station, or ore stockyard. Transportation is also involved in transporting the stripped rocks (wastes) to a waste disposal site. The equipment used to do this is mainly self-dumping trucks, belt conveyors, wheel loaders, track conveyors, and related equipment. Linkage between large-sized mining/loading equipment and transportation equipment (such as the bucket wheel excavator–belt conveyor system, bulldozer–grizzly–belt conveyor system, front-end loader–mobile crushing station–belt conveyor system, and excavator–truck–crushing station–belt conveyor system) provides a seamless connection between mining/loading and transportation, significantly improving mining efficiency and achieving continuous mining. In the last step, waste disposal, a large amount of topsoil covering the upper part of the ore body and surrounding rocks are stripped and transported to a dedicated disposal site.

In underground mining, the first step is development. A series of mine roadways is built from the surface to the mine body in order to set up the necessary mining systems. These systems include traffic, ventilation, lifting, transportation, drainage, power supply, wind supply, and water supply. It is then possible to transport ore, waste stones, exhaust fumes, and sewage to the ground while bringing equipment, materials, personnel, power, and fresh air underground. This step includes adit development, shaft development, slope development, ramp development, and combined development. Construction of auxiliary shafts, ventilation shafts, orepasses, fill shafts, crosscuts, shaft stations, and shaft chambers is also involved in this step. Next is stope preparation. This step determines the producing sections in sections of the ore body where development has been completed, and prepares the necessary conditions for quarrying, including personnel passage, rock drilling, ventilation, and ore loading. The third step is cutting, which opens up the free face and compensating space for large-scale quarrying. The fourth step is quarrying, which involves large-scale mining operations, including rock drilling, ore caving, ore transport, and quarry supporting. The preceding steps are carried out in sequence at the start of mining operations for a particular mine. This sequence, however, does not have to be strictly followed after mining is operational. To ensure sustainable production, further excavation of shafts and drifts, such as extended mine roadways, is required for exploration, stope preparation, and quarrying.

7.2 Digitalization Trends and Challenges

7.2.1 Trends of Digitalization

Owing to various high-tech technologies, mining equipment is becoming large-sized, automated, and intelligent, and mining methods are becoming mass-production-oriented, continuous, and less reliant on human operation. Production equipment and techniques are being optimized; efficiency and safety are being improved; costs and pollution are being reduced. The proportion of open-pit mining increases while the proportion of underground mining declines.

The 14th Five-Year Plan is a critical period in the transformation of China's mining industry. According to the *Outline of the People's Republic of China 14th Five-Year Plan for National Economic and Social Development and Long-Range Objectives for 2035* [3], China will reasonably control the intensity of coal development, promote the integrated development and utilization of energy resources, and strengthen the ecological restoration of mines; it will promote the construction of sustainable development demonstration zones and transformation innovation pilot zones in resource-based regions and implement projects for coal mining subsidence area comprehensive governance and independent industrial and mining area transformation and upgrading; it will improve the level of development and protection of mineral resources, develop a green mining industry, and build green mines; it will strengthen the planning and control of strategic mineral resources, improve reserve security capabilities, and implement a new round of breakthrough strategic initiatives for prospecting; it will strengthen the innovative application of advanced technology and equipment in the fields of deep mining and major disaster prevention and control and promote the substitution of robots for humans in dangerous jobs; it will also promote full coverage of safe production liability insurance in key areas.

In June 2021, the National Development and Reform Commission, the National Energy Administration, the Cyberspace Administration of China, and the Ministry of Industry and Information Technology (MIIT) jointly released the *5G Application Implementation Plan in the Energy Sector* [4]. This plan, paying special attention to the high rate, low latency, and ubiquitous connections of 5G, sees 5G as an important strategic resource and a new type of infrastructure supporting energy transformation. One of its core ideas is that deep integration of 5G with various industries in the energy sector will drive energy production and consumption innovation, providing a powerful driving force for an energy revolution. This plan will support high-quality energy sector development, expanding its 5G application scenarios and exploring new 5G application modes and business forms that can be replicated and easily promoted. For 5G+ smart coal mining in particular, this plan will build basic 5G networks covering coal mines' surface and underground areas, smart platforms converging coal mine management and control, enterprise cloud platforms and big data processing centers, and a cloud-edge-device industrial Internet architecture for the mining industry. Based on the high rate, low latency, ubiquitous connections, and high reliability of 5G, this plan will focus on the development and application of underground inspection, safety protection, and unmanned driving, and explore pilot applications in scenarios such as smart excavation and production control, environment monitoring and safety protection, and virtual interaction, for the purpose of promoting smart mining.

In July 2021, the MIIT and 9 other departments jointly released the *"Set Sail" Action Plan for 5G Applications (2021–2023)* [5]. According to the plan, China will accelerate the development and certification of 5G communication devices that are explosion-proof and otherwise meet the requirements of mining environments; it will advance integrated infrastructure construction of surface and underground mine 5G network systems, smart mining area control platforms, and enterprise cloud platforms; it will promote applications of 5G in all kinds of mines such as energy mineral, metallic mineral, and non-metallic mineral mines; it will expand 5G application scenarios (like remote control, unmanned driving) in the mining industry; it will also advance the remote operation and cluster operation of underground core mining equipment, unmanned operation of mining equipment in deep and high-risk mining areas, smart open-pit continuous operations, as well as unmanned transportation.

2021, the first year of the 14th Five-Year Plan, was also the first year of 5G-enabled digital and smart mining industry transformation. Digital, information, and intelligent technologies help progress toward a number of key mining industry objectives, including high-quality, safe development of mines, mineral resource protection and development, and green and unmanned mines.

7.2.2 Major Challenges in Development and Requirements for Digitalization

The mining industry faces the following challenges and digital transformation requirements:

1. Safety Accidents
   Mining is a high-risk industry. A safety accident may cause significant casualties once it occurs. Mining areas are often dark, closed spaces with low oxygen content and flammable/explosive gases. It is difficult to escape and conduct rescue operations in the case of an accident. Due to high safety risks, more and more people are unwilling to work underground. Labor shortage has become increasingly prominent. To remedy this situation, mining enterprises attach great importance to high-tech technologies, which enable the reduction of underground staff quantity or working hours and smartly perceive and monitor the production environment and process. In addition, mining enterprises work to ensure safe production by increasing underground communication facilities, communication methods, and positioning capabilities, giving them the ability to promptly respond to emergencies and carry out effective evacuation/rescue measures if needed.
2. Environmental Pollution and Ecological Destruction
   Environmental pollution in the mining industry mainly includes air, water, and soil pollution, for example, greenhouse gas emission and dust pollution caused by blasting and mining, soil pollution caused by the release of associated elements, and groundwater/surface water pollution caused by wastewater discharge. Environmental damage may include ground subsidence, mountain fractures or landslides, soil erosion, land desertification, destruction of biological habitats, and seawater intrusion. To better protect the environment while developing mineral resources, there is a need to strengthen mining process supervision and environmental impact monitoring, improve mining automation and intelligence, and reduce the proportion of tailings.

3. Dynamic and Complex Production Environment
   The production environment of mining is complex and dynamic. During underground mining, for example, the production environment needs to be designed and adjusted according to the geology of the working area. The heading face moves 5 to 10 meters forward while the fully mechanized mining face moves 8 to 10 meters every day [6]. As a result, smart mining devices and related infrastructures must support flexible deployment, automatic configuration, mobility, and remote management. In addition, they must adapt to specific production environments including those requiring dust-proof, explosion-proof, shockproof, waterproof, and high-temperature-resistant capabilities.
4. Low Efficiency Despite High Cost
   Recent years have seen a number of trends in the mining industry: slowing down of the demand growth; tightened environmental constraints; higher cost on production and management elements; lower mineral product prices. In 2020, global coal consumption declined by 4.2%, the fourth yearly decrease in the past six years. The coal price (spot price in Qinhuangdao, China) declined by 3.25% in 2020, a 34.7% drop from the 2011 high [1]. In this situation, the integration of new technologies and innovation is a must for the mining industry. The aim is to facilitate an upgrade to a higher industry level, optimize the production process, reduce waste, save energy, minimize costs, improve efficiency, enhance competitive advantages, and accelerate high-quality development.

7.3 Major 5GtoB Large-Scale Replication Scenarios and Typical Cases

5G networks feature high rate, large capacity, low latency, high reliability, high security, high positioning accuracy, and flexible deployment. 5G supports real-time HD video transmission, low-latency remote control, fast and high-accuracy positioning, real-time information exchange, automatic network configuration, and intelligent O&M. In addition, 5G avoids problems common with wired networks, such as long construction periods, high costs, difficult adjustments, and easy cable disconnections. 5G also avoids Wi-Fi/4G network problems such as unsatisfactory coverage, insufficient capacity, and high latency. Given its superior capabilities, 5G is an important enabler for smart mining. Typical scenarios of 5G+ smart mining are:

1. Smart Excavation and Production Control
   5G industrial modules are deeply integrated with mining and transmission equipment to enable key large-sized equipment to support 5G communication. 5G-based real-time production control platforms and dispatching command systems are deployed to achieve real-time information interaction and remote control during mining and production, minimizing the number of onsite staff.

2. Environment Monitoring and Safety Protection
   5G smart helmets, sensing devices, monitoring devices, inspection robots, and rescue robots are deployed to implement visualized underground communications, real-time HD video transmission, environment monitoring data collection, and underground personnel/equipment positioning. These tools help achieve effective safety inspection, pollution monitoring, and disaster warning and rescue.
3. Autonomous Driving of Unmanned Mining Trucks
   5G enables advanced driver assistance systems. When autonomous driving systems are developed and applied to mines, they enable unmanned mining trucks to move in platoons. This yields a number of benefits, including automatic obstacle avoidance, truck following and meeting, and autonomous route planning, greatly improving efficiency and reducing costs.

All of these scenarios require a high-quality 5G network able to operate in a harsh mining environment filled with high-density dust particles and flammable/explosive materials. To ensure safe, efficient, high-quality, and smart mine production, 5G standalone (SA) networking features and enhanced technologies must be deployed. The former includes network slicing, edge computing, and enterprise private networks. The latter includes uplink carrier aggregation (CA), uplink and downlink decoupling, integrated access and backhaul (IAB), low latency and high reliability, high-accuracy positioning, and smart O&M.

Currently, there are three business models for these scenarios. The first one is a turnkey project. In this model, the mining company is the project investor and specifies construction details and standards. The operator/integration service provider is responsible for the construction and delivery. The mining company pays to the operator/integration service provider when the project passes the acceptance. The second model is general contracting by the operator/integration service provider. In this model, the operator/integration service provider provides the equipment, systems, and personnel required for smart mining, and is responsible for construction, commissioning, and operation. The mining company makes regular payments to the operator/integration service provider according to the mine's production volume and previously agreed price per mineral resource unit. The third model is project co-construction with revenue sharing. In this model, the mining company and the operator/integration service provider clearly define their respective responsibilities and jointly complete project investment and construction. When the project becomes operational, they receive regular dividends based on the proportions specified in their revenue sharing agreement.

7.3.1 *Smart Excavation and Production Control*

7.3.1.1 Industry Requirements

Remote control of mining equipment and the production process significantly reduces onsite staff quantity and safety accident risks while improving working environments and production efficiency. Remote control, however, places extremely high requirements on the uplink bandwidth, latency, deployment flexibility, reliability, and security of communication technologies.

First is uplink bandwidth. One of the key reasons uplink bandwidth is so important is video surveillance. Smart excavation and production control require a large number of HD cameras. In an underground coal mine, for example, video surveillance is required at many positions: the middle parts of belt conveyors; conveyor heads and tails; coal drop points; coal feeding points; power distribution rooms, power distribution points, pumping rooms, and drainage points of electromechanical chambers; the front, middle, and rear parts of garages; coal faces; core cutters; heading faces. For 20 channels of 1080p video covering a 300-m coal face to be transmitted concurrently, the required uplink bandwidth is about 120 Mbit/s. For 15 channels of 1080p video covering a 400-m belt conveyor to be transmitted concurrently, the required uplink bandwidth is about 90 Mbit/s [7]. Ever-increasing intelligence places higher requirements on the resolution, frame rate, 3D imaging capability, and quantity of cameras, demanding larger uplink bandwidth. It is important to remember, however, that the transmit power of mining equipment must be controlled within a low level to meet explosion-proof requirements. As such, 5G, able to provide large uplink bandwidth at low transmission power, is a rigid requirement.

Next is network latency. Remote control is the key to unmanned or less-manned operations, placing strict requirements on deterministic latency. For effective and efficient remote control, the E2E latency should be within 50 ms to avoid frame freezing as well as operator dizziness and vomiting after long shifts. 4G, Wi-Fi, and microwave transmission each have a latency of over 100 ms. In contrast, 5G reduces the E2E latency to less than 30 ms.

The issue that should not be overlooked is network deployment flexibility. The heading face and fully mechanized mining face are moving 5 to 10 m every day, meaning that the production environment is changing frequently. As such, network deployment, operation, and reconfiguration should be not only fast, but completely wireless as well.

Last, the need for reliability and safety is paramount. Even a short interruption or a minor safety problem may cause huge economic losses or an irreversible production accident.

The application of 5G technologies in smart excavation and production control is a core application of next-generation information technology to achieve unmanned or less-manned mining operations. 5G features low latency, large bandwidth, and ubiquitous connectivity. It integrates with edge computing, machine vision, and artificial intelligence (AI) to support basic network transmission needs of mines as well as help achieve smart mining: interconnection and interworking of mine equipment and data, smart remote control of the mining processes, real-time control of the core production process, and more. For mining enterprises, operators, and equipment vendors, 5G applications in the mining industry can resolve common problems and present opportunities for future cooperation. As the key to mine safety and production efficiency improvements, 5G can meet enterprise requirements for the return on investment (ROI) and promote close cooperation and collaborative development throughout the whole mining ecosystem.

7.3.1.2 Network Solutions

A 5G private network with user-side access devices deployed is designed for a mine to meet the requirements of smart excavation and production control. Such a 5G network consists of a radio access network, core network, and transport network. The specific solution is subject to specific requirements for large uplink bandwidth, low latency, flexible deployment, high reliability, and high security in smart excavation and production control. The deployed user-side access devices include automated control components such as sensing devices, cameras, controllers, and industrial computers, as well as network access devices such as access routers and 5G customer premise equipment (CPE). The 5G radio access network uses explosion-proof and dust-proof 5G base stations, specially designed for mines, to cover the mining area according to that area's specific conditions. The 5G core network introduces technologies such as network slicing, MEC, and user plane function (UPF) to achieve service isolation, on-edge processing, and local data split, reducing latency and improving security. For the 5G transport network, a ring topology is recommended, with Flexible Ethernet (FlexE) for physical isolation and virtual private network (VPN) for logical isolation to ensure service security.

Super uplink technology is introduced to achieve better network performance. Specifically, the uplink 3.5 GHz TDD spectrum is combined with the uplink 2.1 GHz FDD spectrum to achieve uplink full-timeslot scheduling, increasing data rates near a base station by 20 to 60% and at faraway spots by 2 to 4 times, while reducing air interface latency by 30%. This better meets the needs of service applications on network indicators. In addition, super uplink technology can also be applied together with the active and passive (A + P) ultra-lean site, which can maximize the antenna height without an additional pole, improving the coverage performance by 20%.

Depending on the actual situation of the mining area, a towed smart hydraulic lifting tower can be used to improve the timeliness and accuracy of base station migration, minimizing the impact of migration on onsite operations. In a mine where optical cables cannot be deployed or power cannot be fed, microwave backhaul or storage batteries can be used instead. Microwave transmission replaces optical transmission for data backhaul, and storage batteries reduce the dependency on cable-based power supply.

7.3.1.3 Industry Ecosystems

The 5G application in smart excavation and production control is a convergence of information communication technologies and mining technologies. It involves physical layer applications such as mining processes and methods, equipment adaptation and information system applications, as well as basic technology applications at the network layer and information resource layer such as 5G private networks and edge data centers. The solution architecture and related products are developed through system engineering approaches, with project implementation placing high requirements on the system engineering management capability.

With the growth of the domestic economy and the recovery of the mining industry, high-tech smart mining technologies have seen increasing deployment in China. Research on unmanned and smart mining equipment, however, is still insufficient.

Technical preparations for communication networks are essentially ready, with broad consensus on the basic architecture of 5G private networks for mines. Dust-proof and explosion-proof base stations, industrial gateways, industrial cameras, and CPEs have been put into commercial use. Network coverage solutions addressing the unique needs of mining areas have seen many innovations and breakthroughs. Required network equipment, however, is still expensive, and its combination with mining equipment still needs refinement. Large-scale 5G deployment in smart excavation and production control is expected to solve these problems.

7.3.1.4 Policies and Standards

Smart excavation and production control helps the mining industry achieve mechanization, automation, informatization, and intelligence in accordance with policy requirements regarding unmanned mining areas. The demonstration, commercial use, and replication of smart excavation and production control will deepen the structural reform of the energy supply sector.

Policymakers are highly aware of the importance of smart excavation and production control in the mining industry. Take the coal industry as an example. The *Guiding Opinions on Accelerating the Smart Development of Coal Mines* [8] proposes three objectives: by 2021, by 2025, and by 2035. By 2021, smart coal mines of various types and modes should operate as demonstration sites. A technical system with informatization transmission and automatic operation, covering major links such as coal mine development design, geological assurance, production, and safety, should begin to take shape. Operations at heading faces should have higher efficiency with less manpower, while those at fully mechanized mining faces should require little manpower or be completely unattended. Permanent posts in underground and open-pit coal mines should be transformed into unattended operation and remote monitoring. By 2025, coal mines prone to severe disasters and large-scale coal mines should be essentially smart operations. There should be technical specifications and standards for smart coal mining. Systems such as development design, geological assurance, mining (stripping), transportation, ventilation, washing and logistics should have smart decision-making and automated collaborative operations. Key underground posts should have robot operations. Open-pit coal mines should have continuous smart operations and unmanned transportation. By 2035, various types of coal mines should be essentially smart operations, and there should exist a smart coal mine system that integrates multiple industry chains and sub-systems, featuring smart sensing, smart decision-making, and automatic execution.

According to the *Guiding Opinions on High-quality Development of the Coal Industry for the 14th Five Years*, by the end of the 14th Five-Year Plan, coal mines prone to severe disasters and large-scale coal mines should be essentially smart operations. Systems such as development design, geological assurance, mining (stripping), transportation, ventilation, washing and logistics should have smart decision-making and automated collaborative operations. Key underground posts should have robot operations. Open-pit coal mines should have continuous smart operations and unmanned transportation. The smart mining coal output should reach about 30% of national raw coal production.

Smart excavation and production control involves equipment, network, detection, data, and management technologies, requiring innovation across multiple disciplines and technologies as well as standardization requirements across multiple platforms, devices, services, and categories of data. Current 5G standards in coal mines mainly focus on 5G communications systems. Industry/group standards include *General Technical Conditions for 5G Communications System for Coal Mines*, *Base Stations in 5G Communications for Coal Mines*, *Base Station Controllers in 5G Communications for Coal Mines*, *Terminals in 5G Communications for Coal Mines*, *Use and Management Specifications of 5G Communications System for Coal Mines*, and *General Technical Conditions for the Access of 5G Communications Network Devices in Coal Mines*.

7.3.1.5 Typical Cases

1. 5G Remote Mining in Pangang Mining

   The Panzhihua-Xichang area of Sichuan province is one of the most mineral-rich areas in the world. It is the second largest iron-mining area in China, containing tens of billions of tons of vanadium-titanium magnetite resources. In this area, vanadium reserves account for 52% of China's total reserves for the mineral, while titanium reserves account for 95%. There is also a wide variety of rare and precious mineral resources such as cobalt, chromium, nickel, gallium, and scandium. There is an incredible value in exploiting the resources in this area. The actual work of exploitation, however, is a challenge. Over the years, young people had become unwilling to work in the mining areas of Pangang Group Mining Company due to poor operating environments, extreme working conditions, and difficulty accessing urban areas.

   At present, the open-pit mining process mainly includes piercing, blasting, shovel loading, and transportation. Piercing relies on drilling rigs to drill medium and deep holes. In blasting, explosives are buried in the holes and detonated, breaking ore rocks into pieces. Shovel loading uses electric shovels to load mineral materials onto mining trucks. In transportation, the loaded mining trucks transport the materials to the crushing station, where the materials are further crushed and then conveyed to the storage bin using conveyor belts. Every step of this process relies heavily on workers, who drive mobile equipment such as drilling rigs, electric shovels, and mining trucks to perform exhausting, dangerous operations in high-temperature, dusty environments. Understandably, therefore, it is difficult to recruit workers. The difficulty in recruiting or keeping workers creates another problem: low equipment utilization efficiency, since the effectiveness with which equipment is used depends on the skills of workers. In recent years, part of the transportation work in working face operations has been outsourced, but this is only a temporary solution, which transfers problems outside a mining company. It does not fundamentally address ongoing problems such as safety, labor force aging, and recruitment difficulties. And experienced onsite workers are still needed to operate high-value equipment such as drilling rigs and electric shovels. The mining industry needs a new solution: unmanned or less-manned mining.

The key to unmanned or less-manned mining operations in open-pit mines is unmanned drilling rigs, electric shovels, and mining trucks. Remote control of drilling rigs and electric shovels enables workers to sit in a comfortable office instead of staying in harsh mining areas. This is not only more convenient, but more efficient as well. For mining trucks, autonomous driving frees up drivers and simplifies driving operations, enabling a single driver to manage multiple trucks. With remote control and autonomous driving technologies, stable operation of excavating-loading-transportation equipment plus efficient management ensure safe and efficient production of working face operations in open-pit mines.

Achieving unmanned or less-manned mining operations in open-pit mines is a complex system engineering project, which should be specifically tailored to the actual operating environments and processes of open-pit mining. In terms of system structure, necessary tasks include reconstruction and redesign of equipment such as drilling rigs, electric shovels, and mining trucks, setup of basic wireless transmission networks, readiness of secure and stable computing and storage resources, as well as establishment of reliable and practical software/hardware environments for remote control and autonomous driving. Since existing wired and wireless network technologies cannot meet service requirements, basic wireless transmission networks have become a bottleneck. 5G resolves the issues. 5G applications enable remote intelligent piercing, excavation, and transportation in open-pit mines while supporting requirements of remote control and autonomous driving: large uplink bandwidth required by HD video as well as low latency and high reliability required by control commands. 5G is the natural choice for open-pit mines striving to achieve unmanned or less-manned mining operations.

In March 2021, Panzhihua branch of China Mobile worked with Huawei and other ecosystem partners to develop a 5G+ smart mining solution specific to the mining areas of Pangang Mining. Based on the current state of piercing, excavating, and transportation management as well as the production process of

Pangang Mining's open-pit mines, the solution is designed to achieve intrinsic safety, cost reduction, and efficiency improvement. 5G and edge computing technologies enable this solution to achieve the required large uplink bandwidth for HD videos as well as low latency and high reliability for control commands, which are used for remote control of mining devices (such as drilling rigs and electric shovels) in open-pit mines. In addition, the basic transmission network and edge computing resources lay a foundation for future interconnection of mining equipment and data. The project includes 5G network construction for drilling rigs and electric shovels, 5G edge data center and remote control center, as well as remote control operations.

Based on the equipment reconstruction situations of typical unmanned mines, the project aims to support efficient service operation. Its 5G network effectively reduces the network transmission delay and jitter and provides large uplink bandwidth required by video collection and transmission on the equipment side (including video data collection, encoding, network transmission, decoding, and display refreshing), ensuring efficient operation of 5G-enabled unmanned mines.

Looking at remote mining operations from a broader perspective, the intelligent application of remote, open-pit mining is a system engineering project consisting of four layers: the physical layer, application layer, network layer, and information management layer. As the cornerstone of complex system implementation, the network layer and information management layer supply the 5G network and the 5G edge data center. The application layer is mainly responsible for remote control operations. The physical layer brings about equipment reconstruction and a remote control center. Details are as follows:

(1) 5G private network construction
The 5G network consists of 5G user-side access devices, a 5G radio access network, a 5G transport network, and a 5G core network. The solution is designed based on the requirements of low latency, large uplink bandwidth, high reliability, and high security. 5G user-side access devices include automatic control components (such as sensors, cameras, controllers, and industrial computers) and network access devices (such as access routers and 5G CPEs). The 5G radio access network is composed of 5G base stations and their wireless coverage cells, which cover mining areas to enable real-time connection of mobile devices. The 5G transport network, with a ring topology to achieve high reliability, uses FlexE-based physical isolation and VPN-based logical isolation to ensure service security. The 5G core network consists of the operator's 5G core network as well as a new MEC platform which is deployed in mining areas, specifically in an integrated chassis or cabinet of the 5G edge data center, to support low-latency access of local services. The 5G network lays a foundation for the interconnection of mining equipment and data by setting up the private network environment required by the project, delivering 5G network planning, design, and analysis services, and providing basic commissioning and maintenance services to ensure transmission of related data and control commands.

(2) 5G edge data center construction
The 5G edge data center uses integrated chassis/cabinets in compliance with industry-level standards to integrate basic software/hardware resources for the 5G network and edge computing, meeting the security and real-time communication requirements of production data and supporting the applications related to the intelligent control system for 5G unmanned mines.

- Integrated chassis/cabinet: Compatible with current equipment room conditions in mining areas, it supports seamless deployment and features integration, security and reliability, a lower equipment room footprint, energy saving, quick and easy installation, architecture compatibility, fast and flexible deployment, and comprehensive monitoring.
- Basic hardware for edge computing: It provides the IT service environment and computing capability at the edge of the 5G mobile network, processing services that occur at the edge of the mobile network in real time. Basic hardware for edge computing supports single-user high bandwidth, high throughput, and high performance integration by adopting multi-core and multi-concurrency scheduling. It reduces the forwarding cost per bit by improving the processing capability through a combination of software and hardware.

(3) Remote control system
The remote control system includes an electric shovel remote control system and a rotary blast hole drill remote control system. It collects real-time location, status, and video information about mining trucks, electric shovels, and rotary blast hole drills, and provides control, monitoring, warning, and task scheduling functions. The central control system implements access and monitoring of rotary blast hole drills, mining trucks, and electric shovels. In addition, remote control cockpits are available for real-time remote control of various devices.

The central control system consists of four subsystems: a high-accuracy map, device positioning, safety monitoring, and dispatching. These subsystems cooperate with each other to deliver a stable and reliable smart mining system. In addition, the system provides interfaces for connection to the external mine manufacturing execution system (MES) and the truck dispatching system, supporting specific management and dispatching functions.

The high-accuracy map subsystem automatically creates high-accuracy 2D orthophoto maps or 3D model maps of mining areas using image collection by UAVs. The maps are used for positioning and navigation of intelligent terminals. Based on the 2D high-accuracy maps or 3D model maps, the monitoring screen displays the real-time locations and running status of various devices.

As a positioning and navigation satellite system such as the Global Positioning System (GPS), BeiDou, Global Navigation Satellite System (GLONASS), or Galileo satellite navigation system, the global navigation satellite system (GNSS) is used to locate base stations with the help of real-time kinematic (RTK) positioning. This system provides high-accuracy positioning and orientation data for mobile devices such as electric shovels and mining trucks.

A safety monitoring system displays the real-time status of vehicles, shovels, and drills. The system also sets alarm rules and reports alarms promptly.

(4) Remote control center

The remote control center is equipped with remote cockpits. The remote control and autonomous driving system receives real-time data from all devices, and performs data storage, analysis as well as graphical visualization. A man-machine interface is available for automatic delivery of remote control and mining tasks.

Overall, the remote mining solution integrates and applies mining techniques, network communications, and information technologies. It consists of the physical layer, application layer, network layer, and information management layer, requiring strong integration and collaboration between multiple disciplines, domains, and vendors.

2. 5G Smart Mining of Majialiang Coal Industry (Joint Venture of Datong Coal Mine Group and Zhejiang Energy Group)

In recent years, major mine groups in China have put forward clear requirements for smart mining. One of the key requirements is a central control system, which provides unified coordination and quick feedback control for a device cluster consisting of an HD video backhaul system, a high-density sensor access system, a smart robot inspection system, and a smart coordination control system. Other important requirements include massive data collection and transmission, numerous access devices, as well as ultra-low latency control operations. All these requirements set unprecedented requirements on the quality and capability of network transmission, making a wireless network with large bandwidth, low latency, and high reliability a crucial link for smart mining.

Network is the basis of smart mining. In the industry, currently, a ring industrial Ethernet over optical fibers is chiefly used for data transmission, Wi-Fi is used for wireless coverage in most cases, and 4G is deployed in some demonstration mines. However, the ring industrial Ethernet over optical fibers is easy to damage and difficult to maintain in a complex underground environment, a 4G network cannot effectively support low-latency control signal transmission, and a Wi-Fi network has significant latency in processing data transmission across access point (AP) regions. These problems of the existing underground network are serious obstacles to the development of smart mining.

Majialiang Coal Industry chooses a better way to build its smart mining solution. Based on cutting-edge technologies such as 5G and IP radio access network (IP RAN), Majialiang Coal Industry builds a highly reliable transport network and uses explosion-proof multimode base stations to achieve effective coverage across coal mines. 5G is used to cover working faces and heading faces that require high bandwidth and low latency, while 5G slicing technology ensures stable data transmission of different systems over a single network. This lays an important network foundation for smart mining. The benefits of 5G are significant. It reduces the underground communication latency, increases the transmission bandwidth, and enhances the support for mobile operations. 5G makes the operation processes controllable, and enables the staff to learn in advance about the latest production situations at the working site and identify potential risks in time.

In terms of typical smart applications, a 5G-based high-quality network plus an edge-computing platform promote the development of smart coal mining by enabling such features as smart belt awareness, Narrowband Internet of Things (NB-IoT)+comprehensive awareness, multimedia communication scheduling, HD video collection, smart video analysis, and remote control. Diverse application solutions across multiple scenarios such as underground communication, material management, safety monitoring, centralized control and smart production tools achieve smart interconnection of all production elements, including people, machines, things, environments, and more. The following describes the use cases of smart excavation and production control.

- 5G high-bandwidth transmission enhances the quality of underground video surveillance and enables real-time backhaul of HD industrial videos and images of main roadways, chambers, and working faces.
- HD video-based smart analysis automatically identifies violations in critical scenarios, generates warnings, and implements timely production control.
- Wireless sensors enabled with NB-IoT technology are able to work in a cableless environment where no power cables or network cables are needed. This enhances data collection and network perception capabilities across coal mine scenarios, facilitating production decision-making and control.
- With 5G-based VR/AR technology, connections between frontline personnel and background experts enhance remote device maintenance and commissioning.
- 5G low latency provides a technical foundation for remote control and unattended operation in places such as electric substations, water pump rooms, and gas drainage spots.
- Full-coverage 5G wireless signals support remote control of robot inspection in dangerous areas and timely remote data collection of temperature, gas, and device status.

In the process of implementing its smart mining solution, Majialiang Coal Industry has achieved many network, platform, and application innovations. These include:

1. A Highly Reliable All-in-One Underground Network
   According to underground 5G testing, the upload rate is 200 Mbit/s, the latency is less than 20 ms, and the effective coverage radius reaches 300 m. The basic

characteristics of underground 5G coverage have been identified based on the test data. The key guidelines for underground smart mining construction are:

- Use flame-retardant and impact-resistant optical cables to build a secure, reliable network infrastructure.
- Introduce IPRAN to build a carrier-class 10GE ring network.
- Use 5G base stations to cover critical application scenarios, supporting information backhaul of multiple data sources on the fully mechanized mining face and HD video backhaul on the heading face.

These methods meet all transport network requirements of smart mining and achieve smart management and operation of the all-in-one network.

2. An Edge Computing Core Platform
   As the key technical support for smart coal mining, the edge computing core platform allows local analysis on specific coal mine events and data, directly controls various mining devices, and provides prompt and accurate response. The core cloud is equipped with a smart platform to aggregate and collect mining data and edge cloud analysis results, continuously improving the edge cloud's response capability. This enhances the flexibility, operating efficiency, and intelligence of the entire system.
3. Focus on a High-Performance Transport Network, Exploration of Breakthroughs in 5G Applications
   Based on a cloud-pipe-edge-device structure, all-round technical researches and application deployments are carried out, focusing on gas monitoring, smart video backhaul and analysis, and major device fault prediction. This work aims to meet the requirements of coal mining enterprises in terms of comprehensive awareness, smart control, safe production, and operations management.

   By deploying multiple networks at once, Majialiang Coal Industry minimizes costs and avoids repeat network investment. Now, powerful network assurance is available to enable more wireless applications for smart coal mining, which directly improves the production efficiency and significantly reduces the coal production cost per ton. This improves the economic benefits. In addition, wireless networks require fewer optical/electric cables and devices in underground coal mines. Data of all new systems can be transmitted through nearby wireless networks, reducing the investment in new systems. Furthermore, fewer underground miners mean better overall safety, In conclusion, 5G-powered remote, centralized, and smart control technologies keep miners away from dangerous working environments, reassuring their families and the society. 5G application in the mining industry promotes the development of social productivity and places the enterprise in a better position to attract technical talent.

7.3.2 *Environment Monitoring and Safety Protection*

7.3.2.1 Industry Requirements

Safe production is a top priority of coal enterprises. China, however, so far only ranks in the middle in terms of global coal production safety. In 2019, its mortality rate was 0.083 per million tons of coal produced, 5 times higher than the United States and 11 times higher than Australia. It can do better. As such, reduction of production safety accidents is an important application scenario of 5G in smart mining. With 5G technologies, high-accuracy production and safety data of people, machines, and environments can be continuously and efficiently monitored in real time, ensuring safety while minimizing manual meter reading and onsite problem detection. This approach reduces the number of underground miners while improving the production environment.

Take coal mines as an example. Common sensors used by coal mines for environment monitoring and safety protection include methane sensors, carbon monoxide sensors, carbon dioxide sensors, temperature sensors, humidity sensors, negative pressure sensors, wind speed sensors, liquid level sensors, device startup/shutdown sensors, damper sensors, pipe flow sensors, smoke sensors, and dust concentration sensors.

In addition to safety warning, rescue is a critically important process in the event of an accident. Coal mines should be thoroughly prepared to manage rescues as quickly as possible. Reliable network assurance is especially important here: using surveillance cameras and positioning capabilities, miners equipped with smart helmets and smart handheld terminals can quickly get out of danger. In addition, it can enable rescue by robots, UAVs, and other rescue devices.

Concerning the natural environment, its destruction by mining activities tends to be ignored in many coal mines. This destruction, however, produces very serious effects. For example, the destruction of surface vegetation around a coal mine will seriously affect the local ecology as well as the lives of local residents. It is also inevitable for coal mining to waste or even pollute water resources. As such, it is necessary to monitor and protect the original flow paths of groundwater resources and to discharge the mine's inrush water in time. Strengthening environmental monitoring and safety protection as well as building green mines are absolute requirements for high-quality, sustainable mining industry development as well as for China's carbon peaking and carbon neutrality strategy.

7.3.2.2 Network Solutions

Smart mining aims to achieve unmanned or less-manned mining operations. The achievement of this goal depends on full awareness and in-depth understanding of the underground environment by ground personnel. Environment awareness relies on massive sensor data (such as temperature, humidity, gas, and atmospheric pressure).

5G networks must provide in-depth coverage and use ubiquitous IoT technologies to transmit sensor data to the ground accurately, in a timely manner. In-depth environment understanding depends on static and dynamic backhaul of HD images. The images taken can help ground personnel build a digital twin environment and can be widely used in diverse scenarios, such as coal seam identification, 3D map drawing, unmanned support, and emergency rescue.

To meet the requirements of environment monitoring and safety protection, high-performance 5G network coverage without blind spots must be ensured. Owing to underground environment complexity, such as dust and gas, there are also specific requirements on the transmit power of underground devices in order to ensure explosion proofing: The RF threshold power of wireless devices should not exceed 6 W. The underground environment is generally made up of narrow and long linear areas, with radio propagation markedly different from above-ground circumstances. The low transmit power and special propagation environment under the ground may result in coverage insufficiency. In addition, underground cables are difficult to arrange because the underground space is narrow and the working face is always moving. Additional site deployment may require cable rearrangement. Difficult site selection further aggravates coverage insufficiency.

Such problems can be resolved with signal repeaters, such as ManGeBao, which directly amplifies the signal strength and extends the coverage distance. When base stations are sparsely deployed in an underground mine, signal repeaters can be cascaded and work with directional high-gain antennas to improve signal quality. In addition, signal repeaters only require power supply connections, minimizing cable abrasion. As such, they can be installed onto hydraulic supports and similar equipment. 4G/5G wireless backhaul eliminates the need of wired transmission for mobile base stations. This enables fast movement and deployment of networks keeping pace with dynamic service changes. Beyond that, wireless backhaul is more reliable, without transmission interruptions caused by damaged optical fibers.

Another aspect that can be improved is positioning. On this, environment monitoring and safety protection set especially high requirements: at least meter-level accuracy, and sometimes even centimeter-level. Improved positioning is a key technological development direction for underground areas without GPS signals. Currently, precise 5G positioning can deliver accuracy of 3–5 m by measuring the signal strength or time difference, such as uplink time difference of arrival (U-TDOA) or round-trip time (RTT). If 5G cellular-based positioning cannot fully meet requirements, additional positioning technologies such as ultra-wideband (UWB) can work with 5G to achieve precise positioning. In this case, 5G provides power supply, transmission resources, and sites for UWB devices, effectively reducing deployment and O&M costs. This results in an all-in-one network meeting specific requirements for both communication and positioning.

7.3.2.3 Industry Ecosystems

In terms of mine sensors, there are presently many types. Most of them are connected over cables, Bluetooth, Wi-Fi, Long Range Radio (LoRa), and similar technologies. To accomplish long-distance, large-scale, and fast deployment as well as low-power operation, NB-IoT or Long Term Evolution Category 1 (LTE Cat.1) technology is an option. In the future, 5G reduced capability (RedCap) technology will be a new option for lower transmission latency and higher reliability. With cutting-edge mobile communication technologies, a single base station supports tens of thousands of connections. Additionally, sensors can be deployed far away from gateways, greatly reducing the number of gateways needed and minimizing unnecessary nodes.

For current positioning technologies applied to mining areas, these include GPS (for open-pit mines), UWB, Bluetooth, and ultrasonic wave (UW). A combination of 5G and such technologies can meet the requirements of environment monitoring and safety protection.

About rescue and robots specifically, recent years have seen increasing application of underground support robots, inspection robots, and rescue robots. The operation of these devices requires a high-quality, high-availability 5G network.

7.3.2.4 Policies and Standards

In November 2021, the National Development and Reform Commission, Ministry of Finance, and Ministry of Natural Resources of China jointly released the *14th Five-Year Plan for Promoting the High-Quality Development of Resource-Based Regions* [9]. This plan clearly states that green mines should be vigorously promoted, reconstruction and upgrade of existing mines should be strengthened, and new or expanded mines should all meet the standard requirements. The policy promotes the coordinated development of strategic mineral resources and downstream industries, supports the transformation and upgrade toward low-carbon, green, and intelligent development of resource-based enterprises, and coordinates the necessary steps to achieve carbon peaking and carbon neutrality. The policy urges strict implementation of ecological protection and pollution prevention measures related to resource mining. Ecological problems caused by resource mining should be resolved in a timely manner, such as vegetation damage, soil erosion, goaf subsidence, land salinization, water level subsidence, and heavy metal pollution. The potential pollution risks of closed-pit mines should be prevented. Historical issues such as waste mines, gangue hills, tailings, and extra-large open-pit mines should be resolved as soon as possible. The owner responsibilities of enterprises and local governments should be implemented. To achieve better management and responsibility, the policy also encourages exploring the third-party governance mode and acting in accordance with two principles: "those who destroy something should recover it" and "those who repair something should benefit from it."

In October 2020, the MIIT and Ministry of Emergency Management released the "*Industrial Internet + Safe Production*" *Action Plan (2021–2023)* [10]. This plan proposes to build fast awareness, real-time monitoring, early warning, emergency handling, and system evaluation capabilities. The plan also proposes to carry out "5G+ smart inspection" to implement cloud aggregation and online monitoring of key data for safe production.

In 3GPP Release 17, there is a project about low-power high-accuracy positioning (LPHAP). LPHAP aims to further improve positioning accuracy. Moreover, the battery life of positioning terminals running in long-time sleep mode can reach several months or even one year. In the future, positioning accuracy will be improved in 3GPP Release 18 to decimeter or even centimeter levels.

7.3.2.5 Typical Cases

Xinyan Coal Mine, owned by Lvliang Dongyi Group Coal Gasification Co., Ltd., has recoverable reserves of 160.99 million tons and an approved production capacity of 2.4 million tons per year. It embraced to the national initiative for 5G+ smart mining from the very beginning of 5G development, hoping to achieve safe, less-manned, and effective mining operations and finally accomplish digital transformation with higher profitability through smart safety, production, and management.

The first phase of its 5G smart mining project, which involved implementation of multiple innovative 5G applications, was completed in August 2020, making it one of the first mines in China to achieve full underground 5G coverage. One of those applications was remote coal mining through 5G signals, with the company's solution showcased in multiple national exhibitions. Another was the testing and release of the world's first 5G kite-like solution, which ensured high production system availability. The end of 2020 saw Xinyan Coal Mine selected as one of the first national smart mining demonstration units.

In June 2021, Xinyan Coal Mine started the second phase of the project focusing on convergence of smart subsystems and 5G, based on the phase-1 achievements. In November 2021, the second phase was completed and entered the acceptance procedure. During project rollout, Xinyan Coal Mine worked with ecosystem partners such as China Mobile Shanxi Branch and Huawei to promote 5G network optimization and develop "planning, construction, and optimization" services that adapt to the mining industry. This aids 5G+ smart mining's growing transformation from simply usable to user-friendly.

5G+ smart mining in Xinyan Coal Mine has explored the following 5G aspects:

- Monitoring system for production slope hoists: The type of equipment mainly used to transport materials along the production slope is winches supporting a maximum lifting or lowering weight of 50 tons. The monitoring system uses 5G to control the operation of a hoist and monitor the operational status of each component.
- Electric power monitoring system: An underground power supply system consists of mobile substations equipped with the high-voltage and low-voltage switches. This system uses 5G to read and control the real-time status of the high-voltage

and low-voltage switches, and monitors conditions inside the mobile substations through videos.

- Remote monitoring system for service slope winches: To lift people and objects along the service slope, mines use cages with a maximum lifting weight of 7.5 tons. The remote monitoring system uses 5G to start or stop a winch, lift or drop a cage to a specified position, and monitor the operational status of the winch in real time.
- Unattended system for endless-rope winches: This system uses 5G to perform remote control, positioning, and real-time video surveillance of endless rope winches, which use cyclic steel ropes to pull transportation devices such as mining trucks and flatbeds for long-distance transportation of materials and large objects.
- Remote monitoring system for monkey cars: Monkey cars run steplessly in a cyclical manner through drive wheels and steel ropes. They are a type of equipment used to transport miners over long distances. This system uses 5G to start, stop, and warn monkey cars, as well as monitor their operational status in real time.
- Steel rope flaw detection system: A flaw detection sensor and a speed sensor are used to evaluate the severity and location of flaws in steel ropes and belts. 5G improves this process by collecting and reporting flaw information to the centralized console for manual maintenance. The system greatly reduces the workload on inspections of belts and steel ropes while improving the timeliness of problem identification.
- Unmanned rail electric locomotives: The main function of unmanned rail electric locomotives is transporting materials between the surface and the underground. Here, 5G is used for data backhaul of the surrounding environment and operational status to perform remote start/stop, acceleration/deceleration, honking, and lighting operations. Autonomous driving, remote driving, and mobile terminal-controlled driving are supported.

The deployment of a 5G network plus smart application subsystems reduces manpower and the labor costs of fully mechanized mining teams, improves the mining efficiency, and increases the ore production. Smart application subsystems meet the requirements of less manpower or workload or no manpower in some posts. Centralized monitoring and operations can be performed aboveground in some posts. The status of underground devices can be monitored and managed in a centralized, unified manner aboveground. In this way, the number of underground miners can significantly decline, improving the overall safety. In addition, a comfortable working environment aboveground attracts industry talent, improving company and industry competitiveness. Miners are kept away from dangerous and harsh working environments reassuring their families and the society.

In this typical case, various smart coal mine subsystems over an all-in-one 5G network underground have replaced traditional wired transmissions, achieving a 5G leap from "infrastructure only" to "application integration". While rolling out subsystem services in Xinyan Coal Mine, there are also considerations of 5G network evolution and development of service capabilities in the industry. In the future, Xinyan

Coal Mine will work with partners to continuously improve the "planning, construction, and optimization" services of 5G networks in coal mines and the subsequent O&M system. As an excellent example of a 5G industry application, Xinyan Coal Mine plans to continue showcasing the Xinyan solution to others, building a solid support for the convergence of the coal industry and 5G.

7.3.3 *Autonomous Driving of Unmanned Mining Trucks*

7.3.3.1 Industry Requirements

The operations of open-pit mines face a number of serious challenges. The most serious is safety: due to a complex geological environment, large machinery may fall off at any time. The mining industry is actively exploring ways, including unmanned mining, to free miners from the dangerous operating environment. Another problem is that mining and transportation mainly rely on diesel vehicles, which consume great volumes of oil. This is not only costly, but pollutes the environment as well. A third problem is an increasing labor shortage and continuing high labor cost. Currently, open-pit mines are predominantly run as manual operations. There is an urgent need for new machines, technologies, and smart applications to improve production efficiency. In a word, autonomous driving is an especially important step for the mining industry.

According to surveys, the average age of drivers in China's mining industry is over 45 and rising. Large temperature differences, boring work, and hidden dangers in long-term driving lead to a low willingness of young drivers to take such jobs. Unsurprisingly, labor costs are high. Looking at overall costs, the annual costs of transportation, equipment maintenance, and manpower in mining areas account for 50 to 70% of total production costs. Safety, efficiency, and cost are the three major pain points of traditional mining areas. These pain points urge the autonomous driving application of unmanned mining trucks.

7.3.3.2 Network Solutions

The first requirement of autonomous driving of unmanned mining trucks is a precise 3D map of the mining area and a well-designed driving route. High-altitude cameras and UAVs carry out 3D mapping of the mining area. It is the job of the 5G network to provide not only favorable ground coverage but also low-altitude coverage.

The second requirement is the necessary hardware to support autonomous driving. An unmanned mining truck should support three modes: manual driving, remote driving, and autonomous driving. Manual driving should have the highest priority and autonomous driving the lowest. Autonomous driving works by taking over the driving control system through remote control, which is enabled by a wire-controlled steering system, where the truck receives computer control signals and transmits generated data back to the remote control cockpit through a computer. Implementation of this wire-controlled steering system generally includes several types of refitting: power supply refitting, light refitting, throttle refitting, gear shift refitting, braking refitting, steering refitting, and lifting refitting. It is also necessary to add additional components such as sensors, control motors, solenoid valves, relays, travel switches, and industrial computers. Data transmission, processing, and smart control are achieved by adding cameras, laser radars, millimeter-wave radars, GPS, speed and acceleration sensors, tilt sensors, vibration sensors, vehicle-mounted computing units, and communication modules. Vehicle-mounted computing units and cloud servers work together to implement environment awareness, high-accuracy positioning, autonomous driving, active obstacle avoidance, operation detection, and safety monitoring. All this means the network requires low-latency and high-reliability capabilities as well as an edge computing data center. Moreover, a network running in a complex mining area environment should be well designed and deployed.

The third requirement is an overall dispatching system for the unmanned mine. 5G+ vehicle-to-everything (V2X) technology provides real-time smart control and dispatching of operation groups consisting of unmanned mining vehicles, smart excavators, remote-control drilling rigs, and crushing stations, taking into account the ore grades and mining volume requirements of the mineral processing plant. Smart management and dispatching in real time significantly improves production efficiency and automation.

7.3.3.3 Industry Ecosystems

5G UAV mapping and inspection, 5G remote control, 5G autonomous driving, 5G smart camera, and 5G industrial data collection are all typical 5G application scenarios across industries. Autonomous driving in a closed environment has been applied in ports and logistics transportation. Devices such as laser radars, millimeter-wave radars, and vehicle-mounted computing units have been widely used for autonomous driving. As such, the autonomous driving ecosystem of unmanned mining trucks is mature. The requirements of open-pit mines for 5G base stations are not as strict as those of underground mines, so the autonomous driving application of unmanned mining trucks in open-pit mines is ready for large-scale promotion.

7.3.3.4 Policies and Standards

In March 2020, eight national departments, including the National Development and Reform Commission and the National Energy Administration, formulated the *Guiding Opinions on Accelerating the Smart Development of Coal Mines*. It aims for unmanned transportation in open-pit coal mines by 2025 [8]. In November 2020, CHN Energy issued the *Notice on Further Accelerating the Development of Smart Coal Mining* [11]. It states that, by 2022, there will be 10 national smart coal mine demonstration units as well as 5 open-pit coal mines with autonomous driving. By

2025, CHN Energy plans to complete transformation of all its coal mines into smart mines and take the lead in national smart mining.

7.3.3.5 Typical Cases

The onsite mining process of a coal mine in Inner Mongolia mainly includes perforation, blasting, stripping, coal mining, transportation by mining trucks, crushing, and belt conveyance. Many of the operations require cooperation between core production equipment (such as drilling rigs, electric shovels, hook machines, and mining trucks) and auxiliary production equipment (such as loaders, levelers, bulldozers, cleaning vehicles, and cranes). The planning, construction, maintenance, and optimization of 5G networks are more difficult than those of traditional large networks due to complex and various geographical environments, changing types and quantities of production devices in different regions, and differentiated SLA requirements of service applications on network indicators.

Through cooperation with China Telecom Inner Mongolia Branch, the coal mine applies emerging technologies such as 5G, AI, HD video, big data, and cloud computing to smart mining, achieving smart sensing, ubiquitous connections, and precise control of production. Multiple 5G+ X application scenarios, like autonomous driving, remote control, smart coal mining, and smart inspection, are created to ensure intrinsic safety of mining and improve production and operation efficiency.

5G network construction in this use case converges resources in the autonomous driving industry, breaks through core technical difficulties in autonomous driving refitting, tailors 5G+ technologies to mining scenarios, and sets a benchmark for the industry to follow.

The innovations involved in this use case cover multiple areas such as network solutions and service delivery, building industry-leading advantages. The innovations related to autonomous driving of unmanned mining trucks are as follows:

- "Wing-like" project on a customized 5G network: The edge-deployed 5G core network and stable, secure, reliable signal backhaul of 5G macro base stations meet the requirements of mining trucks, electric shovels, and auxiliary vehicles in terms of low-latency access to the unmanned driving application platform. In addition, the network capacity requirements of smart mining in the next five years are met.
- Digital foundation with a high-accuracy 3D electronic map using UAVs for precise network planning: UAVs collect HD images of mining areas. Then, precise rendering and reconstruction techniques are used to build a 3D model. Beyond that, accurate service capacity estimation and deterministic uplink rate simulation are carried out for precise 5G network planning in mining areas. All these steps avoid the difficulties and low accuracy of traditional manual surveys.
- Super uplink technology used for large uplink bandwidth needed by service applications such as unmanned driving and remote control: To meet the large uplink bandwidth requirements of ToB service applications, China Telecom and

Huawei launched the super uplink technology. It combines the uplink 3.5 GHz TDD spectrum and uplink 2.1 GHz FDD spectrum to achieve uplink full-timeslot scheduling, increasing data rates of user equipment (UE) near a base station by 20 to 60% and UEs at faraway spots by 2 to 4 times, while reducing air interface latency by 30%. This better meets the needs of service applications on network indicators. In addition, super uplink technology can also be applied together with the A + P ultra-lean site, which can maximize the antenna height without an additional pole, improving the coverage performance by 20%.

- Slicing technology used for E2E service isolation and improved experience: Different service flows have different requirements on network indicators. For example, a video service flow has a relatively high requirement on uplink bandwidth, while an instruction service flow is sensitive to latency. Slicing technology maps different types of data flows into network slices, ensuring secure and stable service flow transmission.
- Dual management platforms for efficient O&M of terminals and networks: To ensure efficient O&M of a large number of onsite smart terminals in mining areas, the mining truck health management platform proactively detects faults and security events to ensure secure, stable, and reliable operation of vehicles and unmanned driving systems. The 5G network self-management platform quickly identifies network problems and reduces the production interruptions due to faults. Customized applications and efficient network O&M are supported.

The application scenarios incubated in this 5G+ smart mining case, including remote control, unmanned driving, and smart collaboration, have freed people from dangerous and harsh operating environments and low-value repetitive work. These applications maximize the value of mining equipment, yield significant benefits and improve operation efficiency for mining enterprises.

7.4 Summary and Prospects

"Joint contribution and shared benefits" is the only way to resolve the difficulties in implementing 5G industrial applications. The success of 5G applications in the mining industry depends on a close cooperation among the telecom operators, equipment vendors, mining enterprises, and the mining ecosystem as a whole. A high-quality private 5G network with ground and underground integration is vital for a coal mine, in that it serves as a solid foundation for various applications in the mining areas.

Technological innovations based on a host of technologies such as 5G, AI, and edge computing are launched to explore applications integrating various mining equipment and the existing positioning/sensing approaches. These innovations are all tailored to the actual situations and requirements of the mining areas as well as the production phases. There is also a great emphasis on strictly following safe production regulations in the mining areas. These are the basic principles that practical, efficient, safe 5G+ smart mining should follow.

Smart excavation and production control, environment monitoring and safety protection, as well as autonomous driving of unmanned mining trucks make full use of 5G features including large uplink bandwidth, low latency, flexible deployment, high reliability, and high safety, which have helped mining enterprises improve production efficiency, reduce safety risks, and lower mining and transportation costs. These applications have also improved the mechanization, automation, informatization, and intelligence of the mining industry.

5G has accelerated the transformation and upgrade of traditional mining enterprises by achieving smart mining across the production, operation, and management phases including excavation (stripping), power supply, water supply and drainage, ventilation, main and auxiliary transportation, safety monitoring, and washing. Smart applications designed for mining and transportation collaboration, decision-making management and control, as well as integrated operation have been put into use. Unattended operations in fixed positions and robot operations in dangerous positions have been promoted to achieve unmanned or less-manned mining operations.

5GtoB applications have enhanced the sensing, monitoring, warning, handling, and evaluation capabilities for mine production. Safe production has been transformed from static analysis to dynamic sensing, from post-event response to pre-event prevention, and from single-point prevention to joint prevention by multiple parties. Safety risks have been significantly reduced and even eliminated. The working environment of miners and the intrinsic safety of mine production have been improved.

In addition, environment monitoring and protection have been strengthened. The traditional way of mineral resource development, which features sheer consumption of mineral resources, deterioration of ecology, and high energy consumption, has been fundamentally transformed to a "green mining" approach, by which the coordinated and high-quality development of rational exploitation and utilization of mineral resources together with environmental protection is promoted.

5G applications reduce manpower requirements and mining costs, improve mining efficiency, save energy consumption, and reduce resource waste. All this helps mining enterprises improve their industry levels, optimize their production techniques, and enhance their competitive advantages.

The successful application of 5G in the smart mining field will have lasting positive impacts. By encouraging more enterprises to replicate the success, it will drive the digital transformation of the entire mining industry and promote the adoption of more smart solutions in the industry.

In the future, 5G will be further applied to smart mining. Existing technologies will be continuously optimized and various application scenarios, especially the three major application scenarios discussed in this chapter, will be further promoted as smart mining enterprises play a leading role in the industry. More application scenarios will be explored through a close mine ecosystem cooperation. A deep convergence of 5G technologies into the mining equipment will unlock new opportunities for cross-domain technical innovations, which in turn will lead to innovations in self-developed technologies. All this will enable the industry to realize all-around mining mechanization, informatization, automation, and intelligence, transforming mines to be green, efficient, smart, safe, and unmanned.

References

1. Global energy statistical yearbook (2021)
2. China mineral resources report (2021)
3. Outline of the People's Republic of China 14th Five-Year Plan for national economic and social development and long-range objectives for 2035 (2021)
4. 5G Application implementation plan in the energy sector (2021)
5. "Set Sail" action plan for 5G applications (2021–2023) (2021)
6. 5G Underground mobile communication network (5GDMN) white paper (2021)
7. 5G+ coal mine intelligence white paper (2021)
8. Guiding opinions on accelerating the smart development of coal mines (2020)
9. 14th Five-Year Plan for promoting the high-quality development of resource-based regions (2021)
10. "Industrial Internet+Safe Production" action plan (2021–2023) (2020)
11. Notice on further accelerating the development of smart coal mining (2020)

Chapter 8
5G+ Smart Steel

8.1 Overview

8.1.1 Current State

The iron and steel industry acts as a foundation for numerous industries in China. It relies on the iron ore, coke, and non-ferrous metal industries and heavily impacts the real estate, automotive, shipbuilding, home appliance, machinery, and railway industries.

Iron and steel production also plays a vital role in driving China's economy. According to the National Bureau of Statistics of China (see Fig. 8.1), the national production of pig iron, crude steel, and steel products reached 869 million tons, 1033 million tons, and 1337 million tons, respectively, in 2021. This represents a decrease of 19 million tons, a decrease of 20 million tons, and an increase of 15 million tons compared to 2020, respectively.

After decades of investing in and developing this industry, China has become the world's largest producer and consumer of steel. Since 2018, China has produced more than half of the global crude steel every year. In 2021, China was responsible for 53.32% of global crude steel production, slipping 2.75 percentage points compared to 2020, as illustrated in Fig. 8.2.

Since the 13th Five-Year Plan, major steel enterprises in China have equipped themselves with world-leading technologies and achieved informatization-driven development. Smart manufacturing has been extensively applied to many processes including steel production, enterprise management, logistics and distribution, and product sales. Numerical control has been introduced to over 65% of key manufacturing procedures, and the deployment ratio of Enterprise Resource Planning (ERP) systems has exceeded 70%. Despite becoming a big player in the global steel industry,

P. Sun, *A Guidebook for 5GtoB and 6G Vision for Deep Convergence*, Management for Professionals, https://doi.org/10.1007/978-981-99-4024-0_8

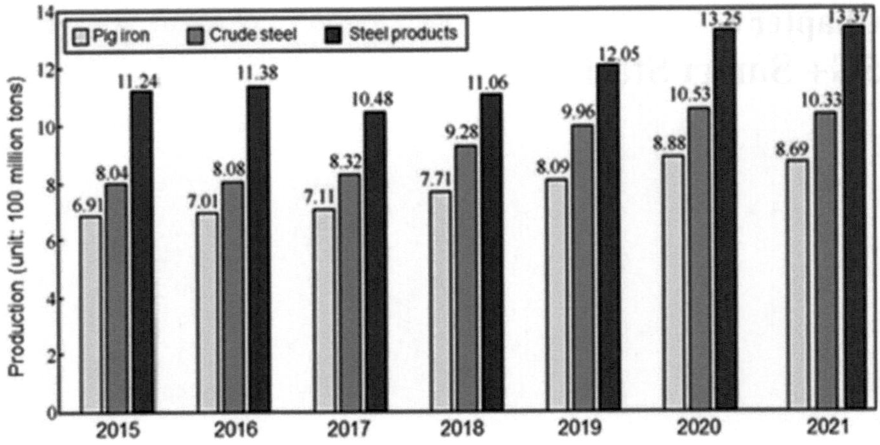

Fig. 8.1 China's iron and steel production from 2015 to 2021. *Source* National Bureau of Statistics of China

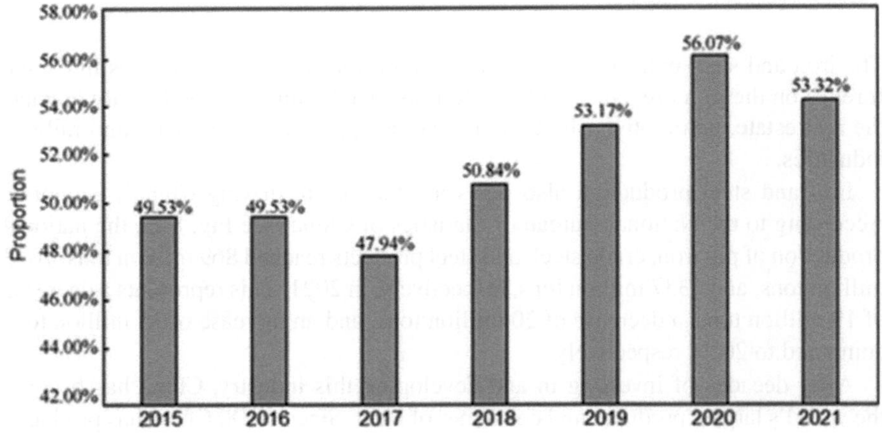

Fig. 8.2 China's contribution to global crude steel production from 2015 to 2021. *Source* National Bureau of Statistics of China and World Steel Association

China still has a long way to go before it can become a steelmaking powerhouse due to the following challenges.

1. Uneven development: China's steel industry is characterized as a fusion of mechanization, electrification, automation, and informatization. However, development varies widely among different enterprises. Leading players like China Baowu Steel Group have reached Industry 3.0 and are now advancing towards Industry 4.0, while a multitude of steel enterprises in China are still at the level of Industry 2.0. In addition, advancement differs greatly among production lines within

individual enterprises. Some branch factories or production lines have implemented remote and unmanned operations, while the majority still rely on manual production.

2. Weak industry foundation: Smart manufacturing is still in the early stages of development and lacks a solid foundation of standards, software, and information security. No true industry standards have been formulated as of yet, and no significant breakthroughs have been made in key technologies.
3. Unmeasurable return on investment (ROI): With soaring labor costs and growing demand for better work environments, the demand for remote and automated production is in turn increasing. However, smart production is still a novel solution, and there is no effective approach to measuring the ROI of smart production as of yet.
4. Insufficient core original innovation: A technological innovation system that features high levels of collaboration between enterprises, universities, and research institutes has still not yet been developed. The entire industry still lacks a cohesive and active pursuit of original research and development. Policy support in this area also needs to increase.

Ultimately, steel production features numerous procedures, long workflows, and specialized technologies. From sintering, coking, chemical processing, and iron-making, to steelmaking, rolling, and forging, the entire production process involves a wide array of materials such as ferroalloys, refractories, and carbon products and an equally broad spread of technologies related to hydrodynamics, electrodynamics, aerodynamics, oxygen dynamics, transportation, machining, and more. These procedures also require a variety of equipment including industrial lime kilns, coke ovens, blast furnaces, rotary/open-hearth furnaces, electric furnaces, and heating furnaces, as well as sintering machines, rolling mills, oxygen generators, large electromechanical devices (such as bridge cranes, cranes, forklifts, air compressors, blowers, motors, shearing machines, and large lathes), and other high-temperature/voltage devices. As such, the steel industry is considered both technology and labor intensive, featuring diverse equipment, large scale, high energy consumption, and massive logistics quantities.

Steel production generally involves mineral processing, sintering, coking, iron-making, steelmaking, continuous casting, and rolling processes. Steel products can be divided into a wide variety of carbon steel and special steel products such as wires, plates, pipes, profiles, and composites. The coking process in steel production also produces by-products such as benzene, phenol, asphalt, blast furnace gas, converter gas, coke oven gas, granulated slag, and steel slag.

8.1.2 Smelting Process

1. Mineral Processing

Mineral processing is a prerequisite for smelting. Freshly mined ore with high metal contents—such as iron, copper, aluminum, and manganese—are selected for smelting.

There are three general operation stages in mineral processing: crushing, grinding, and separation. Crushing, for example, involves coarse, intermediate, and fine crushing, and separation involves magnetic separation, gravity concentration, and flotation.

2. Sintering

In order to even out the iron content in iron ores supplied to a blast furnace and ensure the permeability of the furnace, iron ore concentrates produced from mineral processing need to be processed into 10–25 mm lumps of ore.

Sintering and pelletizing are the two most widely used methods to create iron ore concentrates. Sintering is the process of fusing iron ore concentrates, anthracite powder, lime, blast furnace dust, mill scale, and steel slag according to a certain ratio into solid and coherent iron ore sinters for ironmaking. Pelletizing is the process of adding a small amount of additives to iron ore fines or other iron powder and using pelletizers to agglomerate the fines with moisture into roasted solid iron ore pellets for smelting.

3. Coking

In addition to iron ore (sinters and pellets), coke also needs to be fed into blast furnaces. It serves as the major fuel for blast furnace ironmaking. Coke is burned at the tuyeres to generate a large amount of heat and gas. The ascending gas transfers heat to trigger various physical and chemical reactions in the blast furnace.

4. Blast Furnace Ironmaking

Blast furnace ironmaking is the continuous process of reducing iron ore to pig iron, which is critically important to the steel industry.

During this process, a specified ratio of solid raw materials such as iron ore, coke, and flux are charged via the furnace top to reach a certain level at the furnace throat. They are charged to the furnace in alternating layers of coke and ore. Descending ore burden is gradually reduced and melted into iron and slag that accumulate in the hearth. The molten iron and slag are removed from the furnace through the iron and slag notches at regular intervals.

5. Converter Steelmaking

Converter steelmaking is a process that converts molten iron, steel scrap, and ferroalloy into steel. The converter uses the sensible heat of molten iron and the chemical reactions between the components of molten iron, without relying on external energy.

The primary processes within converter steelmaking are hot metal pretreatment, smelting, and refining.

Hot metal pretreatment removes unwanted components or extracts wanted components from hot metal before smelting. This process can be further divided into common pretreatment (e.g. desulfurization, desiliconization, and dephosphorization) and special pretreatment (e.g. vanadium extraction, niobium extraction, and dechrome).

Hot metal smelting uses the sensible heat of molten pig iron in the converter and the heat generated by chemical reactions between the oxygen blown into the furnace and the pig iron elements (carbon, manganese, silicon, and phosphorus), to achieve a metal composition and temperature ideal for steelmaking.

Hot metal refining refers to deoxidation, desulfurization, inclusion removal, inclusion degeneration, composition fine-tuning, and hot metal temperature control in a vacuum, inert, or controlled atmosphere. Ladle refining operations include argon blowing, vacuum treatment, oxygen blowing, heating, powder spraying, and wire feeding.

6. Continuous Casting

After hot metal treated by a converter is refined within a refining furnace, it is cast into billets of different types and specifications.

Continuous casting is the name of this process for solidifying refined hot metal into billets. A ladle of refined hot metal feeds a holding bath called a tundish via a rotating turret and then flows from the tundish into molds. Molds — one of the core equipment of a continuous casting machine—are then used to freeze and solidify the hot metal directly in contact with them. Withdrawal straightening machines and mold oscillators work together to withdraw the solidified metal from the molds. After cooling and electromagnetic stirring, the solidified metal is cut into slabs of a certain length.

8.1.3 Steel Rolling Process

Semi-finished billets or slabs produced by steel mills must be rolled in rolling mills to become qualified steel products. Hot-rolled finished products are classified into steel coils and ingots with an average thickness of several millimeters. Products of lesser thickness require cold rolling.

7. Hot Rolling

Hot rolling is performed at temperatures above the recrystallization temperature of the metal. Raw materials like slabs or blooms are heated using a walking beam furnace and descaled with high-pressure water before being fed into a rough mill. The rough-rolled materials have their front and tail ends cropped and then are fed into a finish mill for computer-controlled rolling. After laminar cooling, the finish-rolled materials are coiled into straight coils. After front-end cropping, tail-end cropping, edge trimming, and multi-pass finishing processes (e.g. straightening and flattening), the straight coils are either cut into plates or recoiled as hot-rolled steel plates, flat hot-rolled steel coils, slit strips, or other such steel products.

8. Cold Rolling

Cold rolling is performed at temperatures below the recrystallization temperature of the metal. Depending on the specific product requirements, this process generally involves pickling, descaling, cold rolling, degreasing, annealing, flattening, shearing, classification, and packaging.

Cold-rolled materials must go through a pickling line to remove any scale from their surface before rolling. This ensures the strip surface is smooth enough for cold rolling and surface treatments. Pickling, in theory, is followed by rolling. Cold rolling, however, requires the materials with a certain thickness first be softened through an annealing process. Any lubricant on the strip surface will become volatile in the annealing furnace and create dark spots that are difficult to remove from the strip surface. As such, any residual lubricant on the strip surface must be cleared through a degreasing process prior to annealing. The degreased strips are then annealed in a protective gas. Once annealed, the strip surface is bright and does not require pickling for further rolling or flattening. Annealed strips must be flattened to have a flat surface, a uniform thickness, and improved performance. The flattened strips can be sheared according to orders.

8.2 Digitalization Trends and Challenges

8.2.1 Policies and Opportunities

The iron and steel industry is a pillar of China's economic growth and an important part of the country's plans for industrialization, modernization, and green and low-carbon development. In 2021, the Ministry of Industry and Information Technology (MIIT) issued guidelines on promoting high-quality development within the steel industry: "We hope that by 2025, China's steel industry will form a high-quality development pattern, which features reasonable layout, stable resource supply, advanced technology and equipment, prominent quality brands, strong global competitiveness, and green, low-carbon and sustainable high quality."

Accordingly, resource supply for the steel industry is expected to be improved by shutting down outdated production facilities, encouraging energy conservation and emission reduction, and deploying smart manufacturing.

1. Industrial Agglomeration and Legacy Facility Phase Out Requires Innovative Digital Steel Plants.

Industrial agglomeration requires the development of several world-class super-large steel enterprise groups and highly professional enterprises. The overall goal is to increase the concentration ratios of the top 5 players and the top 10 players to 40% and 60%, respectively. The share of crude steel produced in electric furnaces is therefore expected to increase to 15%–20% of the total crude steel production with a scrap rate of 30%.

2. Energy Conservation, Emission Reduction, and Green and Low-Carbon Development Require Reconstruction and Upgrade to Smart Production.

Regarding energy conservation, the total energy consumption and energy intensity of the industry are expected to reduce by more than 5%, while its water consumption intensity is expected to decrease by over 10% and water reuse rate to increase to over 98%.

Regarding emission reduction, the industry's reconstruction completion rate for ultra-low emission is expected to reach over 80%, with all enterprises in key areas with the total emissions of pollutants reducing by over 20%.

Regarding green and low-carbon development, inter-industry coupling is also being promoted to build a cross-resource recycling system to take the lead in achieving peak carbon emissions.

3. Intelligent Upgrade Requires Precise Operations Management Based on 5G, AI, and Cloud.

To achieve smart manufacturing, the industry will need to significantly enhance its manufacturing capabilities, introduce numerical control to approximately 80% of key manufacturing procedures, enable digital transformation for 55% of manufacturing equipment, and have at least 50 smart factories built.

Through equipment upgrade, the production capacity of advanced coke ovens is expected to represent 70% of total production capacity, and the production capacity of both advanced ironmaking and steelmaking is expected to reach over 80% of total capacity.

To facilitate innovation, R&D investment in the industry is also expected to account for 1.5% of the overall business revenue.

Regarding quality and efficiency improvement, current goals are to reach an industry average of 1200 tons of steel, with annual average labor productivity for new common steel enterprises expected to reach 2000 tons of steel. Product quality, performance, and stability are also expected to be further improved.

8.2.2 Challenges Facing Smart Steel

Steel remains a mainstay of China's economy, providing essential materials for the survival and growth of many other industries. However, the steel industry is subject to dangerous and harsh environments, heavy and repetitive work, numerous network silos, outdated monitoring methods, high requirements on the experience and quality of equipment management personnel, and low production and management efficiency due to the large number of onsite spot-check personnel. Smart manufacturing would be the ideal solution to address these challenges.

1. Low Efficiency

Smart manufacturing in steel enterprises would focus on increasing production and reducing consumption. Steel plants are often hindered by high labor costs, high-risk work environments, shift work, and occupational exposure to noise, dust, and high temperature. Workers suffer from this challenging work environment and working hours, reducing work efficiency. Steel plants also typically cover a large area and require large numbers of workshops, warehouses, vehicles, and personnel. This makes traditional management methods, like warehouse stocktaking, very labor- and material-intensive. Logistics management using AI-incapable cameras also makes vehicle identification and dispatch challenging. Reconstruction would be urgently required to improve efficiency.

2. Harsh Environments

Steel plant workers are regularly exposed to high temperature, dust, high pressure, extreme noise, and even toxic and hazardous gases. They are often at risk from thermal radiation and burns, toxic and hazardous gas leaks, 1500°C hot metal splashes, and blast furnace explosions. Such harsh environments make it difficult for enterprises to recruit and retain employees. More investment would be needed to ensure safe and green production.

3. Complex Production Processes

The three key stages of steel production (ironmaking, steelmaking, and rolling) all involve multiple production systems, industrial control systems, and supply chain levels, each of which, in turn, features complex processes and hierarchies. This often wastes resources and limits the production capacity.

4. Network Silos

Separate networks are used for office work, production (industrial intranet, industrial Wi-Fi, and microwave), video capture, energy management, intercom systems (in the 800 MHz frequency band), and emergency handling. They are independent and isolated. Some networks are unstable and incompliant with safe production standards, restricting enterprise process collaboration. Hybrid networks consisting of both optical fiber and Wi-Fi are often used in plants, increasing expansion costs, making O&M more difficult, and resulting in severe Wi-Fi interference. These problems also restrict possible improvements to smart manufacturing capabilities.

8.2.3 Intelligent Reconstruction for 5G Smart Steel

The steel industry involves process-based production, featuring long processes, complex procedures, and strict requirements for safe and green production. Riding the new wave of industrial automation, intelligence, and digitalization initiatives, steel enterprises are already proactively exploring converged applications of 5G technologies in steel production, campus logistics, and safety assurance to promote industry upgrade. This creates a huge market space for smart steel reconstruction.

The high bandwidth, low latency, and massive connectivity of 5G networks are allowing leading enterprises in the information communications and steel industries to explore converged 5G+ smart steel opportunities through pilot applications. As these pilots yield positive results, we have already seen many of these new requirements and application be implemented: 5G smart steel equipment (including upgraded bridge cranes, casthouses, robots, robotic arms, data detection devices, logistics AGVs, locomotives, and AR-based spot-check/inspection devices), smart production applications (including smart steelmaking, hot rolling, cold rolling, and energy environmental monitoring systems), smart operations (including smart operation centers, real-time HD video surveillance, unmanned workshops, and geofence), and smart campus.

8.3 Major 5GtoB Large-Scale Replication Scenarios and Typical Cases

8.3.1 AI Steel Surface Quality Inspection

8.3.1.1 Service Requirements

China has become the world's largest producer of both steel and common non-ferrous metals. When it comes to in-process inspection for steel and non-ferrous metals, the human eye is not capable of inspecting objects at high temperature (usually about 1000 °C). For finished product inspection, manual inspection relies on spot checks due to high production volumes, which reduces inspection accuracy. Manual inspection of objects moving at high speeds (as they often are at many of these production stages) significantly increases missed defect rates. In contrast, machine-vision-based quality inspection systems can automatically identify and classify quality defects on steel surfaces, analyze defect statistics, and assist in diagnosing and solving problems caused by defects. As such, automatic metal surface quality inspection has a large market and a good development prospect in this industry.

Defects on the steel surface include iron oxide scales, holes, folds, wave edges, dents, rust spots, and glues. Steel comes in various types and their unique characteristics have to be taken into consideration during inspection. Due to restrictions on transmission technology, most automatic surface inspection systems in use are standalone. Servers and data processing and analysis software need to be deployed at the edge of each production line. This not only increases costs but also produces subtle synergy effect because it is difficult to share defect data between different production lines. 5G+ AI surface quality inspection uses 5G features to transmit surface images captured by industrial cameras to the MEC platform. The system can be easily deployed, implementing centralized data processing and analysis and reducing a large amount of investment in software and hardware. In addition, AI-based learning is implemented based on the defect data accumulated from multiple production lines, significantly shortening the learning time of each production line and increasing the defect detection rate. According to preliminary predictions, the MEC platform used for centralized processing and maintenance as part of a 5G+ steel surface quality inspection solution can improve O&M efficiency by up to 80%, reducing investment by more than 20% compared to standalone detection systems.

8.3.1.2 Network Solutions

Steel plants often have high ambient temperature, heavy dust, and production lines with the risk of steel breakouts, which is not ideal for optical fiber deployment. This makes 5G wireless networks more suitable for data transmission in steel plants. The device-side detection system enables linear array cameras to connect to a 5G network through industrial computers. Industrial computers are used to compress images, connect with the preprocessing server based on Transmission Control Protocol (TCP), and convert data packets based on the image transmission standard Gigabit Ethernet into TCP packets for transmission. The preprocessing server—connected to the algorithm server through a switch—preprocesses the received images before sending them to the algorithm server for defect detection. It is also interconnected with the customer-oriented manufacturing execution system (MES) through the enterprise's level-1 and level-2 networks to obtain production data and other information. The information can be used to identify defect objects and adjust parameters such as the photographing frequency of device-side linear array cameras.

Fig. 8.3 illustrates the network architecture for 5G+ AI steel surface quality inspection.

8.3.1.3 Industry Ecosystems

AI steel surface quality inspection is one of the pioneering 5G applications in the steel industry.

In terms of software and hardware, network, spectrum, and cloud resources are provided by telecom operators; communications devices and E2E 5G solutions are provided by 5G solution providers or large-scale comprehensive steel solution providers; modules, gateways, industrial cameras, and other quality inspection terminals are required; and vision applications are provided by AI solution providers.

In terms of services, telecom operators and device vendors provide ICT products and services, including networks, cloud services, and edge equipment rooms; and industry terminal manufacturers provide basic hardware, including gateways, industrial cameras, and sensors. AI solutions and cloud resources are essential for steel surface quality inspection, and so AI solution providers offer an AI middle platform

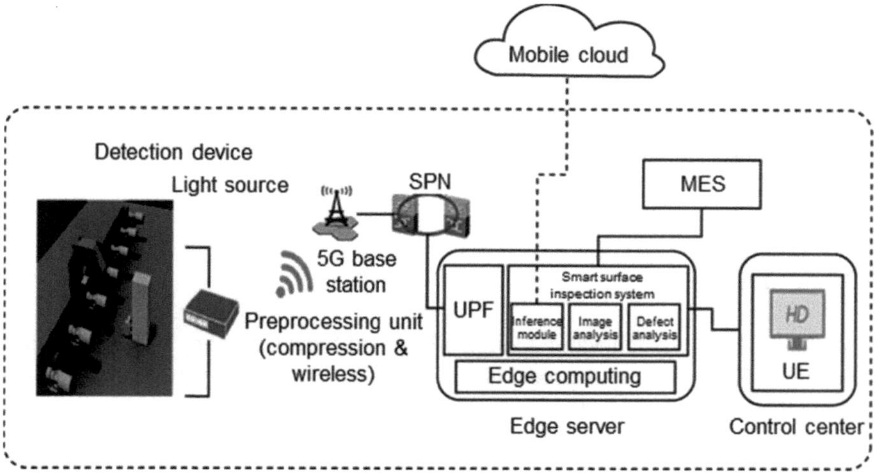

Fig. 8.3 Network architecture for 5G+ AI steel surface quality inspection

to accumulate algorithms, knowledge, methods, and experience to achieve precise prediction and quick troubleshooting through human-machine collaboration. Cloud resource providers also need to deploy cloud service capabilities down to edge nodes on demand to manage resources and services on the cloud while maintaining system and data security in cloud computing for the steel industry. By managing edge nodes, cloud services—such as intelligent video analysis, big data streaming, and machine reasoning—are extended to industrial enterprises as applications. Through two-way data exchange between the edge and cloud, the industrial enterprise applications acquire the intelligent functions of both sides. In addition, enterprise applications need to direct services to the edge through the intelligent edge platform and perform O&M on the cloud, such as unified device/application monitoring and log collection, encompassing a complete edge computing solution for enterprises.

8.3.1.4 Business Models

AI steel surface quality inspection not only improves inspection performance in the steel industry but also further reduces labor and material resource waste by reporting on batch defects and possible process problems in a timely manner. Currently, AI steel surface quality inspection is charged per project. Steel enterprises propose construction content and standards, and telecom operators and integrated service providers quote for the devices, software development, cloud resources, and maintenance fees needed for the project and are responsible for construction and delivery. After acceptance, the project expenses are paid by the steel enterprise. 5G+ AI steel surface quality inspection requires cloud-edge-device collaboration. A large amount of investment is needed to build up the requisite public, private, and edge cloud computing capabilities. In addition, a large number of industrial cameras and laser scanners are required to collect onsite data on the device side. As the clouds and devices are interconnected via the network to complete quality inspection, the initial investment of a project like this is high for industry customers. This is one of the obstacles that must be overcome to popularize machine-vision-based quality inspection. A potential solution to reduce the initial investment and decision-making cycle is charging per month or based on actual 5G data usage.

8.3.1.5 Typical Cases

Founded in 1958, Hunan Valin Xiangtan Iron and Steel Company (XISC) provides more than 400 varieties of products, covering wide and heavy plates, wires, and bars, with an annual production capacity of 16 million tons of steel. As one of major finished products of XISC, bars are processed at high speeds (20–40 m/s), particularly requiring real-time online quality inspection. In this case, a 5G network is used to upload images (with a high uplink bandwidth of 616 Mbit/s, providing surface inspection and size measurement capabilities) and process data together with the MEC to complete quality inspection. The intelligent surface inspection system and algorithms are deployed at the edge cloud server to implement real-time edge-side image processing (including automatic defect detection and classification). Defect images are uploaded to the cloud for training, reducing computing costs of local training. Models are generated through cloud-based training and automatically pushed to the edge cloud server, with algorithms continuously iterated and optimized. With this quality inspection system, XISC has achieved intelligent quality inspection during bar processing and automatic identification of the surface defects of each bar, greatly reducing the work intensity of surface quality inspection personnel. In addition, objects to be inspected are individually recorded to support E2E product quality tracing and problem backtracking. The campus-wide efficiency can be improved by using edge-cloud collaboration to enable rapid algorithm development, local real-time computing and inference to implement online product inspection, and subsequent process guidelines (e.g. grinding and shearing) to improve product quality

8.3.2 AR Remote Assistance

8.3.2.1 Service Requirements

The increasing flexibility and multi-functionality of future steel plants will also impose higher requirements on workshop workers. To quickly meet the requirements of new tasks and production activities, AR technology will play a critical role—especially in device troubleshooting. AR can be used to address challenges such as insufficient technical experts, poor technical support, and high costs, improving device maintenance efficiency.

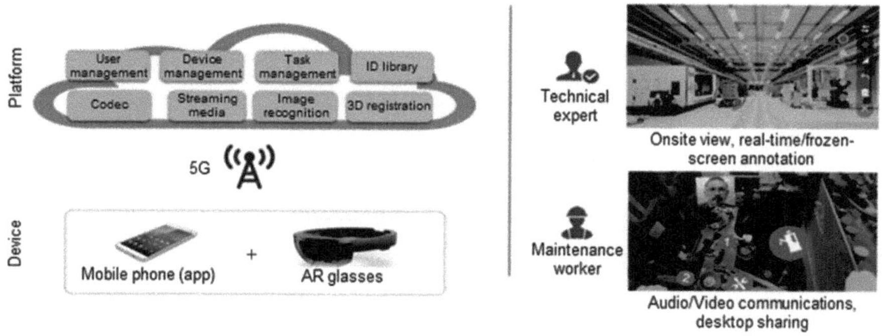

Fig. 8.4 5G AR remote assembly

AR glasses, for example, can help maintenance workers initiate audio/video communications with backend technical experts and collect and share onsite images in real time. Technical experts can also remotely use image annotations while providing voice guidance and sharing desktop to help maintenance workers troubleshoot devices. Figure 8.4 illustrates 5G AR remote assembly.

8.3.2.2 Network Solutions

During equipment assembly in steel plants, 5G network access is enabled for smart devices such as VR/AR glasses, smartphones, and tablets by using built-in 5G modules or deploying 5G gateways. Onsite images, videos, and voices can then be collected and transmitted to the auxiliary assembly system in real time through a 5G network. This system can analyze and process data, generate auxiliary production information, and deliver information to onsite devices through the 5G network. This way, onsite personnel assembling complex or refined equipment can receive superimposition-based augmented images of operation steps and visualized displays showing assembly processes. In addition, expert guidance, equipment operation instructions, drawings, and documents can be synchronized to onsite devices in real time through the 5G network. This shortens the training time of onsite assembly personnel significantly as they need less training and improves their assembly capabilities, bringing intelligence to assembly, and increasing assembly efficiency.

AR remote assistance involves three parts: onsite support (maintenance workers), remote assistance cloud platform, and backend support (technical experts).

1. Onsite Support (Maintenance Workers)

With AR glasses, maintenance workers can collect onsite audio/video information in real time and receive instructions from technical experts, including audio/video information, image annotations, documents, and more.

2. Remote Assistance Cloud Platform

This is a software system used to deploy products, including core function code (e.g. AR annotation), various service modules (e.g. authentication/authorization and audio/video node allocation), various basic service components (e.g. system configuration, account service, and audio/video service), knowledge bases, and more.

3. Backend Support (Technical Experts)

With operation devices, technical experts can obtain real-time onsite images based on audio/video communications and guide onsite maintenance workers through image annotation and file transfer.

8.3.2.3 Industry Ecosystems

1. Networks

Telecom operators provide 5G network services.

2. AR Glasses

AR glasses vendors provide various types of AR glasses (e.g. monocular, binocular, split, and integrated). Customers can select a proper model as required.

3. Servers

Servers are classified into service servers and audio/video servers. They are used to bear core product capabilities and streaming media forwarding capabilities, including user authentication, audio/video node allocation, AR annotation, and video stream codec.

4. Communications Modules

Industrial CPEs are deployed onsite to convert 5G signals into Wi-Fi signals for connected AR glasses to upload and receive data.

8.3.2.4 Business Models

AR remote assistance addresses the challenges such as insufficient technical experts, poor technical support, and high costs, improves device maintenance efficiency, and reduces costs for enterprises. The AR remote assistance system can be integrated into the 5G network reconstructed or built for the enterprise campus. Based on "product + service" cooperation, the system provides paid services such as function expansion, system integration, and system maintenance required by steel plants after product delivery.

8.3.2.5 Typical Cases

In March 2020, XISC began preparing for the installation and commissioning of a new batch of imported production equipment. However, their technical experts in Germany and Austria were unable to come to China for onsite installation and technical services due to the COVID-19 pandemic. To minimize the impact of this restriction on production, XISC's rolling production line relied on the following technical advantages of 5G networks to get the job done.

1. Ultra-high rate: 5G networks overcome bandwidth limitations and provide a downlink rate of 1.2 Gbit/s and an uplink rate of 750 Mbit/s, enabling multiple channels of ultra-HD videos to be transmitted simultaneously.
2. Ultra-low latency: 5G networks provide accurate feedback with a one-way network latency of approximately 10 ms to deliver commands and feedback in real time, enabling remote assistance in maintenance.
3. High stability and reliability: 5G networks outperform in anti-interference and stability, supporting on-demand access anytime and anywhere while providing optimal coverage.

By working with telecom operators, device vendors, and AR vendors, a cross-country private line was quickly provisioned based on the existing 5G network to build a high-speed and low-latency communications network between onsite engineers of XISC and technical experts outside China. With this network, XISC was able to use AR technology and HD panoramic cameras to facilitate remote assembly. With the 5G+ AR remote assembly solution, XISC's onsite engineers were able to push onsite images and environment videos to engineers in Germany and Austria in real time through the 5G network. AR technologies such as real-time annotation, frozen-screen annotation, audio/video communications, and desktop sharing were used by engineers outside China to support the XISC engineers on production line assembly.

XISC took advantage of the 5G+ AR technologies to enable engineers from three countries across the Eurasian continent to work closely and remotely on highly-efficient and high-quality assembly of their rolling production line. By innovatively using the 5G+ AR technologies for remote assembly at an industrial construction site, XISC showed how 5G+ AR-assisted remote collaboration can be widely used in numerous scenarios—from remote collaborative design, remote collaborative assembly, and remote device maintenance, to remote medical diagnosis. This has large implications for enterprises still struggling with post-pandemic work and production resumption.

8.3.3 Remote Control Bridge Cranes

8.3.3.1 Service Requirements

Bridge cranes—an indispensable transportation tool for loading/unloading in various modern steel production scenarios—have been widely and frequently used in raw material warehouses, scrap workshops, ladle transportation, billet warehouses, and finished product warehouses. Iron and steel enterprises use bridge cranes for lifting in various processes, including raw material outbound/inbound, smelting, warehousing, and logistics.

5G remote control bridge cranes are suitable for steel production processes involving intake, output, and adjustment of raw materials, billets, and steel products in steel enterprises. The solution uses 5G networks, automatic control, video/image recognition, and other technologies and reconstructs conventional bridge cranes by installing HD cameras facing different directions on the bridge cranes to build a service platform for remote control and unmanned operations. With this solution, workers in an operation room can remotely control bridge cranes, implementing unmanned and collaborative operation of bridge cranes, helping steel enterprises implement high-precision intelligent transportation of materials and finished products and improving enterprise operation efficiency.

8.3.3.2 Network Solutions

Figure 8.5 illustrates the architecture of a 5G remote control system for bridge cranes.

A 5G network is used to collect and control industrial control data, allowing interconnection between onsite disk operating systems and the MES. Through information convergence, the system enables network communications for integrated management and control and aligns onsite operation plans with bridge crane operations, achieving automated and intelligent bridge crane operations including hoisting, transportation, and scheduling.

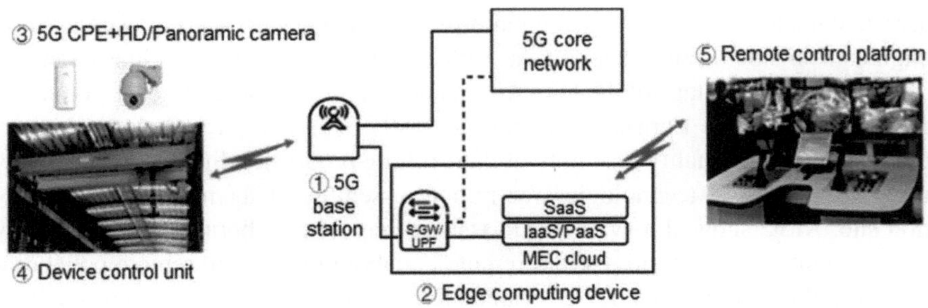

Fig. 8.5 Architecture of a 5G remote control system for bridge cranes

The 5G remote control system for bridge cranes consists of six modules: the bridge crane control module, the video surveillance module, the voice interaction module, the safety protection module, the central control module, and the communications module.

- The bridge crane control module receives and processes remote and local operation instructions and sensor signals through a programmable logic controller (PLC), driving operations of each bridge crane component (e.g. bridge, trolley, hoist, and pendant).
- The video surveillance module collects onsite videos and transmits them to the remote control room for real-time display, providing information that bridge crane operators can use to monitor production sites and control bridge cranes.
- The voice interaction module can be a pager, loudspeaker, Internet phone, wall-mounted speaker, or amplifier.
- The safety protection module provides functions that prevent common rail bridge crane collisions, ground collisions, and incorrect operations, while also supporting system emergency handling, emergency stops, and bridge crane status monitoring.
- The central control module, including software and hardware such as control consoles and control software, is usually deployed in a central control room or a control center, enabling remote control of bridge cranes.
- The communications module consists of various switches, wireless devices, 5G base stations, optical fibers/network cables, fiber optic transceivers, and optical-to-electrical converters.

8.3.3.3 Industry Ecosystems

Existing bridge crane control systems are reconstructed to include 5G network infrastructure, software and servers of the unattended bridge crane management platform, as well as PLC interface adaptation.

Regarding software and hardware ecosystems, network resources (including networks, spectrums, and cloud resources) are provided by telecom operators. Communications devices (including 5G CPEs and industrial gateways) and 5G E2E solutions are provided by 5G solution providers. Regarding bridge crane systems, such as the bridge crane control system, video surveillance system, language interaction system, safety protection system, and central remote control system, large-scale comprehensive steel solution providers provide competitive offerings. Regarding devices, industrial cameras, AI cameras, and other machine-vision-based quality inspection devices are provided by AI solution providers.

8.3.3.4 Business Models

Steel plants are structurally complex and not suitable for cable network deployment, leaving some equipment unable to connect to the network. If connected via industrial Wi-Fi, some equipment (e.g. bridge cranes) is prone to interference, large delay variation, and poor remote operation experience. To ensure production safety and physical health, it is of great significance to use 5G technologies to implement unmanned operations in steel production.

The 5G remote control system for bridge cranes can be quoted as a whole, including equipment/software procurement, customized development fee, and maintenance fee. Maintenance fees vary according to the specifics of the projects.

8.3.3.5 Typical Cases

Guangxi Liuzhou Steel Group, together with China Mobile and Huawei, use 5G technologies to enable remote control of bridge cranes in cold rolling plants at the Fangchenggang Steel Base. Traditional bridge cranes were retrofitted to use 5G networks to transmit control commands and HD videos and enable remote control through a remote control platform. This is great practice for enterprises to reduce labor costs, improve overall efficiency, protect employees' health, and ensure safe production.

8.3.4 Policies and Standards

The Chinese government officially supports the application of 5G in the steel industry and has formulated related policies and standards to promote application demonstrations of 5G and other new technologies in steel plants.

In 2016, the MIIT issued the *Transformation and Upgrade Plan for the Iron and Steel Industry (2016–2020)*, emphasizing that comprehensively promoting smart manufacturing is a key task for the industry. The plan aims to promote new modes of smart manufacturing across the industry and summarize successful experience that can be replicated. Pilot demonstrations of four new modes of smart manufacturing—including process-based smart manufacturing, network collaborative manufacturing, large-scale customization, and remote O&M—are being particularly promoted to improve enterprise capabilities such as efficient product R&D, stable product quality, flexible production organization, and comprehensive cost control.

In December 2021, the Alliance of Industrial Internet, China Iron and Steel Association, and the Chinese Society for Metals jointly compiled the *Industrial Internet and Steel Industry Convergence Application Reference Guide (2021)* to provide useful references for requirement identification, application mode development, key system building, and implementation methods during industrial Internet construction in the steel industry.

On February 7, 2022, the MIIT, the National Development and Reform Commission, and the Ministry of Ecology and Environment also jointly issued a guideline on promoting the high-quality development of the steel industry, proposing an action plan for smart manufacturing in the steel industry to promote application of technologies such as 5G and digital twin.

8.4 Summary and Prospects

Emerging 5G technologies provide strong technical support for the complicated production system of the steel industry and can be used to better meet customized requirements. They are already laying a solid foundation for digital transformation and smart manufacturing for steel enterprises and are significantly stimulating development within the steel industry.

5G has proven to be a strong driving force for the digital economy. Deeply integrated with industries, 5G enables industrial applications and accelerates the exploration of the technical and economic value of data. The convergence of 5G with the steel industry will define convergence standards, high-value scenarios, and objectives. With the implementation of smart manufacturing and Industry 4.0 gaining momentum, the steel industry is advancing toward digital and intelligent transformation.

Chapter 9
5G+ Smart Education

9.1 Overview

9.1.1 Current State

Education is vital to improving the comprehensive quality of the population and promoting the all-round development of individuals, and serves as the cornerstone of national revitalization and social progress. Strengthening education is fundamental to the pursuit of national rejuvenation. Under the background of new development stages, new development concepts, and new development patterns, China must ramp up the modernization of its education system to tackle new changes facing the society. It should shore up the weak links in the education system by using software and hardware resources, as well as adapt to the transformation of educatees in the network era to meet the diverse, personalized, and continuous needs of students.

In recent years, modernization has become the primary goal of China's education reform. In February 2019, *China's Education Modernization 2035* was set out based around eight major development goals, including: (1) Establishment of a modern education system that provides lifelong learning for all; (2) Universalization of quality preschool education; (3) Achievement of high-quality and balanced compulsory education; (4) Complete universalization of upper secondary education; (5) Significant improvement of vocational education services; (6) Marked improvement in the competitiveness of higher education; (7) Provision of the appropriate education for disabled children and adolescents; (8) Formation of a new pattern of education management involving the participation of the whole society. The eight goals cover the entire spectrum of education, and they correspond to the overall strategic goals of national modernization and meet the strategic needs of China's development.

As of 2021, off-campus trainings are regulated more strictly, and the "double reduction" policy is proposed, aiming to cut down homework and off-campus trainings for students in compulsory education. This policy proposes three measures. The first is to improve the after-school service level in schools to meet the diversified needs

P. Sun, *A Guidebook for 5GtoB and 6G Vision for Deep Convergence*,
Management for Professionals, https://doi.org/10.1007/978-981-99-4024-0_9

of students at the primary and secondary education stages. The second is to strictly regulate and govern all kinds of off-campus training activities, especially subject-based training and online training. The third is to encourage diversified activities at spare time to meet students' personalized learning needs. The "double reduction" policy intends to strengthen China's education reform, but poses higher requirements for the development of the education industry.

Informatization and digitalization are key driving forces for education modernization. China values education informatization and has launched a series of major projects and policies in this regard. In March 2012, the Ministry of Education of China issued the *Ten-Year Development Plan for Education Informatization (2011–2020)*, which proposed broadband deployment in every school to make schools more inclusive to everyone. The plan aims to basically create an information-based learning environment where everyone can enjoy high-quality educational resources. In September of the same year, construction of "Three Links and Two Platforms" was named as a priority during the 12th Five-Year Plan in a video teleconference on national education informatization. Three links refer to schools links to information and communications technologies (ICTs), high-quality class resource links to ICTs, and network learning space links to ICTs, and two platforms refer to public service platforms for educational resources and educational management. In 2016, the Ministry of Education issued *China's 13th Five-Year Plan for ICT in Education*, which proposes to build an education informatization system that allows for unrestricted access to education for everyone. Not only that, greater importance is attached to the in-depth application of information technologies in education and teaching to achieve integration of intensive informatization with education and teaching.

With the development of informatization technologies, education informatization is further promoted. In April 2018, the Ministry of Education issued the *Education Informatization 2.0 Action Plan*. This plan announced that, by 2022, teaching application will have covered all teachers, learning application will have covered all school-age students, digital campus construction will have covered all schools, the application level of informatization and the information literacy of teachers and students will have been generally improved, and the platform of "Internet + education" will have been built up.

Digital transformation of the education industry is driven by next-generation information technologies such as 5G, AR, and AI, in accordance with the requirements of education modernization in China. For example, *China's Education Modernization 2035* further proposes to accelerate education transformation in the information era through a wealth of measures, such as building intelligent campuses, building an intelligent platform integrated with teaching, management, and services, and adopting modern technologies to accelerate the reform of talent cultivation modes for combination between large-scale education and personalized training. Given the requirements of competence-oriented education, digital transformation in education has become vital for industry development.

9.1.2 *Major Processes and Operations*

Figure 9.1 illustrates the structure of China's current education system, which is divided into five stages: preschool education, primary education, secondary education, higher education, and continuing education. Each stage of education is developing steadily, oriented towards modernization.

Preschool education is organized for children aged 3–6 years old in kindergartens, which are generally privately owned. It has been made accessible to almost every young child in economically-advanced large- and medium-sized cities, and is being widely promoted to rural areas, resulting in one year of preschool education becoming more and more popular in some rural areas.

Primary education is organized for children aged 6–12 years old in primary schools, which are generally run by local governments, individuals, or civil society groups. Primary education is the key link where the burden of homework for students needs to be reduced in accordance with the "double reduction" policy. Available measures include improving the homework management mechanism, classifying and clarifying the total amount of homework, improving the content quality of homework, and strengthening the guidance of homework completion. At the same time, higher requirements are posed on classroom teaching of teachers.

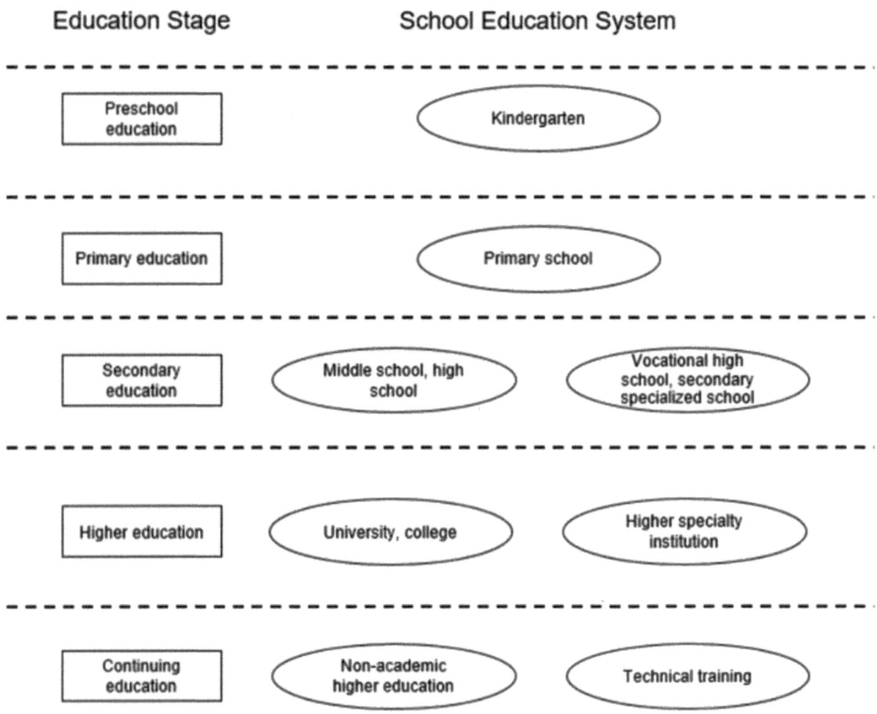

Fig. 9.1 Current education system in China

Secondary education is organized for teenagers aged 12–17 years old in secondary schools, which include middle schools, ordinary high schools, vocational high schools, and secondary specialized schools. Ordinary secondary school comprises three-year middle school and three-year high school. After completing middle school, some students continue into high school, while some enroll in vocational high school and secondary specialized school. Secondary schools are generally run by local governments. The off-campus training services oriented to subject-based middle school education will be further standardized. Improving the vocational education system is a focus of education development in China, where the secondary education system is undergoing a comprehensive reform.

Higher education refers to the specialty, undergraduate, and postgraduate education after secondary education. In China, institutions offering higher education are universities, colleges, and higher specialty institutions, and they perform three major functions: teaching, scientific research, and social service. Improving higher education is crucial to strengthening high-quality education development in China.

Continuing education includes adult technical training, adult non-academic higher education, and literacy education, and has developed quickly under the boom of the Internet.

Teaching resources, teaching modes, and campus management are the three core links of education. Although the teaching content and modes vary across different educational stages, the traditional learning mode in each educational stage in China takes classroom as the core and after-class homework and examinations as the complement. In the learning process, the learning effects are determined by the content of education, teachers' teaching level, and the degree of adaptation. These links together constitute teaching resources and are the core factors that affect teaching quality. In the traditional teaching mode, a blackboard is typically used. Technology has vastly altered the way content is taught—from pure blackboard-based learning to a combination of boards and PowerPoint presentations. In addition, homework check and examinations are becoming more and more convenient through digitalization. As the teaching mode changes, teachers are able to impart knowledge more effectively, and students can better master knowledge through modern scientific and technological means. A campus is the core carrier of teaching. Campus construction covers campus life management, device management, and security management. Higher education campus management started earlier and is closely related to the quality of life of students. After the "double reduction" policy is put forward, the requirements of primary and secondary education campus construction are raised accordingly. The key to realizing education modernization lies in the overall improvement in teaching resources, teaching modes, and campus management.

9.2 Digitalization Trends and Challenges

9.2.1 Trends of Digitalization

9.2.1.1 Scenarios

Digital transformation in education is not only a hallmark of the times, but also a necessity of development. It is an extension to education informatization. Nowadays, the educatees and educational environment are undergoing great changes. With the advent of the Internet information era and the popularization of smart devices, a wider variety of learning methods and more diversified learning forms will open up for educatees to receive education. All these pose a requirement for educational development and reform. With 5G networks and next-generation information technologies such as big data, AI, IoT, and cloud computing, smart development is realized in terms of teaching space, campus space, and learning space. In China, the high-quality development of education is facilitated through comprehensively optimized basic motivation of education and teaching, enhanced resource service supply capability, and integrated online and offline education. Figure 9.2 illustrates the trend of scenarios for digital transformation of the education industry in China.

Currently, digital transformation mainly focuses on distance teaching, smart teaching, and smart campus.

1. **Distance Teaching, a Great Way to Share First-Rate Educational Resources**

Distance teaching, based on online teaching, is vital to popularizing smart teaching. When the COVID-19 pandemic first hit in early 2020, online teaching was widely adopted so that students could continue their studies largely uninterrupted.

However, traditional online teaching is based on optical fiber networks and Wi-Fi and therefore is unavailable in outdoor mobile environments. 5G can effectively solve this problem. More importantly, 5G, with its high bandwidth, can be used

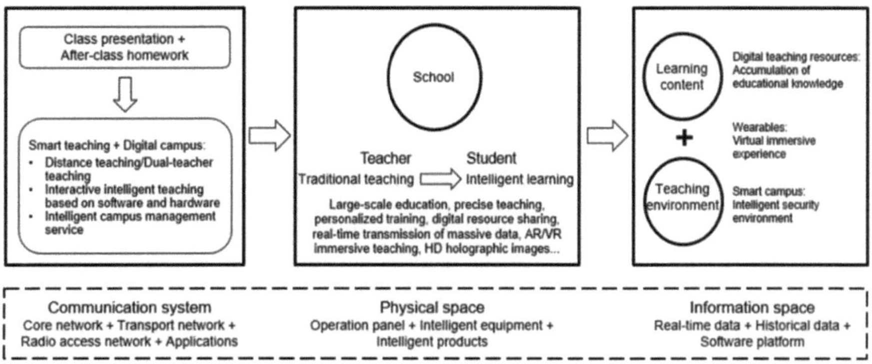

Fig. 9.2 Trend of scenarios for digital transformation of the education industry

together with such technologies as ultra-HD video, virtual reality (VR), augmented reality (AR), and holographic technologies to enrich the content and types of distance teaching, improve the efficiency and quality of distance teaching, and enhance interactions of students with teachers and environments. As such, remote sharing of first-rate educational resources further promotes equitable distribution of basic education services.

2. **Smart Teaching, Building a Smart Classroom That Integrates Virtual and Reality**

Diverse IoT sensing devices and AI technologies are applied in classrooms to deliver real-time sensing of students' learning processes as well as real-time collection of behavior data and appraisal results. Intelligent analysis methods such as classroom interaction identification, classroom focus analysis, and learning track analysis help achieve precision teaching and personalized learning. With these methods, teachers can quickly adjust their teaching progress based on their students' learning status. Intelligent analysis based on 5G and Artificial Intelligence of Things (AioT) improves efficiency of data collection and transmission and facilitates intelligent analysis of students' learning behavior and effects; the analysis results can also be transmitted to teachers in real time.

Based on technologies such as VR, AR, and Mixed Reality (MR), an immersive teaching environment can be built to help deepen students' understanding of abstract theories and natural phenomena or changes that are difficult to observe in real life. The integration of virtual and real environments enables multi-sensory participation in the learning process for students and allows abstract concepts and theories to be presented to students in a more intuitive and vivid manner, thereby making learning more fun and improving classroom efficiency. Despite this, immersive teaching poses high requirements on content and storage. As such, 5G+ XR, which implements cloud-based immersive teaching, is put forward to tackle such issues. With 5G+ XR, high-quality teaching resources can be stored on the cloud and invoked with low latency during teaching. 5G+ XR facilitates professional education and training in various fields, especially in vocational education such as for medicine courses. On the other hand, 5G+ XR helps build immersive training scenarios, in which teachers and students can communicate more efficiently. On the other hand, it eliminates the need for students to stay in a real high-risk environment while undergoing training and reduces training costs by using virtual precision instruments and expensive consumables.

3. **Smart Campus, Delivering an Intelligent Learning Space**

5G is the basis for the Internet of Everything (IoE), with much room for application in the IoT field. It plays a vital role in implementing smart campus applications, which are implemented based on the IoT and typified by campus security, device management, and student health management. 5G+ smart campus drives the build of a more intelligent learning environment. 5G+ AioT-powered devices can implement smart campus management and smart services in all scenarios. With the support of 5G networks, applications such as ultra-HD video surveillance, intelligent video analysis, and intrusion detection and alarm will effectively improve security management efficiency of schools. In addition, 5G+ smart band implements scientific health management of students, and 5G+ smart canteens boost campus service quality.

9.2.1.2 Connotations

Digital transformation in education will transform the relationships between students, teachers, and schools from traditional teaching to intelligent learning.

Through 5G, digital transformation in education integrates various information technologies to realize a smart education environment. It also helps teachers improve their teaching techniques by fully leveraging environmental resources and actively adapting to students' needs. Ubiquitous communication networks and sensing devices are used to intelligently sense the learning scenarios and learners' characteristics, and proactively create learning environments, plan learning paths, and push adaptive learning resources for the learners, shifting from "information search by persons" to "information search for persons". Data analysis is the basis for implementing intelligent education services. With education informatization construction, standard models will be established for collecting and analyzing various types of data. Meanwhile, data aggregation and cross-space transmission are implemented based on how the physical environment is perceived, and the regulation of education services is enhanced. This breaks the restrictions of time, space, content, and media, and realizes seamless transmission of education information.

Intelligent learning brings more possibilities to teaching settings. With digital transformation, traditional class-based education will transform to large-scale education, precision teaching, and personalized training. High-quality teaching resources will be further popularized, students can select teaching resources based on their needs, and personalized training solutions can be proposed in schools based on students' advantages. Moreover, intelligent learning streamlines the sharing of educational resources, of which digital resources will become an important channel for equitable distribution of resources.

Making education accessible to everyone has always been a challenge, but with digital education, this is no longer a major concern. Digital education can promote the interconnection of high-quality educational resources and services through networks, expand the time and space for learning, and transform China's education to a situation where learning opportunities are equal for everyone in a world of ubiquitous technologies. This form of education can significantly improve the capabilities of on-demand and equal education.

In traditional network and technical environments, a multitude of learners are offered the same learning resources. However, the network transmission technology in the intelligent era makes learning resource transmission among learners possible, so that an intelligent learning service system can accurately analyze individuals and provide high-quality resources and services according to the specific needs of individuals. This also helps stimulate students' enthusiasm towards their studies.

Digital education management not only makes campus and student learning management more professional through smart applications, but also solves the "management silos" by streamlining campus networks for intelligent collaboration. With intelligent technologies and ubiquitous high-speed communications, various education businesses can be conveniently, quickly, efficiently, and intelligently coordinated anywhere, anytime, and in any way. This means that many education businesses are no longer provided independently; rather, the management, teaching, training, and service businesses in the education field are intelligently coordinated. This reorganizes business processes and innovates service forms.

9.2.1.3 Expected Effects

Digital transformation in education will enable new learning content and teaching environments. In digital teaching, the ducates and environments are undergoing radical changes. With widespread Internet and smart devices, the future ducates will grow in a brand new social environment, and a gradual transformation to the network-based, digital, and personalized learning mode is underway. In response to this trend, campuses and training institutions will undergo a comprehensive reform.

Teaching resources, teaching equipment, and campuses will be comprehensively innovated through digitalization to streamline the accumulation and sharing of teaching resources. Teaching resources can be further upgraded to achieve education modernization objectives and promote the overall development of the industry. Teaching equipment will be further developed. Traditional and modern devices will be further integrated. Devices that focus on providing virtual immersive experience will also enrich classroom education methods. Software and hardware facilities further improve the campus management level and enable a smart campus to create an intelligent and secure learning environment.

The integration of communication systems, physical space, and information space will reshape the education industry ecosystem and promote all-round digitalization of the education sector. Upgrades of communication systems, represented by 5G, are the basis of digital transformation. The comprehensive development of core networks, transport networks, wireless networks, and applications supports the comprehensive upgrade of physical and information space to form an operation environment for digital education. The development of digital education relies on software and hardware resources. In the physical space, the operation panel, intelligent equipment, and intelligent products are comprehensively innovated. In the information space, operating systems serve as the base, and data centers and innovative applications are integrated with hardware to improve the work efficiency of classrooms.

9.2.2 Network Upgrade

9.2.2.1 From "Three Links and Two Platforms" to "New Educational Infrastructure Construction"

Currently, a further upgrade of educational infrastructure is an important step to promoting digital transformation, and the 12th Five-Year Plan and 13th Five-Year Plan have brought about rapid informatization to China's education sector, with a strong focus on "Three Links and Two Platforms". In July 2021, the Ministry of Education and other departments in China jointly issued the *Guiding Opinions on Promoting the Construction of New Educational Infrastructure and Building a High-Quality Education Support System* (hereinafter referred to as "Opinions"). The "Opinions" pointed out that the new educational infrastructure will be guided by new development concepts, led by informatization, and oriented to the needs of high-quality education development, focusing on information networks, platform systems, digital resources, smart campuses, innovative applications, credible security, to name but a few. As an important part of the country's new infrastructure construction, the new educational infrastructure construction is a driving force for educational reform in the information age and a strategic measure to accelerate the modernization of education and build an educational power, focusing on information networks, digital resources, smart campuses, innovative applications, and credible security.

Based on the requirements of digital transformation and the development path of new educational infrastructure construction, digital transformation in education should create an intelligent learning space that is student-centric and covers teachers, parents, and administrators, implementing education and teaching as well as campus management in a comprehensively digital, network-based, and intelligent manner. 5G plays an important role in digital transformation of the education industry. It improves network connectivity to enhance the level of intelligence, and accelerates the development of various application scenarios with 5G network capabilities. 5G can promote the development of various applications in multiple scenarios to provide personalized education. Centered on an education space, the large-scale promotion of 5G in the education field has resulted in infrastructure of four layers: terminal layer, network layer, platform layer, and application layer. Information security and related standards also play a vital role in the development of each layer.

9.2.2.2 Terminal Layer: Important Support for Personalized Education

The terminal layer consists of various intelligent devices such as VR, holographic platforms, and intelligent terminals. At this layer, the smart education scenario information can be comprehensively sensed, and smart devices can be used to facilitate personalized education for teachers and students. 5G+ education will adopt a large number of new intelligent terminals, covering mobile devices such as mobile phones and tablets, cameras, XR devices like VR/AR, wristbands, sensors, and even service robots. The terminal layer constitutes an overall interface for data collection and service provisioning, and lays an important foundation for the development of 5G+ education applications.

9.2.2.3 Network Layer: New Education Network

Besides 5G networks, the network layer for 5G+ education supports traditional networks such as 2G/3G/4G, broadband, and Wi-Fi, as well as various IoT networks such as Bluetooth and ZigBee. Compared with the traditional education network, the new network integrates multiple networks by leveraging diversified technologies and offers ubiquitous basic network access to implement various intelligent applications.

A 5G+ education network generally includes the core network, transport network, radio access network (RAN), and edge nodes. The core network serves as an information processing center to implement functions such as local data split and flexible routing. The transport network is an E2E network that connects the RAN and the core network. Responsible for wireless network access, the RAN provides coverage/capacity separation by separating the control and data planes, and implements centralized coordination and management of radio resources by means of cluster-based centralized control. With cache and computing capabilities, the edge nodes are deployed at the network edge and connected to mobile devices, sensors, and users to reduce the load on the core network, thus reducing the data transmission delay.

9.2.2.4 Platform Layer: Core of Data Processing and Decision-Making Support

Various system platforms deployed on the cloud can be used with the 5G converged network and other intelligent technologies to support education applications. With data processing and decision-making support as the core, the platform layer provides multiple functions, such as user management, device management, security management, big data analysis, and intelligent decision-making, as well as cloud computing and storage service capabilities. It delivers efficient and real-time intelligent data analysis and decision-making support to implement various 5G+ education applications by leveraging databases such as the campus basic database, individual education database, teaching database, class management database, and school management database, as well as algorithm libraries oriented to different applications.

9.2.2.5 Application Layer: Engine for Transforming Learning Content and Teaching Environment

With the comprehensive transformation of the terminal layer, network layer, and platform layer, 5G will create a series of applications to comprehensively transform the learning content and teaching environment, bringing comprehensive improvements in teaching quality, management efficiency, and service level. Such applications involve 5G+ distance teaching, 5G+ interactive teaching, 5G+ immersive teaching, 5G+ security management, 5G+ device management, 5G+ campus services, and others.

9.2.3 *Major Challenges*

9.2.3.1 Interconnection Between Digital Technologies and Teaching Requirements

As for digital education, China is facing the challenge of improving coordination between the technical level and teaching requirements, and is in need of an intelligent environment with information technologies to break the "information silos" in the original education informatization process. Currently, the education information system needs optimization in terms of resource sharing. Information systems (such as teaching, scientific research, management, technical service, and life service) are silo-like, which hinders service and process integration. 5G communication and data analysis technologies can be adopted to facilitate data sharing inside and outside of classrooms and advance the seamless integration of online and offline courses. Alternatively, edge computing technologies can be utilized to intelligently manage and control specific requirements and businesses in education management. To achieve this, a comprehensive reconstruction is required for digital education.

Although there is a great focus on promoting education informatization, teachers' teaching skills are still not fully improved yet. In response to digital education, teachers must change their teaching mindset, adapt to the characteristics of perceptual learning scenarios centered on communication networks and sensor devices, and proactively create learning environments, plan learning processes, and match learning resources for students. Teachers also need to adapt to the seamless flow of information and master the use of related tools. During practice, teaching scenarios should be developed based on user requirements to facilitate intelligent collaboration. Moreover, teachers need to seek converged innovation on the basis of traditional teaching modes and comprehensively improve their own teaching levels to meet the requirements of digital education.

9.2.3.2 Industry Challenges Facing Digital Education

The business model of the digital education industry remains immature. It is still held back by expensive devices, few buyers, and high operation costs. Comprehensive upgrades of application development and teaching resources are needed to realize commercialization of digital education. On the one hand, due to a lack of technical and industry standards, teachers cannot effectively perform education and teaching upgrades, limiting the development of industry business models. Smart application teaching requires teachers to change their teaching mindset from knowledge-oriented to learner-oriented, develop their own teaching content, and stimulate proactive learning. However, teachers are short of resource development standards, hindering the business model development. On the other hand, there is a lack of teaching content resources related to digital education, slowing down the development progress of commercial applications. New teaching methods require new teaching content, most of which needs to be produced by professional technical personnel, for example, 5G+ XR teaching content. How to meet the teaching requirements of courses is another challenge. In addition, the teaching content like XR is costly and time-consuming, and most education content needs to be customized, which restricts the commercialization of related applications in a short term.

From a long-term perspective, ethical and security problems during the development of digital education need to be highly concerned and appropriately handled. Some applications involved in digital education require comprehensive monitoring, analysis, and evaluation of students' learning processes and teachers' teaching processes. Some researchers believe that such applications should be prohibited because they infringe on students' and teachers' privacy and do not deliver the expected application effects. As smart education develops, the data protection system needs further improving for the purpose of privacy protection and data security. This is because a large number of smart education applications involve collection and analysis of data that mainly comes from students, of whom the students in primary and secondary education are mainly minors. In addition, great concerns have been raised regarding the supervision requirements for applications in the education sector.

9.2.3.3 Problems Facing the Digital Education Development in China

Due to insufficient resource distribution and unbalanced development of China's education, digital transformation will also be affected. The technology maturity of digital education and network coverage need to be improved. 5G networks are steadily advancing in China, but for various schools, in-depth coverage of 5G networks has not been implemented to fully facilitate digital education applications, and key technologies such as network slicing are not mature. As such, there is not enough capacity to bear new education services. New services (such as 4K/8K live classroom, VR/AR classroom, holographic education, 4K HD surveillance, and mobile patrol vehicles) pose higher requirements on network bandwidth.

In China, there are obviously more educational resources available in cities and eastern regions compared with rural areas and western regions. As digital education continues to develop, educational resources may be rolled out online so that similarly high-quality educational resources can be shared with students in western regions or rural areas. However, the uneven distribution of infrastructure may further widen the gap between education levels and magnify the imbalance of educational resources. To solve these problems while developing digital education, digital education applications should be developed to facilitate equitable distribution of resources such as teachers and teaching materials, and transfer payments should be used to shore up weak links in resources and facilities in related regions as soon as possible.

What's more, in China's educational ecological structure, vocational education (excluding university education) is not developing very well. China's vocational education lags behind many developed countries. The very best vocational educational resources are typically allocated in the eastern regions of China, while labor resources are relatively concentrated in the central and western regions. The labor resources receiving vocational education in the central and western regions fall below the superior employment requirements in the eastern regions.

9.3 Major 5GtoB Large-Scale Replication Scenarios and Typical Cases

9.3.1 Distance Teaching

9.3.1.1 Industry Requirements

Distance teaching is a form of online education that realizes teaching and learning over the Internet. Currently, online education is considered an important supplement to classroom education in schools to open up access to education for all. However, online education has long been mainly used in fields such as extracurricular training and coaching. In 2020, online education was popularized at a faster speed and became applied in a large-scale manner across schools for the first time under the background of "suspending classes without suspending learning" during the COVID-19 pandemic.

Distance teaching is of great significance for realizing equitable distribution of educational resources. China is hindered by long-lasting uneven and insufficient distribution of educational resources. Compared with the eastern regions, the central and western regions of China generally lack high-quality educational resources. Likewise, teachers who are experts in certain fields are also lacking in some regions.

The large-scale commercial use of 5G networks will significantly improve the learning environment of online education from the following perspectives: First of all, 5G supports online education in long-distance mobile environments, meeting distance teaching requirements in more scenarios. Secondly, 5G is integrated with AI technologies to conduct real-time analysis on online education. As such, while teaching online, teachers can quickly analyze how their students are progressing with their studies and provide personalized coaching accordingly. Thirdly, 5G features high rates and low latency and supports technologies that require high bandwidth, such as audio/video streams and XR. This enriches the content of remote classrooms, brings immersive communication experience for teachers and students, promotes the co-construction and sharing of high-quality educational resources, and ultimately improves the situation of insufficient and unbalanced distribution of high-quality resources in basic education.

Dual-teacher classroom and holographic teaching are two target scenarios for large-scale replication of distance teaching.

1. **Dual-Teacher Classroom**

The dual-teacher classroom integrates online and offline teaching, with the teacher and assistant working together. The teacher explains the course content through a live broadcast, and the assistant cooperates with the teacher to carry out teaching and class activities. In dual-teacher classroom mode, students must watch live content in the classroom and interact with the teacher through devices such as the quiz feedback tool.

2. **Holographic Teaching**

Holographic technology can reproduce an object by means of diffractive light emitted from the object. The position and size of the reproduced object are the same as those before, but the images of the object observed from different positions are different. Holographic technology can be used for optical storage, reproduction, and information processing. Holographic projection is currently the most widely used holographic technology, a technology for recording and reproducing a three-dimensional image of a real object by using interference and diffraction principles. It implements virtual imaging, and its applications in performance activities have been explored a lot.

In distance teaching, holographic projection can be used to project the teacher's real-life image, the courseware, and teaching appliances to students using three-dimensional technology. Remote synchronization and real-time interaction are also supported to create a fun and immersive classroom environment for students, bettering the teaching quality of distance teaching.

Distance teaching can function as a remote class observation/evaluation system or a remote class inspection system. Different from the traditional recording-based system, 5G-based remote class observation/evaluation uses 5G-based recording terminals. Based on the 5G mobile network, teachers can give lectures in local classrooms, and interaction, class joining, and commenting are allowed in remote classrooms, promoting teaching reflection and improving teaching levels.

9.3.1.2 Network Solutions

1. **Dual-Teacher Classroom**

In dual-teacher classrooms, 5G expands coverage while enhancing interaction through performance improvement and construction flexibility. Wired networks are hindered by a long construction period, high costs, and poor flexibility, whereas wireless networks such as Wi-Fi are prone to audio and video delay and frame freezing.

5G, however, comes with a high bandwidth and low latency to enable flexible, on-demand, and mobile teaching, 4K HD video transmission, and low-latency immersive interaction. 5G effectively improves the experience of interaction in traditional practices, guaranteeing long-term development of the dual-teacher classroom.

This can be achieved by adding only 5G communication modules to the existing dual-teacher classroom to enable 5G network access at any time, greatly shortening the service provisioning time.

2. **Holographic Teaching**

Holographic teaching poses higher requirements on the rate and latency of 5G. Holographic teaching is an augmented reality (AR) application in essence, which requires a 20–40 ms E2E latency. It implements low-latency transmission of high-bandwidth content such as audio/video streams and applications, supports communication between teachers and students with lower latency in remote classrooms, and further enhances immersion that reflects face-to-face classroom teaching, making virtual classrooms feel authentic.

9.3.1.3 Industry Ecosystems

The development of dual-teacher classroom and holographic teaching depends on education network construction. Therefore, telecom operators often act as the core promotion party to manage network and device upgrades. 5G technology providers offer the technical support for telecom operators' solutions to promote the development of distance teaching. Multimedia classroom suppliers provide devices and solutions. Based on the dual-teacher classroom and holographic classroom, relevant 5G technology providers and software/hardware developers can continuously promote application upgrades.

1. **Dual-Teacher Classroom**

For the dual-teacher classroom, multimedia classrooms with data transmission and receiving functions are the core. Except for network reconstruction and upgrades, other multimedia devices are relatively mature. The dual-teacher classroom integrates online and offline teaching, with the teacher and assistant working together.

In dual-teacher teaching, multimedia classroom devices are relatively mature and are usually procured in a centralized manner by telecom operators.

2. **Holographic Teaching**

Similar to a standard green screen studio, a holographic live teaching area is used to collect audio and video data of teachers. Teachers can learn how students are doing through HD displays and interact with them in real time, without the need for adding special equipment.

In the classroom, a dedicated podium needs to be deployed for holographic teaching. A holographic screen displays teachers' images in the naked-eye 3D projection effect. HD cameras and microphones are deployed in the classroom to streamline real-time interaction between teachers and students.

The holographic classroom adopts technologically advanced devices, and enterprises are more motivated to provide hardware solutions and continuously offer solution upgrades to boost revenues.

9.3.1.4 Business Models

The implementation of dual-teacher classroom and holographic teaching involves two phases: deployment and application. Investment in new educational infrastructure construction is the main source of income for relevant entities.

In the deployment phase, telecom operators manage the network and hardware construction solution, determine the device and hardware investment in bidding mode, and submit related information to schools for approval. The schools pay through the new infrastructure construction fund. Telecom operators pay a fee to 5G technology providers and multimedia classroom suppliers.

In the application phase, telecom operators charge the channel fees in traffic charging mode, and other entities obtain profits through application upgrades. In holographic teaching, green screen studios are often leased to provide services. In this case, the holographic classroom provider charges the corresponding single use fee. Related application developers can continuously upgrade the network and hardware to improve the distance education effect and gain revenues.

9.3.1.5 Typical Cases

1. **5G+ Dual-Teacher Classroom**

In April 2019, dual-teacher teaching for an open biology class was carried out using an E2E 5G network between Guangdong Experimental High School and its branch. The 5G network with high bandwidth and low latency allowed the teacher to communicate with students in both the local and branch schools in real time [1].

2. **5G+ Holographic Classroom**

On February 28, 2019, China Unicom, together with the National Engineering Research Center for E-Learning (NERCEL) at Huazhong Normal University, the National Engineering Laboratory For Educational Big Data, and other organizations, held the "5G+ Intelligent Education" industry application conference in the First Affiliated Secondary School of Huazhong Normal University. At the conference, holographic signals were transmitted over the 5G network to deliver an open physics class across Wuhan and Fuzhou. Thanks to the 5G network, a teacher at the Fuzhou campus was holographically projected to the students based at the Wuhan campus. In the future, the wide application of 5G+ holographic projection technology in distance teaching will help further promote the cross-temporal allocation of high-quality educational resources, undertaking research and innovation to balance out educational resource allocation [2].

9.3.2 *Smart Teaching*

9.3.2.1 Industry Requirements

In the years to come, 5G communication modules will be installed on various types of smart teaching terminals (such as XR devices in recording rooms/remote classrooms, smart libraries, experimental devices, monitoring devices, and portable smart devices). These terminals will be used in various application scenarios in the teaching environment to comprehensively improve education capabilities, placing higher demand on communication networks. Interactive teaching and immersive teaching are the two main teaching forms.

1. **Interactive Teaching**

Interactive teaching is a teaching mode implemented in smart classrooms consisting of smart devices such as intelligent interactive large screens, tablets, and quiz feedback tools. This mode of teaching is capable of providing intelligent and interactive instruction such as exercise delivery and submission, and is an important application for promoting the digital, networked-based, and intelligent development of classroom teaching. Currently, intelligent interactive teaching relies primarily on wired networks and local wireless networks such as Wi-Fi, Bluetooth, and ZigBee. 5G with high bandwidth and low latency can be integrated with interactive teaching to make interactive teaching smoother without drastically changing the existing classroom environment. However, its application into the classroom will significantly improve teaching quality due to its improved network capability.

The basis of interactive teaching is a smart classroom that comprises paper-pen interactive classroom applications. Interactive teaching in smart classrooms is centered on smart boards. Mobile terminals (such as tablets, mobile phones, and quiz feedback tools) and other devices like cameras, lights, and air conditioners are connected to implement interactive teaching between one teacher and many students as well as intelligent sensing and control of the classroom environment. During teaching, multi-screen group-based cooperative learning is allowed, and classroom data can be automatically collected by using intelligent recording devices and intelligent terminals. More specifically, during interactive teaching in smart classrooms, the teacher can synchronously send class content and questions to students' intelligent learning terminals such as tablets. Students then use intelligent terminals to synchronously send answers to the teacher in real time, and the system automatically reviews the answers based on image recognition and big data analysis. This reduces the burden on the teacher and improves classroom efficiency. Furthermore, all classroom Q&A data is collected to support subsequent intelligent analysis and learning evaluation.

5G+ interactive teaching enhances data transmission efficiency even more as well as space flexibility of interactive teaching. It supports mobile teaching in a larger area and real-time transmission of HD videos and AR- and VR-based teaching content. For example, teachers can invoke teaching applications on the 5G edge cloud platform as they wish, and push and distribute various teaching content in groups based on the 5G network to tablets of different student groups. Regular recording and live broadcast in interactive teaching are also supported. What's more, 5G can be used to combine distance teaching with interactive teaching to enlarge the coverage area of the latter. This further improves the coverage of high-quality teaching resources, and achieves teaching effects beyond traditional teaching.

2. **Immersive Teaching**

In various teaching applications, technologies such as AR, VR, and MR can be used to create a virtual-physical learning environment for students so that they can visually learn about abstract concepts and theories or stay in scenarios and activities that are difficult to experience in the real world. This breaks through the limitations of time, space, and even real environments, creates an immersive teaching environment, and improves teachers' teaching efficiency and students' learning experience.

5G+ cloud-based XR interactive teaching deepens the virtual+real teaching mode and helps students visually and acoustically participate in learning. Abstract concepts and theories are presented to students visually and intuitively, which makes learning more fun and improves teaching efficiency.

High-quality applications of various XR technologies pose higher requirements on transmission bandwidth and latency, for the purpose of good image quality, fast interaction, and high immersive experience level. Take the VR technology as an example. According to the research in *5G Application Innovation and Development White Paper* and *Virtual Reality/Augmented Reality White Paper*, in terms of image quality, some immersive phases require a 100 Mbit/s bandwidth, which can be achieved by using 5G. 5G delivers a rate 10 times higher than that of 4G and supports transmission at 100 Mbit/s or even 1000 Mbit/s [3]. In terms of interaction response, the interval from when a user rotates the head to when the corresponding image is displayed should be controlled within 20 ms to avoid dizziness [4]. If all these are processed at terminals locally, the terminals will be complex and expensive. If visual computing is deployed on the cloud, terminal complexity can be significantly reduced, but an extra network transmission delay is introduced. Currently, compared with 4G, which suffers from an air interface delay of dozens of milliseconds, 5G can control such a delay to 10 ms or less, meeting the delay requirement for interaction response.

As such, 5G will make XR terminals represented by VR easy to use. The balance between user experience and terminal costs is a key issue that affects the application of XR technologies. 5G+ cloud-based VR is expected to accelerate promotion of large-scale VR applications. It implements the content processing and computing capabilities required by VR applications on the cloud, cutting terminal costs and configuration complexity and ensuring smooth, immersive, and cordless VR services.

Based on the high rate, low latency, and massive connections of 5G, XR teaching content is uploaded to the cloud which then runs, renders, presents, and controls XR applications. XR images and voices are efficiently encoded into audio/video streams and then transmitted to terminals in real time through the 5G network. XR cloud platforms are built to develop cloud-based XR applications, including virtual experiment courses and virtual popular science courses, enhancing immersive learning experience of students and enabling operable interactive learning of digital learning content.

9.3.2.2 Network Solutions

1. **Interactive Teaching**

Compared with immersive teaching, interactive teaching does not pose specific network upgrade requirements. However, 5G significantly improves network experience and optimizes application service capabilities through integrated network reconstruction and platform system deployment. To be more specific, 5G has the following advantages:

1. Unified network bearer: Various networks are no longer required in schools. All audio-visual education terminals can access 5G networks like mobile phones, and they are ready to use upon power-on. All teaching background applications can be securely carried on the 5G edge cloud platforms of telecom operators or schools, without the need for maintenance. The integrated network gives schools access to multiple applications, such as information-based teaching, teaching evaluation, class culture evaluation, and course management and control.
2. Ultra-high bandwidth: Ultra-high bandwidth ensures that interactive display terminals as well as signal transmission and processing terminals in smart classrooms can perfectly reproduce 4K images and support the upcoming 8K interactive terminals. This brings clearer and natural images to teachers and students and ensures leading smart classroom technologies.
3. Low latency: Paper-pen interaction data collection is greatly improved in terms of efficiency, stability, and security. With the 5G network, classroom devices can collect students' real-time writing data so that teachers can quickly identify issues and master the learning process of their students. With high reliability, the 5G network ensures the stability and reliability of data collection and transmission and provides more stable, rich, efficient, and valuable data services for students and teachers.

2. **Immersive Teaching**

The high bandwidth and low latency features of 5G are the fundamental support for immersive teaching experience. This support depends on the rendering capability provided by edge computing. VR/AR teaching content is uploaded to the cloud which then runs, renders, presents, and controls AR applications. VR/AR images and voices are efficiently encoded into audio/video streams and then transmitted to terminals in real time through the 5G network. The edge cloud architecture allows the rendering function that requires low latency to be deployed close to users so that service data is directly processed on the edge rendering platform and then transmitted to users

Table 9.1 Network bandwidth requirements for different VR experience

Standard	Early Phase	Entry-Level Experience Phase	Advanced Experience Phase	Ultimate Experience Phase
Continuous experience time	<20 min	<20 min	20–60 min	>60 min
Video resolution	Full-picture resolution: 3840 × 1920	Full-picture resolution: 7680 × 3840	Full-picture resolution: 11,520 × 5760	Full-picture resolution: 23,040 × 11,520
Equivalent traditional TV screen resolution	240p	480p	2 K	4 K
Typical network bandwidth requirement	25 Mbit/s	100 Mbit/s	418 Mbit/s	1 Gbit/s
RTT requirement	40 ms	30 ms	20 ms	10 ms
Packet loss requirement	1.4×10^{-4}	1.5×10^{-5}	1.9×10^{-6}	5.5×10^{-8}

Source Technical White Paper on Virtual Reality Experience Standards

without passing through the core network. The 5G-based edge cloud solution effectively solves the problems of insufficient connections, low speed, and high latency of cloud services in traditional solutions. Table 9.1 lists the network bandwidth requirements for different VR experience.

9.3.2.3 Industry Ecosystems

Currently, two ecosystem modes are available in smart teaching.

The first mode is smart network+smart classroom reconstruction, centered on telecom operators. In this mode, telecom operators closely interwork with schools, and work together with 5G technology providers and hardware solution and software service suppliers to deliver a smart teaching platform based on smart classroom reconstruction. Due to the particularity of smart classroom software and hardware devices, an integrator is needed to integrate various subsystem applications into classrooms or device platforms before providing them to telecom operators. This mode is suitable for large-scale replication in regions and used in regional smart classroom and smart campus reconstruction projects managed by telecom operators.

The second mode is overall smart classroom construction, centered on smart classroom and VR/AR enterprises. In this mode, smart classroom and VR/AR enterprises manage network reconstruction and various subsystems to provide customized classroom construction solutions for campuses. Originated from VR/AR, this mode is closely related to software and hardware development, and therefore equipment suppliers are the core of the solution while 5G communication technology providers and telecom operators provide support. This mode is more common for exploratory smart classroom construction in a single school.

Interactive teaching and immersive teaching have different requirements for ecological devices.

1. **Interactive Teaching**

The interactive teaching ecosystem can be divided into two parts: hardware devices and software applications.

Hardware devices cover the following: interactive large screens used for teachers' electronic teaching, such as smart tablets, e-blackboards, and integrated blackboards; smart learning pens; regular recording and playing kits, including recording hosts, cameras, and microphones; class bulletin terminals that display culture values, examination schedules, and class attendance; smart teaching terminals that improve classroom interaction efficiency; high-speed document scanner and mobile booth terminals that display teaching materials; student feedback devices are used in student-centric flipped classrooms and group discussion scenarios.

Software devices cover the following: (1) Information-based teaching applications: Before classes, teachers can prepare courses on 5G mobile terminals, and during classes, they can use the quiz feedback tool to obtain students' test data and adapt their lessons in real time. After classes, teachers can use portable 5G terminals to deliver lectures. (2) Streamlined teaching effect evaluation: With 5G, data generated by students throughout a class is recorded and transmitted to the 5G edge cloud platform for big data and AI analysis to output more comprehensive and objective analysis results for different users (students, parents, teachers, school administrators, and education department personnel). This way, targeted teaching and scientific policies can be provided for students based on impersonal evaluation, thereby ensuring education equality. (3) Centralized application management and control: A smart classroom with cloud, pipe, and device capabilities implements unified access control for all 5G terminals, centrally manages device status, service applications, and log data, and intuitively displays data for stable, secure, and efficient operation.

Generally, the smart classroom platform is provided by telecom operators. Therefore, the smart network+smart classroom reconstruction mode centered on telecom operators is more common in smart classroom construction.

2. **Immersive Teaching**

In addition to smart teaching functions, immersive teaching requires VR devices with certain interaction functions. Such VR devices must be able to play content based on pre-customized rules and allow for interactions between users and the virtual environment. In immersive teaching, the interaction between the virtual environment and entities is implemented so that users can feel the changes of the virtual environment. As such, the VR devices should support 360° videos, free-view videos, and computer graphics simulation.

VR/AR delivers platform features. Device development enterprises are more likely to participate in the construction in overall smart classroom construction solution mode centered on smart classrooms and VR/AR enterprises.

9.3.2.4 Business Models

Based on the business ecosystem of smart teaching, telecom operators and 5G technology providers generally provide solutions to gain construction benefits. Telecom operators obtain continuous revenues through traffic and platform operation. Related equipment suppliers can obtain single-time benefits through equipment sales or continuous benefits through platform operation.

When acting as an integrator in the smart teaching process, telecom operators provide smart classroom solutions for campuses with network construction as the core. After the construction is complete, they can continuously operate related platforms and promote application updates to generate revenues.

Smart classroom and VR/AR device vendors can balance device sales revenues and subsequent operation revenues. Due to the continuous updates of smart classroom and VR/AR applications, relevant enterprises can choose to obtain long-term platform operation at the cost of certain device sales discounts to gain profits.

9.3.2.5 Typical Cases

1. **Smart Classroom in Yangpu Primary School in Shanghai**

In April 2019, a smart classroom activity was held in Yangpu Primary School in Shanghai. In the classroom, students answered questions assigned by the teacher in the system. The teacher randomly selected students' notes to know about how they were progressing and provided targeted teaching guidance based on their learning difficulties. This brings brand-new teaching experience compared with traditional classrooms and makes interaction in classrooms easier and more convenient without altering the way they teach. Automatic accumulation of classroom data provides valuable reference for subsequent teaching guidance.

2. **VR-based Cosmos Education Course Series in Youdao Premium Courses**

On July 24, 2020, Youdao Premium Courses of NetEase Youdao reached an agreement with China Telecom's Tianyi cloud VR to launch a series of VR-based cosmos education courses, aiming to deliver fun classes and more immersive learning through VR technologies. With all-in-one VR headsets and Youdao intelligent learning terminals, viewers can dive into simulated environments to learn about aerospace and the universe [5].

9.3.3 *Smart Campus*

9.3.3.1 Industry Requirements

As the basic environment that supports smart teaching, the smart campus provides smart management services for schools, teachers, students, and parents, and provides a communication platform and teaching space. A smart campus refines the management and service work of schools, making it more human-centered. IoT is the key foundation for smart campus construction. The construction and application of 5G networks will enable more efficient ubiquitous IoT sensing in the campus environment, creating a secure, intelligent, and convenient campus environment. The 5G network is used to upload the data collected by HD cameras and intelligent sensor terminals (including the data of campus environments, people, behavior, and teaching facilities) to the smart campus management platform in real time. The collected data is comprehensively analyzed based on technologies such as AI and big data, and the analysis results and decision-making instructions are delivered to various management and service terminals in real time.

Currently, due to the campus security requirements and impact of the COVID-19 pandemic, the campus security platform with video surveillance and security robots as the core has become the benchmark for large-scale 5G development. At the same time, information-based campus construction adds complexity to asset management. As such, campus asset management becomes another important development direction of 5G applications.

1. **Campus Security Management**

Video surveillance is the main way to realize campus security management. Currently, security management in most campuses still relies on traditional video surveillance of a passive and lagging defense system. Cameras deployed in schools transmit surveillance images to surveillance rooms. Manual processing makes it difficult for security personnel to promptly obtain valid information in video images.

The intelligent security protection system implements facial recognition, image recognition, cross-border tracking, and behavior recognition and helps detect exceptions and security risks in time and report alarms. Based on video and image information integration, the system turns "post-event O&M" into "fault forecast" through video-based patrol and intelligent deployment, significantly enhancing campus security protection.

5G networks implement HD video surveillance to strengthen real-time monitoring and intelligent identification of video streams, and also facilitate the construction of a smart, campus-wide security monitoring system. A unified platform with centralized access of various IoT sensors and data storage, analysis, and sharing can be built to effectively integrate campus videos and IoT surveillance resources, meeting various security application requirements.

Based on 5G networks and 5G network slicing, technologies such as IoT, cloud computing, edge computing, AI, satellite positioning, and geographic information system (GIS) are used to integrate daily campus entry and exit security control as well as campus internal environment data for comprehensive analysis, thereby effectively monitoring the security of key public areas and campus borders.

Take school bus security management as an example. Cameras installed on school buses are used to detect the vehicle's operation environment, passenger information, and internal environment in real time. Massive data is transmitted to the school management center through the 5G network featuring high rate and low latency so that school administrators can learn about the school bus situation in real time. This implements remote monitoring and management of school buses. On top of this, alarms are generated as a response to abnormal events such as thefts, speeding, and diverting off route. School bus security management improves incident feedback speed and ensures safe school bus operation by means of preventing children from being left in vehicles, tracking students' locations throughout the journey, monitoring dangerous driving habits of the driver such as speeding and fatigue driving, preventing school bus overload, and protecting the safety of students on the school buses.

Moreover, the 5G education private network mode prevents video surveillance data from being transmitted off campus and integrates prevention and control data to facilitate data viewing by management departments. With unified management, prevention, and control, personnel of prevention and control departments can accurately and quickly respond to campus emergencies. Security data analysis helps determine the crowd flow model and dangerous areas on the campus, formulate scientific security defense plans, and improve the security management level of campus video surveillance.

Security robots can be used to strengthen campus security. Based on the intelligent ontology of the robot, a security robot adopts the intelligent brain technology of the cloud robot to improve its cognitive capabilities, including object recognition capabilities (such as facial recognition and vehicle recognition) and natural language interaction capabilities. In challenging situations, Human Augmented Robotics Intelligence (HARI) is introduced to assist robots with walking (for the purpose of outdoor navigation, obstacle avoidance, or others) and audio/video interactions in terms of listening, speaking, and monitoring. The security robot system must also be able to backhaul and receive UAV video data, and send UAV start/stop control instructions, start/stop coordinates, as well as destination area and navigation path data.

2. **Equipment Management**

With the continuous development of information-based education, numerous informatization devices of various types appear on campuses. However, due to the lack of training, supervision, and awareness, the usage of the purchased expensive devices is low in some schools. Not only that, but because there is no unified device management tool, teachers cannot utilize these devices in an orderly and effective manner, and authorities of education bureaus and schools therefore cannot obtain the device usage in real time. This means there will be no basis for subsequent investment decision-making on information-based education.

A unified 5G device management platform manages and controls 5G audio-visual education terminals in terms of classrooms, schools, and regions in a unified manner.

This platform enables remote operation and visualized display of the terminals in subsequent education informatization construction, provides teachers with guidance on how to use audio-visual education equipment, and lets school administrators understand the informatization application level of teachers as well as the usage of audio-visual education equipment. This provides the analysis basis for decision-making by the authorities of regional education bureaus.

9.3.3.2 Network Solutions

1. **Campus Security Management**

Edge computing and AI-powered exception analysis are the core of the network solution.

A secure campus management platform is built to enable unified access to the monitoring videos of all campuses. Service subscription based on the schools or classes of students is supported through multiple channels such as apps or websites.

The MEC processes and intelligently analyzes surveillance videos, identifies exceptions, and promptly generates alarms when necessary. Moreover, the MEC allows for video stream distribution based on rights and regions, reducing network bandwidth requirements. The edge-computing-based campus surveillance video aggregation and processing gateway aggregates, preprocesses, identifies, and forwards videos collected by surveillance cameras on the campus to the centralized platform, and provides the videos for playback clients on the campus.

The application of intelligent robots poses higher requirements on the campus security management network. More specifically, such functions as HD video surveillance, positioning image, and thermal imaging must be simultaneously required, and the 5G transmission rates in the uplink and downlink as well as the latency must meet specific standards.

2. **Asset Management**

The network requirements of 5G equipment management depend on managed objects. The 5G equipment management platform matches each piece of equipment because it needs to monitor the running status of each application network.

9.3.3.3 Industry Ecosystems

Currently, smart campus applications are mainly promoted and implemented in university campuses, where a dedicated informatization and asset management department is set up for building the campus platform and integrating the network and information system. Therefore, telecom operators and information system enterprises should closely work with such departments to launch long-term operation solutions.

The campus management platforms in other education phases are generally built, applied, and managed by education departments in a unified manner.

9.3.3.4 Business Models

On campuses, the unified procurement mode is usually used to promote the construction of related systems. In most cases, the procurement involves both construction expenses and operation expenses for a certain period of time, avoiding path dependency of telecom operators and information system suppliers.

When responding to a bidding, telecom operators and information system suppliers should formulate construction schemes as well as subsequent service plans so that they can obtain service benefits by upgrading applications after the system is constructed and put into operation.

9.3.3.5 Typical Cases

On May 27, 2019, China Unicom Shanghai Branch and Shanghai University held an opening ceremony simultaneously on Baoshan, Jiading, and Jing'an campuses based on the 5G high-speed network. The university is the first university in Shanghai to implement 5G-based multi-campus synchronous live broadcast and interaction.

In November 2019, China Unicom Shanghai Branch helped Shanghai Luwan Senior High School become the first 5G+ smart high school in China. It explored and promoted the construction of virtual labs and holographic live classroom applications based on the 5G network, and helped better integrate information technologies into teaching so that information tools can become an important means for teaching. This effectively promotes information sharing and exchange within the school, expands the channels for teachers and students to obtain knowledge and information, and cultivates students' creative thinking.

In March 2020, Shanghai Jiading Luwan No.1 Central Experimental Primary School hosted an online open day using the 5G network and advanced interactive live broadcast/recording technologies, in compliance with COVID-19 pandemic prevention and control requirements. Students and their parents could visit the school virtually by scanning a QR code and were able to communicate with one another in real time in the interactive area.

In June 2020, Donghua University carried out a cloud-based graduation ceremony through the 5G network and advanced interactive live broadcast and recording technologies. During the ceremony, 360-degree panoramic cameras were used for VR live broadcast across two campuses, and graduates gained immersive experience in the ceremony through remote interaction.

9.3.4 Policies and Standards

9.3.4.1 Policies

In July 2021, the Ministry of Education and other departments jointly issued *Guiding Opinions on Promoting the Construction of New Educational Infrastructure and Building a High-quality Education Support System*, which pointed out that by 2025, an intensive, efficient, secure, and reliable new educational infrastructure system with an optimized structure will be basically established, and long-term, comprehensive development will be achieved through iterative upgrades, improvement, and continuous construction. An education private network and an "Internet + education" platform will be built to provide a digital foundation for high-quality education development. High-quality resources are converged to promote structural reform on the supply side. A new campus that integrates physical space and cyberspace will be built to expand the new education space. Innovative education applications are developed to support education process redesign and model reconstruction. Improvements to all-round and all-weather security protection protect the safety and security of teachers and students.

On September 26, 2021, the Ministry of Industry and Information Technology (MIIT) of China and the Ministry of Education jointly issued a notice on organizing the application of "5G+ smart education" pilot project to further promote 5G+ smart education development. Ten departments, including the MIIT, also jointly issued the *5G Application Yangfan Action Plan (2021–2023)*, which took "5G+ smart education" as one of the key application fields and proposed the objective of building over 100 5G application benchmarks. The "5G+ smart education" pilot project promotes collaborative innovation in all links of the industry chain to form a batch of technologically advanced typical applications that integrate 5G with education/teaching and deliver obvious effects. It helps explore development models that can be replicated, promoted, and applied on a large scale, facilitates commercial use of "5G+ smart education" applications, and offers a path for the high-quality education development in the new era.

9.3.4.2 Standards

Currently, the 5G+ education application standards have not been specified, but the establishment of the 5G+ smart education industry alliance lays a foundation for setting up such standards. On April 29, 2019, China Mobile, together with more than 50 enterprises and institutions in the communications, Internet, and education fields as well as universities and scientific research institutes, announced the formation of the 5G Smart Education Cooperation Alliance in Hangzhou. For this alliance, the initial members include Beijing Normal University, Huawei, iFLYTEK, TAL Education Group, NetDragon, Dell, and Talkweb, and China Mobile serves as the first director unit. Aiming to create a cooperative system that integrates "teaching", "learning", "production", "research", and "investment" in the 5G network and attains the goal of promoting the development and fusion of 5G and smart educational technologies, the alliance collaborates regarding the enactment of smart education standards in the context of 5G, research of key technologies, business pilot demonstration, exchange and cooperation, and startup business cultivation to realize cooperative innovation, integration, and win-win results among all partners. Technical standards for smart devices represented by VR are also being steadily established.

9.4 Summary and Prospects

In 2021, China's education industry underwent major transformations, opening up new doors for innovation and development of the education industry. This has laid the path for developing 5G+ education applications. The requirements of students, teachers, parents, and multi-party administrators have been preliminarily determined. The transformation of smart education may be fulfilled through the following phases:

- Initial application phase of 5G: Currently, 5G networks are at the initial deployment phase in the education industry, and there are no standards available for 5G network applications and convergence. Therefore, the convergence mode between 5G networks and traditional networks as well as 5G education application standards need to be explored in this phase.
- Mature phase of 5G standard development: In one to two years, 5G, featuring high rate and low latency, will be a mature technology for the development of basic services in the education field. For example, 5G will improve the transmission efficiency of HD video rebroadcasting and change the supervision mode through remote video surveillance. It also supports the transmission of massive data thanks to the high bandwidth. Therefore, it is possible to explore interactive applications for interactive classrooms.
- Mature phase of 5G terminal modules: With the maturity of 5G network standards, telecom operators will develop terminal modules for 5G network transmission and sensing, such as physical environment sensors and human body feature sensors, to support intelligent sensing, identification, and data collection in different scenarios. Such terminal modules will be widely used in scenarios such as intelligent security protection.
- Mature 5G education network: With mature 5G terminals, a comprehensive education service network will be established based on different teaching, research, and education management scenarios to meet service requirements in the education field. With powerful context awareness, data processing, and analysis functions, the network can provide adaptive services for different users.
- 5G-powered education model transformation: This is the ultimate goal of 5G + education development and is based on the comprehensive improvement of multiple technologies, such as 5G education networks, big data, and cloud

computing. Continuous changes of the basic education model promote significant improvement of teaching efficiency. In this phase, large-scale data computing is required to explore teaching models that meet the cognition requirements of learners and teachers. Technologies like VR, AR, and holography help implement immersive learning simulation and interactive holographic services.

In the future, as related standards and terminals gradually mature, the new educational infrastructure construction will be closely combined with multiple new ICT technologies represented by 5G through various development paths to form the core power that supports China's education modernization.

References

1. People's Daily Online. China Unicom Guangdong Branch. Guangdong Experimental High School Attempts "5G Intelligent Education" Classes [R]. 2019.
2. China Unicom. "5G+Holographic Projection" Enables Remote Allocation of High-Quality Educational Resources [R]. 2020.
3. 5G Application Innovation and Development White Paper [R]. 2021.
4. Virtual Reality/Augmented Reality White Paper [R]. 2021.
5. China Telecom. Youdao Premium Courses of NetEase Youdao and China Telecom's Tianyi cloud VR Jointly Promoted VR-based Cosmos Exploration Courses [R]. 2020.

Chapter 10
5G+ Smart Healthcare

10.1 Overview

10.1.1 Major Processes and Operations

The healthcare industry is an aggregation and integration of health-related sectors, including hospitals, medications, medical devices, and health management. It involves not only sectors directly related to medication and healthcare—specifically pharmaceuticals, medical devices, and healthcare services—in a narrow definition, but also peripheral sectors—such as pharmaceutical devices, packaging materials, and talent services—in a broader definition. Generally, the industry consists of various healthcare institutions (such as hospitals) and their upstream and downstream services and can be divided into three primary branches: pre-hospital, in-hospital, and out-of-hospital.

10.1.1.1 Pre-hospital

Pre-hospital services include disease screening, first aid, and pharmaceutical cold chain logistics.

Disease screening refers to specific examinations provided by primary care institutions or checkup centers on people who are vulnerable to diseases but have yet to suffer any symptoms of such diseases. The goal is early warning, detection, intervention, and treatment, to improve efficacy, save lives, and reduce disease burdens. Such early intervention can be performed through inspection, palpation, percussion, auscultation, blood and secretion tests, imaging tests, pap smears, in vitro diagnostic reagents, genetic testing, and others.

First aid refers to the immediate care and ambulance services provided to patients with acute illness or injury before they reach hospital. It is essential for helping hospitalists pre-assess patients' conditions and prepare necessary treatment. The entire process includes emergency call handling, ambulance dispatching, on-scene

P. Sun, *A Guidebook for 5GtoB and 6G Vision for Deep Convergence*,
Management for Professionals, https://doi.org/10.1007/978-981-99-4024-0_10

medical care, and continuing care during transport, which imposes requirements on emergency dispatch, quality control, centralized monitoring, and first aid data transmission.

Pharmaceutical cold chain logistics refers to the storage and transportation of refrigerated and perishable pharmaceutical products, which is a systematic project from production through to consumption. It keeps temperature-sensitive medications refrigerated from the warehouses of pharmaceutical companies to pharmacies, placing high requirements on temperature monitoring.

10.1.1.2 In-hospital

In-hospital services are classified into two categories: medical diagnosis and treatment, as well as hospital operations management. Medical diagnosis and treatment covers the whole process from appointment/registration right through to guidance, interrogation, examination, consultation, payment, medication dispensing, surgery, and nursing. In this process, guidance involves not only assisting patients' inquiries but also escorting patients to various areas and departments for tests, examinations, payment, medication pickup, inpatient admission, etc. Medical diagnosis depends on medical examinations (e.g., radiological imaging, ultrasound imaging, nuclear medicine and molecular imaging, endoscopic imaging, and pathological imaging) and the advice of healthcare professionals. Medical examinations are able to visualize most medical data and cover a majority of clinical departments in a hospital.

Hospital operations management includes nursing care, ward rounds, and logistics management. Nursing care—one of the core hospital services—has extremely high labor and time demands. As such, mobile applications for ward rounds and mobile nursing stations have been developed to implement closed-loop management based on bar coding throughout the entire life cycle of medical orders, standardizing care processes and services. The use of personal digital assistants (PDAs) helps personnel execute medical orders and record patient information, preventing human errors from arising. Logistics management involves managing consumables, medications, waste, and personnel, requiring appropriate arrangement of human and material resources for hospitals. Smart hospital management systems are available on the market to help hospitals increase management efficiency and reduce resource costs.

10.1.1.3 Out-of-hospital

Out-of-hospital services include household health management, rehabilitative care, and community healthcare.

Household health management refers to monitoring patients' health and nursing them in their homes. This is possible because vital sign data is usually monitored by a portable health monitoring device at home and uploaded to a health management system through a household gateway, so as to obtain remote guidance and nursing care. In an emergency, a patient can call medical support or community healthcare providers for quick response and treatment.

Rehabilitative care refers to promoting and restoring the health of patients with chronic diseases, injuries, or disabilities. Unlike traditional rehabilitation—which is often slow and repetitive, the use of intelligent rehabilitation devices (such as rehabilitation robots and balance function testers) assists in rehabilitation training and intelligently evaluates the effect, which can greatly facilitate patients' strength and cognitive recovery.

Community healthcare usually uses medical-grade health indicator detection devices and the health management platform to provide various healthcare services for residents, including comprehensive health status monitoring, chronic disease screening, age-related disease prevention and treatment, as well as preoperative and postoperative prevention, prediction, monitoring, and rehabilitation. This information can be shared with community healthcare stations and residential health management stations.

10.1.2 Current State

10.1.2.1 Growing Healthcare Needs with Aging Population

An aging population has become a global issue. According to the World Bank's open data, the proportion of the population aged 65 and above had grown from 7.78% in 2013 to 9.32% at the end of 2020. Fertility decline is affecting global economic growth. "The global fertility rate, which fell from 3.2 births per woman in 1990 to 2.5 in 2019, is projected to decline further to 2.2 in 2050," as stated in *World Population Prospects 2019: Highlights*. With falling fertility level and increasing life expectancy, population aging will remain a global trend in the next century, which in turn places great pressure on the healthcare industry.

China has the largest elderly population in the world. According to the seventh national population census conducted by the National Bureau of Statistics of the People's Republic of China, the number of Chinese people aged 60 and above exceeded 264 million in 2020, which is equal to 18.7% of the population, up by 5.44 percentage points from 2010. Among them, there are 191 million people aged 65 and above, comprising 13.5% of China's population. 149 cities across the country have entered the stage of deep aging, among which 11 are experiencing super aging. It is predicted that China's population of elderly people aged 60 and above will reach 300 million, becoming a moderately aging society by the end of the 14th Five-Year Plan period, according to the statistical releases from the national authorities. In the face of accelerating population aging and increasing life expectancy, China aims for long-term, balanced population development.

The soaring elderly population increases the expenditure on healthcare and pensions, posing significant challenges to resource sustainability. The elderly usually have great healthcare needs, require long-term care, suffer chronic diseases (a majority of the elderly aged over 60 have at least two chronic diseases), and lack self-care ability. These present unprecedented challenges to nursing homes, sub-health management platforms, and healthcare service assurance.

Population aging also creates great opportunities for industries related to traditional Chinese medicine, innovative medicine, and medical devices. For example, as the elderly have reduced self-care ability, this opens up a market for medical devices and assistive products. In particular, the demand for medical devices for diagnosis, monitoring, and treatment is expected to witness an explosive growth. Moreover, the elderly will likely value healthcare experience and services more, stimulating a growing demand for healthcare devices and wellness products.

10.1.2.2 Overwhelming Shortage of Resources with Rising Prevalence of Chronic Diseases

Although China's economic growth has improved the standard of living, rapid urbanization, sedentary lifestyles, changing dietary habits, and rising obesity have aggravated the rising of chronic diseases—especially cancer, diabetes, and hypertension. Chronic diseases have become the top killers in China, featuring low awareness, treatment, and control rates and leading to high morbidity, mortality, and disability rates. According to the report on nutrition and chronic diseases of Chinese residents (2020), deaths caused by chronic diseases accounted for 88.5% of the total deaths in China in 2019, including 80.7% of deaths from cardiovascular and cerebrovascular diseases, cancer, and chronic respiratory diseases. The prevalence rates of hypertension, diabetes, and high cholesterol among Chinese residents aged 18 and above reached 27.5%, 11.9%, and 8.2%, respectively. The prevalence of chronic obstructive pulmonary disease among residents aged 40 and above reached 13.6% [1].

The rising prevalence of chronic diseases will induce long-term medication use and disease management costs, raising China's healthcare expenditure. On the one hand, chronic disease treatment requires high communication costs. Without frequent communication between patients and doctors, disease management is often ineffective. On the other hand, chronic disease management usually requires patients to adjust their lifestyles, such as eating a healthy diet, giving up smoking, and undertaking regular exercise, all of which depend heavily on the will of patients. As such, chronic disease management cannot achieve the expected effects in most cases.

Traditional chronic disease management mainly involves four processes—diagnosis, treatment, rehabilitation, and follow-up—and three players—patient, hospital, and community healthcare institution. As dominant players, hospitals and community healthcare institutions are responsible for chronic disease prevention, primary care, treatment, rehabilitation, education—all of which depend on people. However, quality resources are usually concentrated in large healthcare institutions. Primary care institutions lack not only general practitioners and preventive care specialists for chronic disease management but also sufficient service facilities and medical devices. Relevant policy support and financial investment are inadequate. This means that patients suffering from chronic diseases cannot fully trust primary care workers and institutions, placing even more burden on healthcare resources.

10.1.2.3 Uneven Distribution of Healthcare Resources

Considering economic and social development as well as ever-increasing service requirements, the total amount of healthcare resources in China is relatively insufficient. Unavailable, unaffordable, and troublesome healthcare are prevailing problems in China. According to the statistical bulletin on China's healthcare development in 2020, the number of healthcare institutions in China reached 1,023,000 by the end of 2020, an increase of approximately 15,400 over the previous year—with a year-on-year growth rate of 1.5%. There had been approximately 971,000 primary care institutions (accounting for 94.9% of the total number of healthcare institutions), 35,000 hospitals, and 14,000 public healthcare institutions in China by the end of 2020. From 2014 to 2020, the number of hospitals and primary care institutions had been increasing, but the number of public healthcare institutions had been declining [2].

On the 2020 list of China's top 100 hospitals, at least one hospital in each province (except for Inner Mongolia, Qinghai, Tibet, Guizhou, and Hainan) is ranked—with Beijing, Shanghai, and Guangdong (in descending order) boasting most of the top-ranked hospitals. For example, 11 hospitals are located in Shanghai, 10 in Guangdong, 6 in Zhejiang, 5 in Hubei, 5 in Jiangsu, 5 in Shandong, 4 in Fujian, 4 in Hunan, and 4 in Liaoning. This highlights the uneven distribution of quality healthcare resources across the country [3].

As China's quality healthcare resources are mainly concentrated in major cities—such as Beijing, Shanghai, Guangzhou, and Shenzhen, there exists a significant divide in the quality of healthcare resources between rural and urban areas as well as between different provinces. The central and western regions of China in particular have far fewer healthcare professionals than the eastern region, highlighting their need for further improved diagnosis and treatment skills and hospital capabilities. Furthermore, inadequate capabilities and inefficient utilization of China's primary care institutions impel patients to visit grade—A tertiary hospitals for any illness—no matter serious or minor, chronic or acute. This consequently overwhelms large public hospitals and wastes primary care institutions in China.

10.1.3 Challenges

10.1.3.1 Lack of Healthcare Providers

According to statistics by the end of 2020, there were 10.678 million healthcare providers, 792,000 village healthcare workers, and 530,000 other healthcare workers in China. The number of healthcare providers (including 4.086 million practicing/

assistant physicians and 4.709 million registered nurses) increased by 524,000 (up by 5.2%) compared with 2019. There were 2.9 practicing/assistant physicians per 1000 people, 3.34 registered nurses per 1000 people, 2.9 general practitioners per 10,000 people, and 6.56 healthcare workers in public healthcare institutions per 10,000 people [2]. However, healthcare providers are still in short supply, especially during the COVID-19 pandemic, inducing a growing imbalance between supply and demand in China's healthcare industry. An annual growth of approximately 500,000 healthcare providers still cannot meet China's all-round and multi-tiered healthcare requirements.

Not only that, as public hospitals attract the majority of physician resources, primary care institutions are short of healthcare providers and service capabilities due to low payment and insufficient policy support. By the end of 2020, 609,000 village clinics had been established in 509,000 administrative villages in China, which is 7000 fewer than the number in 2019, causing a decrease in the total number of healthcare workers. As such, there is an urgent need to increase investment in primary care institutions, improve the healthcare environment, and replace aging and outdated medical devices in primary care institutions. By constructing an integrated healthcare system, quality healthcare resources of upper-tier hospitals could be fully utilized by primary care institutions.

10.1.3.2 Weak Localization of High-End Devices

With continuous innovation and development of healthcare technologies, ever-emerging diverse medical devices—especially high-end devices—become an increasingly vital input to the healthcare service system and an important guarantee for healthcare institutions to provide healthcare services. Medical devices can be divided into diagnostic, treatment, and auxiliary devices. However, a large share of China's high-end medical devices are not made domestically, meaning they primarily rely on imports, which in turn drives up their purchase cost and complicates after-sales service.

Import substitution has been substantially completed in medical devices for patient monitoring and ultrasound imaging but not in devices for computed tomography (CT), magnetic resonance imaging (MRI), and digital subtraction angiography (DSA) in China. As a leading domestic medical device provider in China, Mindray's domestic market shares in monitoring and ultrasound devices have reached approximately 50% and 10%, respectively. For example, DSA—an indispensable technique for interventional diagnosis and treatment—can produce real-time and accurate images. It is the industry-recognized "gold standard" for interventional treatment and one of the most effective methods for detecting vascular diseases. However, 92% of DSA devices in China were purchased from multinational enterprises such as GE, Philips, and Siemens over a long period.

As China lags behind developed countries in terms of medical device development, Chinese hospitals are limited to where they can purchase high-end medical devices.

The monopoly of foreign enterprises on after-sales service of medical devices is largely related to unsolved internal problems of domestic enterprises and healthcare institutions.

Currently, most Chinese enterprises lack competitiveness in enterprise scales and self-owned brands. One of the reasons why Chinese residents pay for expensive medical treatment is because high-end medical devices are imported and their after-sales service always costs exponentially more than domestic ones.

China has only a very small number of international-certified high-end medical device manufacturers and products, significantly lagging behind developed countries in terms of production management and quality assurance. Furthermore, the export of Chinese high-end medical devices is facing manifold non-tariff barriers including certification and green requirements.

10.2 Digitalization Trends and Challenges

10.2.1 Trends of Digitalization

10.2.1.1 Rapid Advancement of Hospital Informatization

With the thorough development of China's social economy and the gradual improvement of the medical system, high-quality medical services have become an urgent need in the new era. Nowadays, China is facing great opportunities to deepen the reform of the medical system and promote the combination of medicine and high-tech. The traditional medical mode is embracing new and intelligence, and the medical informatization is gradually promoted.

So far, despite the digital transformation of China's healthcare industry starting early, it is still relatively slow. The previous digital transformation of hospitals revolves around service processes. As big data, AI, and 5G technologies develop, the transformation of separate services may create barriers between them. New digital transformation focuses on leveraging information technologies such as the medical IoT, medical cloud, and medical big data to break "information silos" between hospital departments in the traditional medical mode, streamline coordination among departments, and improve the efficiency of medical personnel. As information technologies continue to evolve, smart healthcare is the trend in modern society. Traditional hospitals will transform into smart hospitals capable of active thinking, perception, and execution.

In a smart healthcare ecosystem, hospitals will develop a full understanding of its patients through devices, systems, and processes to sense, measure, capture, and transfer patient information in real time. In this way, hospitals can realize such intelligent functions as all-round automatic information collection (for IoT), in-time information transmission (for interconnection), as well as intelligent decision support. These three core functions can be integrated to facilitate information resource sharing and dependency, and therefore are key to smart hospitals. These intelligent functions help hospitals to better serve people, reflecting people-oriented ideology.

At the same time, emerging technologies such as cloud computing, IoT, and 5G are leveraged to build new ICT integration platforms and solutions. Together with partners, the healthcare industry will promote full-process and full-data digital transformation. Information that is open and linked not only helps doctors and patients communicate with each other, but also streamlines communication between doctors and nurses, between different departments, as well as between healthcare institutions and other institutions (such as management organizations and insurance units) to achieve service collaboration. In this way, the healthcare data is fully open, shared, and exchanged. This not only improves the service efficiency of healthcare personnel and provides higher quality services, but also opens up premium health services to patients. Moreover, hospital management and policy monitors can obtain more accurate monitoring information to improve the management level and decision-making of the government.

10.2.1.2 Innovation and Iteration of New Medical Equipment

Owing to favorable policies, the national medical equipment industry has entered a stage of rapid growth, one where next-generation information technologies, such as 5G and AI, can effectively improve the processing and application capabilities of medical devices, enhance the depth and efficiency of data processing, and continuously drive new medical devices and products. For example, new medical devices such as medical robots and wearable devices have developed rapidly, and are widely used in medical and health fields such as health monitoring, disease management, rehabilitation, and traditional Chinese medicine.

The application of new medical equipment changes the service model of traditional medical care. For example, new medical imaging devices can assist in implementing real-time diagnosis, improving efficiency while ensuring diagnosis accuracy. Medical robots replace manual operations to a large extent, making medical operations more convenient and accurate, and thereby reducing risks of medical misoperations. Additionally, the living standards and healthcare awareness of residents in China are increasing and their demand for smart medical equipment continues to grow. With the improvement of technologies such as wearable materials and sensor technologies, especially the development of 5G communications technologies and big data monitoring, technical platforms are available for the development of wearable medical monitoring devices, greatly increasing the storage capacity of the devices and improving communication quality and data transmission speed. The personalized development of wearable medical devices will be a major trend in the future. These devices are expected to be applied to more fields such as family diagnosis, sports health monitoring, and major disease prevention and treatment, with the purpose of providing universal medical services for the elderly, opening up medical care to anyone whenever or wherever they need it.

10.2.1.3 Wide Attention to Patient Privacy and Data Security

Digital healthcare collects and receives all records of the entire health process in real time, dynamically, and continuously. Medical information includes patients' health status and personal privacy data. For a long time, medical data has been in a closed environment. Digital transformation requires different platforms of medical institutions to be interconnected, and this has aroused heated social discussion on how the security of medical data can be protected.

Patient data from online medical consultation to offline medical registration and treatment is automatically uploaded to the cloud of mobile medical applications. Leaked patient information caused by improper technical operations or subjective factors such as business purposes may cause physical and psychological harm to patients and their families. For example, leaked medical records of mental behavior diseases or acquired immunodeficiency syndrome (AIDS) may fuel discrimination towards certain patients. As time goes by, patients' medical records accumulate more and more personal information, including medical diagnosis history, treatment and medication history, dietary habits, and psychological archives. As the number of users increases, more and more data will be stored on the server. For such a large database, there is a pressing need to ensure data security and protect patient privacy.

10.2.1.4 5G: A Critical Support for Healthcare Digital Transformation

Thanks to 5G networks, the healthcare industry will undergo unprecedented improvements in development and digital transformation. Similar to the information highway, 5G provides possibilities for the transmission of massive data and information, as well as higher transmission speeds. 5G features flexible and configurable new network architecture, ultra-high bandwidth, millisecond-level transmission latency, million-level connection capability, and high reliability. It can effectively compensate the drawbacks of wired networks, Wi-Fi, and 4G, and promote comprehensive collaboration of in-hospital and out-of-hospital care.

First, 5G has become a key exploration direction for the new infrastructure construction of hospitals. 5G capabilities are fully utilized to provide real-time computing and low-latency edge cloud services for healthcare to provide future healthcare, such as remote surgery, emergency first aid, remote ultrasound, VR visit, and smart ward, as well as other high-value application scenarios. 5G healthcare can improve inter-hospital information exchange and service collaboration. Upper-tier central hospitals with medical expert resources and comprehensive medical facilities can remotely guide the medical services of lower-tier hospitals in the medical consortium over a wireless network to improve medical diagnosis performance.

Second, 5G networks are deployed and maintained by telecom operators, significantly reducing hospital operation costs. Previously, hospitals needed to purchase a large number of communication devices and servers to establish physical private networks for secure and reliable communication of medical services. On top of this, dedicated operation teams are required for routine maintenance. After the 5G medical network is introduced, telecom operators provide and deploy communication devices and are responsible for O&M, slashing investment costs on the hospital side.

Third, 5G can facilitate healthcare convergence and innovation and develop new smart healthcare services. A 5G medical network has smooth evolution capabilities, and when combined with cloud computing, big data, digital imaging, and AI technologies, it can penetrate into all phases of healthcare services, helping healthcare achieve wireless and intelligent development.

Fourth, 5G has become the core foundation for the innovation of high-end devices. It will drive the development of multiple fields, such as intelligent medical devices, traceability of high-value consumables, intelligent wearable devices, and medical robots. According to KPMG's estimation, the potential value of the global market in major vertical industries of 5G technologies could reach USD4.3 trillion. Healthcare is a key 5G application field for profitability.

10.2.2 Trends of Technologies Related to 5G Application Scenarios

10.2.2.1 Terminal Layer—5G+ Intelligent Medical Equipment and Terminals

Intelligent terminals, such as tablets used for ward-round, video conference terminals for teleconsultation, video collection terminals, and wearable devices, can be integrated with 5G universal modules to connect to 5G networks. By utilizing 5G mobile communications technologies, test and examination devices and mobile medical workstations can be integrated to implement mobile test and examination services. Through 5G, patients' vital signs can be monitored in real time for a long period, and vital sign data and critical alarm information are transmitted to medical personnel to help them understand the status of patients and make decisions.

Traditional medical devices, such as large medical equipment and robots, are complex. It is difficult to reconstruct them by directly integrating 5G universal modules. Rather, they can support 5G by using data transfer units (DTUs) or USB dongles. Based on 5G network slicing, private networks can be provided for medical detection and nursing devices that bear large traffic to ensure stable and smooth data transmission. In this way, a large number of medical sensor terminals and video-related devices can be remotely used to sense, measure, capture, and transfer patient information in real time, achieving all-round perception of patients. Not only that, intelligent medical terminals can constantly and accurately monitor disease information, without time and space restrictions, and break the technical bottleneck for the wide replication and promotion of remote monitoring.

10.2.2.2 Network Layer—Three 5G Network Capabilities Adapted to Scenario Requirements

5G features high rate, low latency, and massive connectivity, which correspond to enhanced Mobile Broadband (eMBB), ultra-reliable low-latency communication (URLLC), and Massive Machine-Type Communications (mMTC), respectively. eMBB features ultra-high bandwidth and rate, and is used for continuous wide area coverage and hotspot high-capacity environment. In wide coverage scenarios, the user-perceived rate can reach 100 Mbit/s and a high mobility of 500 km/h is supported. In hotspot high-capacity scenarios, the user-perceived rate can reach 1 Gbit/s, the cell peak rate 20 Gbit/s, and the network traffic density 10 Tbit/(s·km^2). eMBB can provide continuous wide area coverage for ambulances and help to realize the "hospital-like treatment for patients while onboard" vision. HD videos, along with a large amount of vital information such as patients' vital signs and conditions can be sent back to the background command center in real time through 5G networks. Additionally, data of wearable devices of patients can be collected to detect vital signs in real time, around the clock.

URLLC supports connections with a minimum one-way air interface latency of 1 ms, and 99.999% reliability in high-speed mobility scenarios. URLLC is mainly used in low-latency scenarios such as wireless monitoring, remote detection, and remote surgery in hospitals. In wireless monitoring, vital sign information of a large number of patients is collected in a unified manner in the background, facilitating unified monitoring and management and greatly improving the efficiency of medical personnel in existing intensive care units (ICUs). Remote B-mode ultrasound and remote surgery have high requirements on detection technologies because real-time feedback is required. Remote B-mode ultrasound and remote surgery can eliminate the physical distance between doctors and patients and allow for real-time detection and surgery thousands of miles away.

mMTC supports a connection density of 1 million/km^2. Devices feature lower power consumption and lower costs, achieving connectivity of everything. mMTC is mainly used inside a hospital. Existing hospitals have thousands of medical devices and they are eager to manage and monitor the medical devices in a unified manner. In the future, 5G can act as a solution and allow all devices to access networks.

Although the bandwidth, rate, and latency of 5G can meet the current demands of the healthcare industry, a 5G private network is required to satisfy more requirements. In actual network deployment, the following items still need to be considered:

First, public network spectrum of the operator dedicated to the local area. The virtual private network and physical private network can be provided. The virtual private network enables the healthcare industry and public users to share the spectrum resources of the operator. The physical private network provides dedicated frequencies for hospitals to build 5G networks.

Second, hierarchical isolation. Existing hospitals have urgent requirements for medical data security. Therefore, data isolation requirements of the medical industry must be fully considered during 5G network construction. Currently, hospitals are sensitive to medical data leaving hospitals. As such, data needs to be transmitted inside the hospitals.

Third, customized services. Currently, there are a large number of uplink high-bandwidth services in hospitals, such as remote ultrasound and remote B-mode ultrasound, and patients' vital sign data uploaded by a large number of IoT devices. The existing networks of telecom operators cannot meet those high-bandwidth requirements in the uplink. Therefore, customized flexible frame structures and differentiated wireless services are required. In addition, various base station types need to be developed to meet the deployment requirements of various scenarios in hospitals.

Fourth, smart operations capabilities. With the networks, devices in existing hospitals can be managed, services can be controlled and visualized, and faults can be easily located.

10.2.2.3 Platform Layer—Healthcare Informatization and Telemedicine Platform Reconstruction and Upgrade

In the future, smart healthcare will benefit from the high rate, low latency, and big data analysis capabilities of 5G to provide convenient healthcare services for everyone and improve the effects of existing healthcare methods. 5G MEC capabilities can be fully utilized to meet the growing demands that people have for future healthcare, such as real-time computing and low-latency healthcare edge cloud services, mobile ambulances, AI-assisted diagnosis and treatment, VR teaching, and imaging device enablement.

In view of the urgency and importance of mobile healthcare development, new technologies and capabilities must support modeling and prediction of various diseases to facilitate service application. Medical imaging can be used to identify and classify lesions. Based on mobile terminals and wearable devices, people's requirements for daily health management and chronic disease rehabilitation and treatment can be met, supporting self-health management. AI-based intelligent triage, diagnosis assistance, and electronic medical record (EMR) writing are supported. IoT application architecture based on sensor networks can be achieved. Data collection and utilization of various medical terminals are feasible. Big data distributed computing frameworks are supported, such as MapReduce, Spark, and Tez. The blockchain technology, as the underlying technology, can encrypt the underlying data to protect the privacy of patients' data. Many functions are available, such as multiple algorithm libraries, big data storage and access, and distributed computing task scheduling. A large number of services that are explored and practiced in clinical medicine lay a foundation for providing patients with digital, intelligent, and personalized diagnosis and treatment services. The whole process of health management is involved, including prevention, diagnosis, treatment, and nursing care.

10.2.2.4 Application Layer—Intelligent and Personalized Patient Experience as Two Development Paths

At the end of 2008, IBM proposed the concept of "smart hospital" for the first time, involving medical information interconnection, sharing and collaboration, clinical innovation, and diagnosis science. Advanced information communications technologies, such as mobile communication, Internet, IoT, cloud computing, big data, and AI, are used to establish a medical informatization platform centered on EMRs to connect patients, medical personnel, medical devices, and medical institutions. This platform enables highly information-based, personalized, and intelligent diagnosis, treatment, rehabilitation, payment, and health management, providing high-quality mobile healthcare services for people. Driven by national policies, social economy, and industry requirements, mobile healthcare is developing rapidly.

Technological progress enables smart hospital construction. With the development of the IoT, big data, cloud computing, AI, and sensing technologies, the data processing capability of computers increases exponentially, and many decision-making and medical assistance means become possible. The application of 5G smart healthcare will promote hospitals to work with medical insurance and social service departments to streamline patients' medical treatment and hospital service processes including pre-hospital assessment, in-hospital treatment, and post-hospital aftercare. Medical information can be shared among patients, medical devices, hospital information systems, and medical personnel, allowing the medical personnel to obtain medical information anytime and anywhere. This mobilizes medical services and greatly improves the working efficiency.

10.2.3 Major Challenges

Currently, the technical system, business model, and industry ecosystem of digital transformation in the healthcare industry are still evolving and being explored. Therefore, top-level architecture, system design, and implementation mode need to be continuously improved. The early exploration of digital healthcare produces good demos, seeing wide application in a variety of scenarios such as remote surgery, emergency rescue, middle-end control, medical robot control, mobile ward round, remote monitoring, remote training, surgery demonstration, and indoor positioning. Despite this, the digital transformation of the healthcare industry has not reached maturity, and there are still many problems in popularizing applications, as detailed in the following.

10.2.3.1 Network Construction Still Faces Many Challenges

Different from other network services, telemedicine networks have higher requirements on network speed, stability, and security. Establishing service quality standards for telemedicine networks and monitoring and controlling telemedicine network quality are inevitable for improving telemedicine network service quality.

The 5G healthcare network needs to carry medical device services and patient management services, and extend to inter-hospital and out-of-hospital scenarios. Technically, the network needs to be isolated from and collaborate with the public Internet in hospitals. As such, sinking of the UPF based on healthcare industry slicing will be the focus of deployment. However, the E2E industry chain of 5G standalone (SA) networking is not mature enough. New 5GC deployment requires simultaneous user migration and network reconstruction. System joint commissioning is also complex and the implementation period is long. Currently, 5G network deployment in the healthcare industry is still slow. Major challenges are as follows:

> First, the quality of medical networks varies. Various telemedicine services do not have unified network service quality requirements, so there is no basic assurance for standard telemedicine services, and the network stability is poor. Second, healthcare network systems are isolated. In existing telemedicine systems, networks, protocols, and platform systems are difficult to interconnect. Furthermore, 5G smart healthcare services have varying functions in different hospitals, hindering cross-region and cross-network telemedicine collaboration. Third, medical network security risks exist. Standards for networks, devices, systems, data, and applications are not unified, and security risks exist during interconnection.

10.2.3.2 Industry Support for Common Technologies Needs to Be Strengthened

Most 5G+ healthcare applications are leading attempts at the early stage of application exploration. There are few technical verification and solution feasibility studies. Likewise, there are also no mature technical solutions in technology R&D, outcome conversion, product manufacturing, and application deployment. What's more, there are numerous 5G+ healthcare application scenarios and different application scenarios have varying network performance requirements. There are no specific standards and specifications for network indicators of different applications. Standardization issues exist in network construction mode, device access, data interconnection, network service quality, and data security.

10.2.3.3 Business Model of Convergent Applications Are Still Under Exploration

Healthcare involves multiple disciplines and high levels of professionalism. Deeper convergence and collaboration between the healthcare industry and the information communications field is needed. 5G healthcare applications are set back by large investment in the startup period and unclear market returns. Industry customers are simply not motivated enough to innovate. Most of those applications are invested by governments or telecom operators. There is no clear business model and no business closed-loop.

5G healthcare is a convergence solution. Currently, medical institutions maintain service quality by themselves, involving many industry chains. A stable industry ecosystem has not been formed. For now, cooperation is mainly led by telecom operators. The leading capability of healthcare informatization solution providers is insufficient, and multi-party collaboration is desperately needed. Furthermore, 5G smart healthcare services demand high service quality, but there is no test and verification solution for the healthcare industry.

10.2.3.4 Medical Data Security Faces Challenges

Medical information resources have huge data and value. However, challenges such as information asymmetry, credit in-transparency, and difficult supervision exist. First, most medical institutions focus more on procurement but less on device use and management. Non-transparent device use and management leads to idle and wasted medical device resources. Second, medical device data lacks support from industry standards and industry master data. As a result, data disorder makes it difficult to implement data-based intelligent supervision. Medical device distribution data is incomplete, inaccurate, and not dynamic. Lack of device usage data and data quality problems make it difficult to evaluate healthcare technologies of related applications. There is no standard for reporting adverse events of medical devices, making it difficult to collect and analyze statistics. Without the support of precise data, it is difficult to achieve reasonable allocation of medical devices.

In addition, hospitals need to collect patient data in multiple scenarios, platforms, and time segments for in-depth analysis. There are many data transmission and circulation links, making it difficult to prevent personal information from being spread or abused. New cyber security risks caused by the converged application of AI technologies in the healthcare industry are spreading and intensifying. As data privacy draws close attention, people are more and more vigilant about the security of personal health data. How to improve data security assurance capabilities through technical means and standards, and seek a balance between openness and privacy of medical big data will become another challenge for medical digital transformation.

10.2.3.5 Policies and Regulations for Digitalization Need to Be Improved

5G+ healthcare still faces a number of challenges such as commercial license issuance, national law and regulation support, frequency clearance, and electricity usage charge discount policies. To achieve large-scale commercial use of 5G, cooperation from multiple parties is required to overcome these challenges. The government has issued national laws and regulations on communications infrastructure planning, construction, sharing, and protection. The laws and regulations specify the public attributes of communications infrastructure and stipulate that communications infrastructure construction must be included in urban and rural planning, providing public site resources.

Regulatory policies for digital applications need to be further optimized. 5G+ healthcare applications still face challenges including patient distrust, weak profitability of enterprises, and low participation of doctors and medical insurance institutions. On top of this, the legal responsibilities, charging standards, and network tariff policies for 5G+ healthcare services have not been specified, making it difficult to form a sustainable business model in a short term. 5G applications accelerate the data circulation of various applications in the healthcare field, posing risks in healthcare quality and data security. Innovative supervision methods are urgently needed to ensure secure promotion of 5G healthcare applications.

10.3 Major 5GtoB Large-Scale Replication Scenarios and Typical Cases

10.3.1 Teleconsultation

Teleconsultation uses communication, computer, and Internet technologies to complete medical diagnosis and provide medical information and services. In 1988, the United States took the telemedicine system as a whole and proposed a widely-recognized concept of an open distributed system, which uses computer and communication technologies to provide medical services for specific groups of people. China's telemedicine started later and began to be explored in the 1980s. 5G+ teleconsultation is a medical consultation service that allows remote medical experts to provide real-time video guidance for doctors at primary care institutions to check and diagnose diseases. Based on the 5G+ teleconsultation platform, small diseases can be diagnosed and treated in community hospitals while mysterious, difficult, urgent, and serious diseases can be treated by experts through the teleconsultation system by implementing resource sharing.

10.3.1.1 Industry Requirements

China has a vast territory and uneven distribution of medical resources. Top medical resources are often concentrated in first-tier cities such as Beijing, Shanghai, Guangzhou, and Shenzhen. Medical resources in counties and rural areas are relatively inadequate, and there is a huge demand for medical care. Residents in these areas, especially the elderly, cannot obtain timely and high-quality medical services. Thanks to teleconsultation services, patients can receive diagnosis and treatment locally or at home from experts of grade-A tertiary hospitals. The teleconsultation services can provide convenience for the elderly and ease inadequate medical services in remote areas, helping handle challenging and urgent diseases. Moreover, teleconsultation can help alleviate overcrowding in hospitals to some extent.

The outbreak of COVID-19 and shortage in pandemic prevention materials have greatly increased the demand for telemedicine. Through remote devices, doctors can communicate with patients without direct contact. Unlike the video connections of common mobile phones, remote diagnosis and treatment devices integrate booths and supports required by medical care, letting doctors observe patients more carefully. Patients can upload medical materials generated during consultation, such as CT images or medical records, to doctors for diagnosis. With auxiliary services, doctors can also perform online preview and comment to record and track the diseases, which cannot be achieved by common videoconferencing devices.

With the 5G telemedicine service platform, hospitals can build bedside mobile medical carts based on the 5G wireless network. The carts integrate doctor and nurse workstations, as well as teleconsultation platform applications, and are equipped with 5G CPE wireless network adapters for 5G network access. A 5G private network is used to access the hospital information system and connects teleconsultation experts. Fig. 10.1 illustrates the bedside teleconsultation architecture.

For example, in Wuhan huoshenshan hospital, medical resources were in great shortage. Experts who went to the pandemic area were skilled at tackling respiratory diseases and treating severe patients. Many patients at the frontline had complex conditions, which required the diagnosis and treatment opinions of multidisciplinary experts. Based on the 5G network, mobile medical carts were used for patients at bedsides to conduct video consultation with experts. Through the telemedicine service platform, experts could learn about patients' medical history in advance or ask such information online, and listen to patients' chief complaints. The high rate of 5G network and the minor streaming function of the consultation platform enable

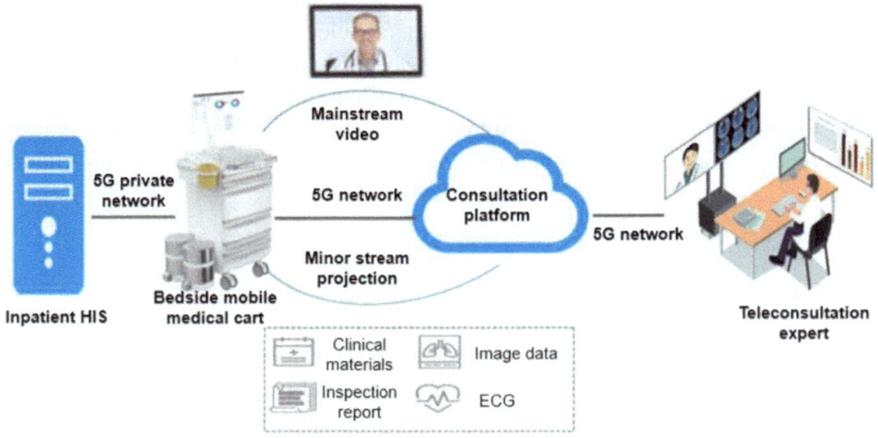

Fig. 10.1 Bedside teleconsultation architecture

high-speed transmission and sharing of massive medical data such as vital signs and medical images. Experts can make judgments and analysis based on medical records and images, and provide diagnosis and treatment suggestions in a timely manner. With the high rate, high stability, and low latency of 5G wireless network technologies, multiple parties can participate in a consultation at the same time. Experts from multi-disciplines can carry out consultation and discuss treatment solutions anytime and anywhere, improving diagnosis and treatment accuracy and efficiency. This also promotes easy access to high-quality medical resources.

5G teleconsultation has been widely used across the country and has demonstrated obvious advantages: First, 5G teleconsultation is free from geographical restrictions. It can quickly gather high-quality medical resources, and improve diagnosis and treatment efficiency. The teleconsultation platform can quickly gather experts from multiple hospitals, bypassing geographical restrictions, and use communication devices to provide multi-disciplinary telemedicine services for clinical treatment. In this way, medical resources can be quickly optimized, the cure rate can be improved, and the mortality rate can be reduced. Second, 5G teleconsultation reduces the risk of cross infection. For infectious diseases, teleconsultation can reduce close contact with patients while enhancing diagnosis and treatment. Frontline medical personnel in the clinical isolation area can output medical records through the teleconsultation platform without taking clinical materials out of the isolation area, reducing the chance of infection. Medical experts can use the remote platform to view patients' diagnosis and treatment records, provide high-quality medical services for infectious areas such as fever outpatient departments and isolation areas for suspected and confirmed patients, and provide diagnosis and treatment suggestions through remote audio and video technologies. Third, 5G teleconsultation improves the efficiency of diagnosis and treatment. The outbreak of the COVID-19 pandemic led to an increasing shortage of doctors as the number of patients increases exponentially, especially the shortage of medical personnel for severe and critical patients. The teleconsultation platform allows one-to-one, one-to-many, two-party, or multiparty consultation. Multiple patients from different hospitals can be treated at the same time in a relatively fixed period. Fourth, 5G teleconsultation helps enhance the qualification of medical personnel and knowledge sharing. During teleconsultation, multiple medical personnel can participate in discussions and therefore learn more. This allows medical personnel to improve clinical diagnosis and treatment thinking and enhance clinical professional experience under the guidance of consultation experts.

10.3.1.2 Network Solutions

On a 4G network, teleconsultation supports 1080p HD videos at most for both doctors and patients. However, the problems such as poor real-time performance, low definition, and frame freezing persist. The high rate and low latency features of 5G networks are vital to the healthcare field which has high reliability requirements. 5G networks support high-speed synchronous transmission and sharing of remote 4K/

8K HD videos, VR/AR consultation, and medical image data, allowing experts to carry out online consultation, improving diagnosis accuracy and guidance efficiency. This promotes easy access to high-quality medical resources.

10.3.1.3 Industry Ecosystems

Teleconsultation involves complex multi-party collaboration, covering medical institutions, telecom operators, integrators, and medical informatization enterprises. In most cases, teleconsultation is mainly implemented by medical institution groups. The medical institutions of all parties closely cooperate with each other to form a consultation group.

Operators play an indispensable role in healthcare transformation. They are not just 5G network providers, but also service promoters and creators. Operators integrate 5G technologies and capabilities of healthcare informatization enterprises to form an integrated teleconsultation solution.

Currently, teleconsultation is still in the initial phase. Medical institutions propose teleconsultation center construction plans based on their own requirements, which will be driven by the coordinated efforts of telecom operators, 5G technology suppliers, and healthcare informatization enterprises.

10.3.1.4 Typical Cases

1. 5G+ MEC Smart Healthcare of Medical Consortium in Futian District, Shenzhen

   Futian district is one of the first pilot areas in China and the only pilot area in Shenzhen for basic public service standardization, with the aim of building "a world-class resilient and healthy urban area and medical highland". Cooperation between NHC and Ministry of Industry and Information Technology (MIIT) in the Futian district has seen the Eighth Affiliated Hospital of Sun Yat-sen University lead efforts to overcome a number of problems including uneven medical resource allocation, limited informatization budgets of primary care institutions, and an inability to track and manage patients' rehabilitation activities. 5G technologies were used to innovate fully-connected hospitals, helping to build a private and shared regional 5G healthcare network system that serves as a benchmark for the country. 5G modules were embedded into medical devices, and basic capabilities including 5G, MEC, and AI were used to empower primary care and clinical departments to realize "low-cost, intensive, and replicable" innovation.

   Eight application scenarios including 5G mobile ward round, 5G teleconsultation, 5G pre-hospital first aid, and 5G public health emergency response, were built to implement remote, mobile, and informatization upgrades and the reconstruction of medical consortium services. During the COVID-19 pandemic, the 5G+ MEC-based healthcare private network implements hierarchical diagnosis and treatment in application scenarios such as bedside consultation, teleconsultation, and first aid for community health, helping with pandemic prevention. Through optimized medical resource allocation, the goal of "treating minor diseases in communities, handling serious diseases in hospitals, and letting people recover at home" can be realized, together with a shift from disease treatment to disease prevention. This case showcases the 5G medical system of Shenzhen, to act as a benchmark for the healthy China Program.

2. Teleconsultation Platform of the Wuhan Huoshenshan Hospital

In February 2020, the Wuhan huoshenshan hospital officially started accepting COVID-19 patients, becoming the very first hospital in Wuhan dedicated for COVID-19 sufferers. The first teleconsultation platform of this hospital, supported by Huawei technologies, was officially put into use [4].

This platform connected medical experts based in Beijing with frontline medical staff of huoshenshan hospital to hold teleconsultations with patients. Not only did this improve the efficiency and accuracy of patient diagnosis and treatment, it also helped relieve the pressure of frontline medical workers in Wuhan, while preserving the health and safety of medical experts outside Wuhan.

The teleconsultation platform used by huoshenshan hospital has several technical highlights, including the integrated HD videoconferencing terminal TE20 and management platform donated by Huawei, which could ensure smooth telemedicine even in an extreme network environment with a bandwidth of 512 kbit/s. The system also supports 1080p HD videos. In telemedicine scenarios, medical experts in two places might need to share medical records such as CT films of patients using auxiliary streams to help with patient diagnosis. Therefore, HD images are particularly important.

The first teleconsultation platform of huoshenshan hospital uses gigabit optical cables and this system uses a 5G network for backup. In later phase, the 5G network could fully support teleconsultation based on actual medical requirements. Currently, huoshenshan hospital has purchased medical carts that are fitted with mobile cameras, which are used to remotely perform ward rounds. The images captured by the carts can be used on the teleconsultation system.

10.3.2 Remote Ultrasound

Based on communications, sensor, and robot technologies, 5G+ remote ultrasound can be used to remotely control robot arms and ultrasonic probes over communications networks, facilitating remote ultrasonic services. Ultrasonic experts can use the HD audio/video system to communicate with doctors and patients in lower-tier hospitals in real time. Moreover, they can control the robotic arms of lower-tier hospitals using a controller to perform remote ultrasound examinations.

10.3.2.1 Industry Requirements

Ultrasound is a "super rigid demand" in all clinical examinations. Compared with CT and magnetic resonance (MR), ultrasound largely relies on a doctor's expertise as different doctors operate the device differently. This in turn may affect the result.

Many hospitals in China have been hindered by a shortage of medical resources. Although the government's investment in ultrasonic diagnosis equipment in primary care institutions has been increasing in recent years, these institutions still face other challenges such as a shortage of medical personnel, undeveloped medical and diagnosis technologies, and limited diagnosis and treatment services, which may fail to meet patients' needs. Many patients that can be examined in primary care institutions choose large hospitals for ultrasound examinations because the quality control of ultrasonic diagnosis and treatment in primary care institutions is loose, degrading patients' trust in these institutions.

Therefore, an HD remote ultrasound system without latency needs to be established to fully utilize expert resources from high-quality hospitals, enabling service guidance and quality control across regions and hospitals. Such a remote ultrasound system is helpful for ensuring standardized and reasonable ultrasound examinations in primary care institutions. Experts can remotely control robotic arms to perform ultrasound examinations on patients in primary care institutions. Remote ultrasound systems can be used in different hospitals of medical consortium and remote rural areas to help improve primary care.

5G+ remote ultrasound is used in the hierarchical diagnosis and treatment system and is generally carried out by upper-tier and lower-tier hospitals in medical consortium and community. For example, the Fifth People's Hospital of Ningxia Hui Autonomous Region has set up a close city consortium consisting of 25 members. The 5G remote ultrasound solution enables upper-tier hospitals to provide clinical consultation and teaching guidance for primary care institutions, and strengthens service/academic exchanges between hospitals. In this scenario, upper-tier hospitals enhance their influence and primary care institutions improve their ultrasonic diagnosis proficiency. An experience sharing platform is provided for doctors to exchange cases, reducing the time and transportation fees caused by onsite visits. This platform helps primary care institutions to provide high-quality diagnosis and treatment for patients in critical conditions.

The remote ultrasound system consists of the software system, hardware system, and peripherals, of which the software runs on the PC and can be accessed by the peripherals through a PC. The software system realizes the dynamic and orderly operation of the whole remote ultrasound system. The workstation is connected to the ultrasonic equipment through the high-definition multimedia interface (HDMI) cable to collect images. At the same time, cameras are connected to the workstation to capture how doctors perform scans while a microphone collects sound. After being processed by the workstation, multiple channels of signals are sent to the consultation end to obtain real-time and dynamic ultrasound images, as well as synchronous gesture images and sound.

With the 5G+ remote ultrasound system, real-time quality control experts/directors can view multiple channels of real-time operations of each ultrasound machine simultaneously. They can also choose to view one channel of video in full screen and provide voice guidance. Doctors in the consulting room can request real-time help from experts. This solves the problems of insufficient ultrasound teaching and training methods, high entry thresholds, and gets rid of traditional training methods that are based on static ultrasound images or segmented video files. The ultrasound platform facilitates remote ultrasound training and teaching, helping to improve the diagnosis proficiency of doctors.

10.3.2.2 Network Solutions

In remote ultrasound applications, ultrasound image quality is affected by contrast sensitivity, image resolution, noise and contrast definition. Sonographers need to perform professional diagnosis based on the image definition, image uniformity, ultrasonic tangent plane type, color blood flow display, and organ detection depth. During ultrasound diagnosis using convex array, linear array, and phased array, high-quality ultrasound images are of great significance. Since dynamic ultrasound images are captured in real time, there is a large amount of data to be transmitted, which has high network requirements. For example, a cardiopulmonary ultrasound examination of several minutes generates an amount of data up to 2 GB. In this case, the high-speed, stable, and low-latency 5G network guarantees stable, secure, and real-time remote control [5].

Compared with traditional private networks and Wi-Fi, 5G networks can overcome the problems of difficult and costly private network construction in remote areas such as rural areas and islands, insecure Wi-Fi data transmission in hospitals, and large remote control latency. 5G networks have a peak rate 10 times that of 4G networks and millisecond-level latency. Remote ultrasound based on 5G network technologies allows doctors in upper-tier hospitals to control robotic arms and perform remote ultrasound examinations in real time. High-resolution ultrasound image data and real-time HD audio/video consultation images can be transmitted in a wireless environment. Medical record analysis, ultrasound image diagnosis, and remote video consultation can be completed for patients to further determine specific treatment solutions.

10.3.2.3 Industry Ecosystems

The construction and maintenance costs of 5G medical private networks are beyond the capacities of hospitals and social health centers in some districts. In "Public network for private use", telecom operators are responsible for the construction and O&M of 5G public networks. This can help the healthcare industry quickly build private networks at low costs. With new technologies such as 5G slicing and MEC, telecom operators can securely isolate services of private network users from those of public network users. Operator network devices strictly comply with 3GPP security standards. E2E physical isolation and security assurance are achieved for private networks: 1. Medical devices use dedicated cards to access the network. 2. Wireless devices, transmission devices, and core network devices must pass strict network access authentication. 3. Operators' IDCs housing MECs comply with level-3 security protection standards. 4. MECs are connected to hospital data centers through private circuit lines.

10.3.2.4 Typical Cases

On February 2, 2020, the health of a suspected patient with COVID-19, treated in Zhejiang Provincial People's Hospital Tongxiang branch, quickly degraded. A further, more detailed scan of the lung and heart was needed, meaning there was an urgent need for an immediate ultrasound examination. Considering the nature of the patient's illness and the safety of medical personnel, the branch hospital sought help from Zhejiang Provincial People's Hospital, hoping to use remote ultrasonic robots to examine the patient [6].

After an emergency connection was set up, doctors in Zhejiang Provincial People's Hospital immediately came to the remote ultrasound medical center inside the hospital. Through 5G, a sonographer remotely controlled the ultrasonic robotic arms beside the patient's bed in the hospital branch, 60 km away from Hangzhou. With the free and flexible movement of the robotic arms, real-time ultrasound images were sent back to Hangzhou. A few minutes later, the doctors issued an ultrasound diagnosis and the clinical doctor in charge had a clear understanding of the patient's condition, helping determine the next stage of diagnosis and treatment.

Zhejiang Provincial People's Hospital successfully used 5G technology for the first time in China to apply ultrasound robots in the evaluation of COVID-19. This not only avoided infection risks caused by onsite visits, but also provided diagnosis in a quick, efficient, and timely manner. Zhejiang Provincial People's Hospital, together with China Telecom Zhejiang branch and 5G Smart Medical Innovation Laboratory founded by Huawei, provided technical support for this 5G remote ultrasound.

10.3.3 Emergency Rescue

Emergency medicine is a multi-disciplinary science that deals with and studies various acute pathologies and acute traumas, and refers to the science of emergency care measures taken in a short period of time against accidents, injuries and diseases that threaten human life. Emergency medicine does not deal with the whole process of injury, but focuses on the first aid phase, involving heart, lung, and brain resuscitation, shocks caused by circulatory functions, acute trauma, multi-organ failure, and acute intoxication. Moreover, emergency medicine needs to study and design on-site rescue, transport, and communication and so on, among which, pre-hospital treatment (first aid center) is a crucial part of emergency medicine.

"Time is life" and therefore pre-hospital first aid is demanding on the timeliness of treatment. For patients in critical or severe conditions, every second counts. The ideal first aid is to establish information interconnection and sharing among patients, emergency centers, ambulances, and hospitals as quickly as possible to identify patients' locations and conditions, dispatch an ambulance to the site, quickly provide professional first aid, and select a suitable hospital. Patient information and audio/video data inside the ambulance are sent to the hospital simultaneously, so that doctors in the hospital can get ready to receive the patient. If necessary, experts in the hospital can join in multi-party consultation to help pre-hospital doctors determine a course of treatment.

As such, an efficient pre- and in-hospital informatization system needs to be established. However, a traditional 4G network cannot support such a system. Thanks to the 5G era, the high rate, high bandwidth, and massive connection features make it possible to efficiently connect patients, pre-hospital doctors, and hospital experts. 5G technologies not only ensure a high-speed and stable network for the informatization system, saving "time" and reducing "distance" in pre-hospital first aid, but also optimize the pre-hospital first aid service mode, improving the treatment rate of critical and severe patients.

10.3.3.1 Industry Requirements

Currently, emergency medicine in China is still in the initial stage and faces challenges such as unbalanced development between rural and urban areas, improper configuration of emergency medical personnel, and insufficient equipment. In existing first aid, slow information transmission greatly hinders information sharing between pre-hospital personnel and first aid personnel in the hospital. After a patient is sent to the ICU, the patient's basic personal information, physical status, and current vital signs need to be re-collected and re-evaluated, which severely affects the efficiency and effectiveness of first aid.

With 5G, if there is no specialist or general practitioner on site, the wireless network can transmit patients' vital signs and critical alarm information to remote experts for help, which is critical to saving patients' lives. In addition, remote monitoring enables hospitals to quickly learn about patients' conditions, determine suitable first aid methods, and prepare resources, achieving seamless interconnection between pre-hospital first aid and in-hospital treatment.

Specifically, the 5G healthcare private network streamlines the networks inside and outside the hospital. Hospitals can provide "hospital-like treatment for patients while onboard": Patients can register and create archives immediately after getting on the ambulance. Information such as patients' vital signs, electrocardiograms (ECGs), HD videos, and vehicle locations is transmitted to the emergency command center in the hospital in real time. This helps emergency doctors learn about patients' conditions and prepare for necessary first aid in the hospital. Not only that, pre-hospital data is interconnected with the in-hospital pre-examination and triage system to open green channels for first aid and reserve beds in advance. This reduces the time required for in-hospital examination and handover.

The 5G intelligent first aid information system consists of four parts: smart first aid cloud platform, vehicle-mounted first aid management system, first aid teleconsultation system, and first aid assistance system. 1. The smart first aid cloud platform consists of the intelligent first aid dispatching system, integrated first aid platform system, and structured pre-hospital first aid EMR system. It provides main functions of emergency dispatch, background O&M management, and first aid quality control management. 2. The vehicle-mounted first aid management system includes the vehicle management system, medical device information collection and transmission system, AI intelligent image-based decision-making system, and structured pre-hospital first aid EMR system. 3. The first aid teleconsultation system refers to a guidance system based on HD video and AR/MR. It can transmit real-time HD audios/videos and provide hypermedia medical records, first aid maps, as well as large-screen bulletins. 4. The first aid assistance system includes smart medical backpacks, first aid recorders, in-vehicle mobile workstations, and in-hospital mobile workstations.

10.3.3.2 Network Solutions

5G networks can transmit medical equipment monitoring information, real-time vehicle locations, and video images inside and outside vehicles, facilitating teleconsultation and remote guidance. The collection, processing, storage, transmission, and sharing of pre-hospital first aid information can fully improve management and treatment efficiency, enhance service quality, and optimize the service process and mode. Big data technologies help to mine and use the value of medical data, and apply, evaluate, and assist in first aid management and decision-making.

The 5G edge healthcare cloud provides secure and reliable medical data transmission, facilitates information resource sharing and system interconnection, and provides powerful technical support for pre-hospital first aid and smart healthcare. Fig. 10.2 illustrates the network architecture of the emergency rescue solution.

10.3.3.3 Ecosystems and Business Models

The core of the emergency rescue system is the 5G smart first aid platform, which is generally purchased by local first aid centers in a centralized manner. Services will be provided by telecom operators and information system enterprises. One or both of them will continue to participate in the O&M of the smart first aid platform.

Government procurement for emergency centers generally includes equipment and subsequent O&M expenses. Telecom operators and information technology enterprises provide corresponding services and can continuously generate revenue through O&M and application upgrades after the first aid platform is constructed.

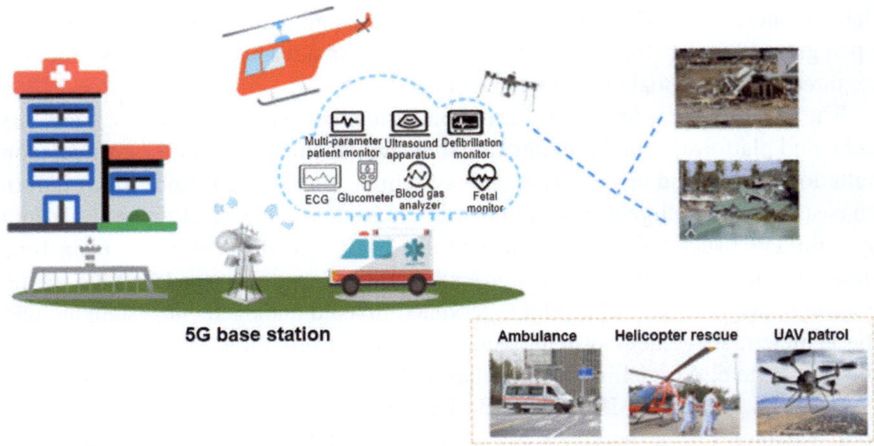

Fig. 10.2 Network architecture of the emergency rescue solution

10.3.3.4 Typical Cases

1. 5G Remote Emergency Command Center of SAHZU
   The Second Affiliated Hospital Zhejiang University School of Medicine (SAHZU for short) created the first multi-dimensional 5G smart first aid green channel in China. The 5G remote first aid command center in the Binjiang campus of SAHZU integrates multiple subsystems including 5G remote ultrasound, 5G first aid command platform, immersive and real-time panoramic VR experience, remote HD audio/video interaction, and UAV monitoring. This command center uses the 5G network to establish a multi-dimensional "runway" for real-time data interconnection [7].

   In the command center, doctors can use VR glasses to virtually immerse themselves in a 5G ambulance and monitor vital sign data of patients in the ambulance in real time. Information about patients on ambulances, such as ECG, ultrasound images, blood pressure, heart rate, blood oxygen saturation, and body temperature, is transmitted in real time through the high-speed 5G network and displayed on the large screens in the command center. Medical personnel in the hospital can quickly identify patients, obtain medical records, monitor patients' status in real time, and assist medical personnel on ambulances, saving valuable time for diagnosis and treatment.
2. 5G Remote First Aid in Expo
   On May 25, 2019, a middle-aged tourist in the Beijing Expo Park suddenly had an acute myocardial infarction. He suffered from chest pain and was sweating in a medical station located inside the park. The 5G remote first aid system built by Huawei and China Unicom played a key role in saving this patient. Multiple parties collaborated with each other, taking only 1 hour and 38 minutes

to complete the surgery. The ultra-low latency of 5G truly demonstrated its power in assisting time-sensitive first aid [8].

3. 5G+ Emergency Water Rescue in Fuzhou
In January 2020, Fuzhou launched a "5G emergency water rescue" drill. Near a wharf of Minjiang River, an old man was shouting help in the river. A red emergency rescue hovercraft arrived quickly. The driver carried out correct rescue according to instructions from the headset. At the rear command center on the shore, the rescue process was transmitted by UAVs over 5G and videos were displayed on the screens back in the command center. The commander could make rescue deployment based on the site conditions [9].
Water rescue is a challenging, urgent, and difficult rescue project full of risks and has high demands on rescue equipment and personnel. In terms of drowning rescue, issues such as slow response, difficult positioning, and a challenging search and rescue environment exist. In the emergency water rescue drill in Fuzhou, a modern emergency rescue hovercraft was used. It was equipped with a 5G "sea-land-air" integrated real-time communications system and could cruise on land, water, sand, mud, swamp, snow, desert, ice, salt lake, or freshwater lake, making it easier to quickly reach areas that ships and vehicles could not. With high bandwidth, low latency, and massive connection features of 5G networks, multiple HD videos could be simultaneously transmitted, the latency of which was as low as 1 ms. This enabled fast and stable information transmission, saving more time for emergency water rescue.

10.3.4 Policies and Standards

10.3.4.1 Policies

Related policies and measures have been released to facilitate the development of 5G+ healthcare applications. In March 2020, the MIIT released *Notice Concerning Promoting the Accelerated Development of 5G*, urging 5G network deployment. In addition, it calls for promoting the innovation and development of "5G+ healthcare", specifically, building 5G smart healthcare systems, setting up 5G smart healthcare demonstration networks and healthcare platforms, promoting 5G applications in COVID-19 pandemic warning, pre-hospital first aid, remote diagnosis and treatment, and intelligent image-assisted diagnosis, optimizing and promoting excellent applications of 5G in combating the pandemic, promoting services such as remote health examination, consultation, and medical assistance, and promoting medical resource sharing.

In the same period, the National Development and Reform Commission (NDRC) and MIIT jointly released *Notice on 2020 New Infrastructure Construction Project (Broadband Network and 5G)*, proposing seven 5G innovative application improvement projects, including the construction of 5G smart healthcare systems for major public health emergencies. This project proposed to carry out R&D into smart healthcare technologies based on the new 5G network architecture, build 5G smart healthcare demonstration networks, set up an evaluation and verification environment, and promote the industrialization of key network devices and prototype systems that meet smart healthcare collaboration requirements, accelerate 5G applications in COVID-19 pandemic warning, pre-hospital first aid, real-time teleconsultation, remote surgery, wireless monitoring, and mobile ward rounds, aiming to provide important support for major public health emergencies while effectively ensuring the health of medical personnel.

In March 2021, the Outline of the People's Republic of China 14th Five-Year Plan for National Economic and Social Development and Long-Range Objectives for 2035 was released, which proposed "building 5G-based application scenarios and industry ecosystems, and carrying out pilot projects in key fields such as smart transportation, smart logistics, smart energy, and smart healthcare". Furthermore, a column was set up for 10 digital application scenarios, such as smart transportation, smart energy, smart manufacturing, smart agriculture and water conservancy, smart education, smart healthcare, and smart culture and tourism.

5G+ healthcare involves cross-domain cooperation in wireless communications and healthcare technologies. Therefore, coordination at the national level is required for developing policies and mechanisms related to the digital economy that can be replicated as soon as possible through pilot projects. The government needs to accelerate the formulation of policies supporting the construction of telemedicine private networks, and the extension of telemedicine services to primary care institutions, promoting nationwide medical resource sharing. In terms of application, pilot projects will be promoted. Application researches based on 5G networks, such as mobile first aid, teleconsultation, robotic ultrasound, robotic ward rounds, medical private wireless network, and telemedicine teaching, will be carried out to promote innovative integration of 5G and the healthcare industry.

10.3.4.2 Standards

A lack of established standards is a major challenge to 5G+ healthcare. 5G not only has a faster rate and larger connection scale than previous generations of technologies, but it will also create a series of new service forms and integrate a series of technologies including machine to machine (M2M), audio and video technologies. These technologies use a much higher spectrum range than previous generations of networks. Five kinds of requirements: terminal, network architecture, platform, security, and ethics drive the evolution of standards for the 5G+ healthcare converged application. In terms of medical terminals, the service requirements are diversified, and unified standards can provide support for terminals' technical requirements. As for network architecture, the 5G healthcare network architecture, key network devices, and network interfaces need to be standardized. In terms of the platform, the platform architecture, construction guide, service functions, and service evaluation must be standardized. In terms of security and ethics, 5G accelerates the data

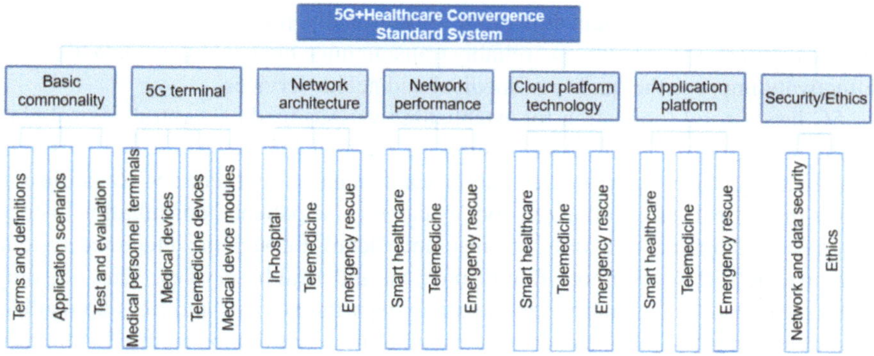

Fig. 10.3 The 5G+ healthcare convergence standard system

flows of various applications in the healthcare field. In this process, strict technical standards and refined ethical standards are required to combat potential healthcare data security risks.

In November 2020, the NHC officially began researching into 5G healthcare industry standards. It would work with hospitals, research institutes, and leading enterprises in the 5G technology industry to research 5G healthcare industry standards, promote 5G healthcare application innovations, and standardize the development of the 5G medical terminal industry to support the construction of smart hospitals. To further support large-scale development of 5G+ healthcare applications, a 5G+ healthcare convergence standard system should be established as soon as possible to standardize the convergence of healthcare industry applications and 5G technologies in the top-level design. The system covers basic commonality, 5G terminals, network architecture, network performance, cloud platform technologies, application platforms, security, and ethics. It standardizes the implementation and promotion of 5G applications throughout the process and enables technologies and medical applications to better serve the public. Fig. 10.3 illustrates the 5G+ healthcare convergence standard system.

10.4 Summary and Prospects

5G will enable a brand-new healthcare ecosystem that can accurately, conveniently, and efficiently meet the requirements of patients and hospitals in a cost-effective manner. 5G networks will revolutionize all key links in the healthcare industry and high-speed 5G networks can be added to the existing hospital architecture to make full use of the high bandwidth and low latency features of 5G, meeting the network requirements of medical diagnosis systems. 5G will significantly improve the effect of telemedicine, effectively promote the accessibility to high-quality medical resources in China, and facilitate talent cultivation. Additionally, it can meet increasing bandwidth requirements, quickly and reliably transmit huge medical image data files, and effectively improve the quality of medical and health services.

5G lays a solid foundation for establishing various medical central control platforms in the healthcare industry. It supports medical IoT devices so that hospitals can gradually transform to provide public health management services. Regardless of where the patients are, authorized doctors can browse patients' consultation history, diagnosis and treatment records, and insurance details through the integrated system, and provide different remote family medical and health services. This not only effectively reduces the number of outpatient department visits and frees up medical resources, but also prevents diseases through routine health monitoring. In this way, a patient-centric medical data network can be formed. All the preceding measures can help streamline health data input and implement personal health management through cloud-based big data analysis, such as real-time monitoring and evaluation, disease warning, chronic disease screening, and proactive intervention. In this manner, the healthcare industry can enter the smart healthcare era.

References

1. Report on Chinese Residents' Chronic Diseases and Nutrition (2020)
2. National Health Commission of the People's Republic of China (2019) Statistical bulletin for China's health care development in 2018 . Chin Pract J Rural Dr 26(7):2–13
3. Zhuang YQ (2021) Blue book of hospitals: annual report on china's hospital competitiveness (2020–2021). Social Sciences Academic Press, Beijing
4. Global Network (2020) The first "teleconsultation platform" of Huoshenshan Hospital supported by Huawei will be put into use (in Chinese)
5. 5G Smart Healthcare Full-process service white paper (in Chinese) (2020)
6. China Telecom 5G helps complete remote ultrasound diagnosis and treatment for COVID-19 patients (in Chinese) (2020)
7. 5G smart ambulance first aid of SAHZU. It's so fast! (in Chinese) (2019)
8. 5G makes telemedicine more effective (in Chinese) (2019)
9. "Sea-Land-Air" integration achieves emergency rescue in complex environments—"emergency water rescue." In: Fuzhou enters the 5G Era (in Chinese) (2019)

Chapter 11
5GtoB Standardization Progress

11.1 5GtoB Standardization Progress in 3GPP

The commercial deployment of 5G has led to the large-scale implementation of convergence application scenarios. 5G has made great strides across a wide range of industries, including industrial Internet, healthcare, electric power, vehicle-to-everything (V2X), steel, and ports. Nonetheless, 5G technologies need to follow specific standards in order to truly penetrate the core business of such industries. Research on 5G convergence industry standards is underway in leading industries, and their adoption will undoubtedly usher in a new era of digitalization.

3GPP SA6 activities provide a common platform for new verticals (to 3GPP), which are ready to use the 5G system for their own needs. 3GPP SA6 stands for the 3GPP Technical Specification Group Service and System Aspects Working Group 6. It studies application enablement for vertical markets, with the goal of providing application layer architecture specifications, including those related to architecture requirements, functional architectures, procedures, information flows, interworking with non-3GPP application layer solutions, and deployment models. Generally speaking, the main focus of 3GPP SA6 can be categorized into service frameworks suitable for all verticals, and enablers that ensure the efficient use and deployment of vertical applications. Service frameworks include the common application programming interface (API) framework, service enabler architecture layer, and edge application enablement. Enablers for vertical applications include solutions for railways, V2X, and factories of the future (FF).

11.1.1 Developing the Common API Framework

To avoid duplication and inconsistencies between different API specifications, 3GPP has considered developing a common API framework (CAPIF) that is applicable to

P. Sun, *A Guidebook for 5GtoB and 6G Vision for Deep Convergence*,
Management for Professionals, https://doi.org/10.1007/978-981-99-4024-0_11

any northbound APIs. There is a single and harmonized approach for API development, with a number of 3GPP specifications at work on specifying a framework to host APIs of public land mobile network (PLMN), and on enabling verticals to leverage CAPIF to host their APIs.

In order to apply a unified northbound API framework across several 3GPP functions, CAPIF was delivered in 3GPP Release 15, with enhancements coming in Release 16, and was integrated with the northbound APIs developed by 3GPP SA2 (SCEF/NEF) and 3GPP SA4 (xMB).

Figure 11.1 illustrates the functional model of CAPIF. The CAPIF architecture has three main specified entities: CAPIF core function (CCF), API exposing function (AEF), and API invoker.

The CCF is the central repository for all of the APIs, taking care of authentication, onboarding, and aspects related to logging and charging.

The AEF wraps around the 3GPP functions to expose the APIs. One of the example functions is the NEF APIs that can be exposed to the AEF function—with the API invoker on top, which has access to the CAPIF core functions for onboarding and discovery and the availability of services provided by the AEFs.

The following key features covered in 3GPP TS 23.222 [1] indicate what CAPIF could offer:

- Onboarding and offboarding API invokers
- Publishing, unpublishing, retrieving, and updating APIs

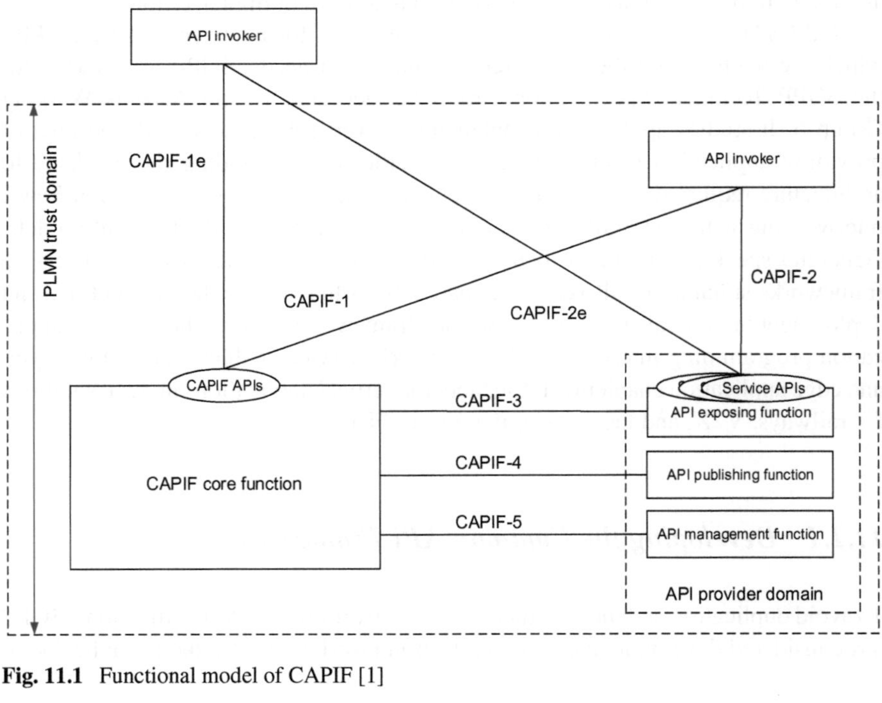

Fig. 11.1 Functional model of CAPIF [1]

- Discovering APIs
- Authenticating APIs
- Supporting access control
- Supporting topology hiding
- Supporting third-party domains
- Supporting the federation of CAPIF in support of distributed deployments.

11.1.2 Developing the Service Enabler Architecture Layer

Many of the industries considering running applications over 3GPP systems may require similar common capabilities as others, and thus benefiting from work that has already been done. 3GPP TS 23.434 [2] specifies a functional architecture for the service enabler architecture layer (SEAL) over 3GPP networks to support vertical applications (e.g., V2X applications). This functional architecture will include common application plane and signaling plane entities. It allows a set of common services (including group management, configuration management, location management, and network slice capability enablement) to be shared among different verticals.

11.1.2.1 Functional Architecture for SEAL

SEAL services are supported for UEs both on-network and off-network in a B2B model. To allow a UE to create a group based in a particular location, the interconnection between multiple SEAL servers to enable distributed SEAL deployments and inter-service communication for interaction between SEAL servers are also specified.

Figure 11.2 illustrates the generic on-network functional model for SEAL. The model includes the SEAL client for client-side functionality and the SEAL server on the network. The SEAL client entity consists of all of the various clients related to services, like group, location, key, and identity management, and the same concept is used on the network side.

Above SEAL is the vertical application layer (VAL), which uses SEAL to realize its services. SEAL supports all of the common functions, that provide capabilities to VAL—which can be built by reusing what is there. V2X is an early example, but all new verticals can now leverage what has been done.

The vertical applications consume the SEAL services in the form of APIs. Each SEAL service offers these APIs on a service-based interface to all of its consumer entities.

Figure 11.3 illustrates the service-based interface representation of the functional model for SEAL services.

The SEAL function(s) exhibit the service-based interfaces, which are used for providing and consuming SEAL services. The service APIs are specified for each SEAL function enabled over the service-based interface.

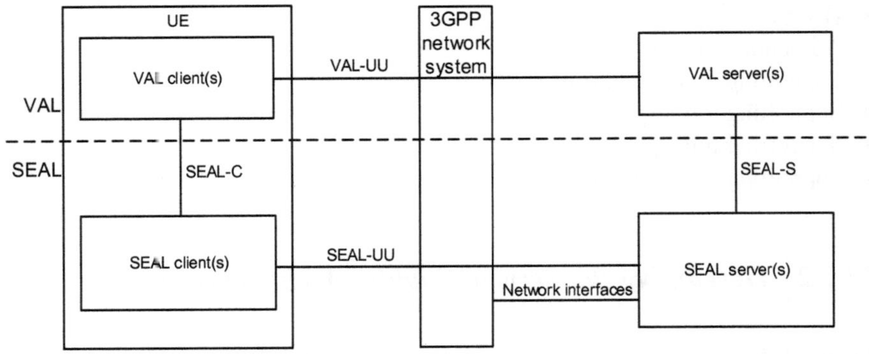

Fig. 11.2 Generic on-network functional model [2]

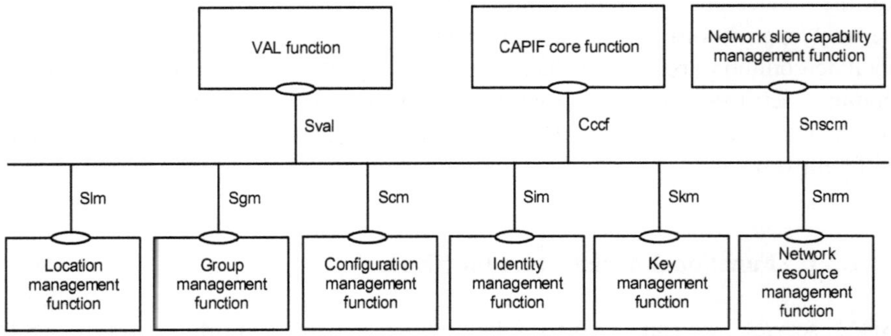

Fig. 11.3 SEAL generic functional model representation using service-based interfaces [2]

The service APIs offered by the SEAL function(s) are published and discovered on the CAPIF core function, as specified in 3GPP TS 23.222 [1].

11.1.2.2 Common Services for SEAL

The common services described below can be regarded as optional value-added functional models for SEAL. Each common service server acts as a separate application function (AF).

- Network slice capability enablement (NSCE)

  Network slicing is a key technology for differentiated network requirements of vertical industries. NSCE is a SEAL service that offers network slice capability enablement capabilities to one or more vertical applications. The capabilities include application layer network slice management, communication service management exposure, discovery of management service exposure, network slice fault management, network-slice-related performance and analytics exposure, slice API configuration and translation, as well as network slice application policy management. Figure 11.4 illustrates the service-based interfaces for providing and consuming network slice capability enablement services.

Figure 11.5 illustrates the network slice capability enablement architecture in the non-roaming case, using the reference point representation to show how various entities interact with each other.

The detailed functionalities and procedures that support the network slice capabilities are detailed in 3GPP TR 23.700-99 [3] and 3GPP TS 23.435 [4], respectively.

- Location management

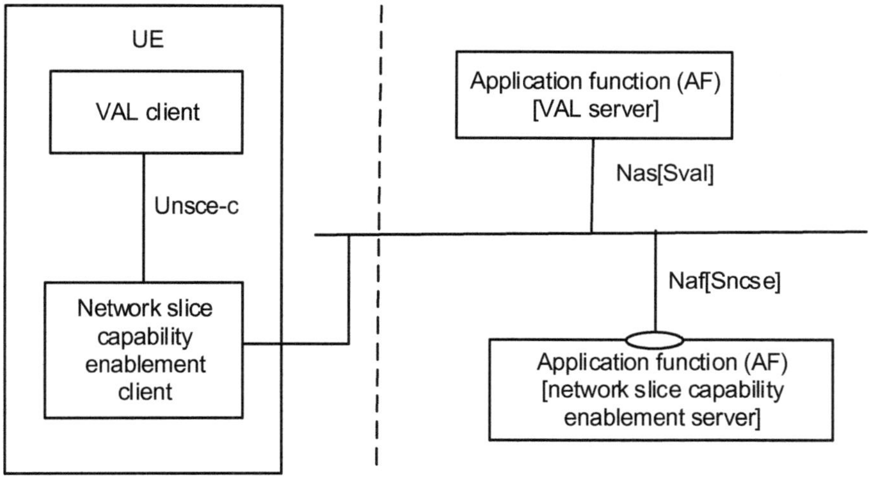

Fig. 11.4 Architecture for network slice capability enablement—service-based representation [3]

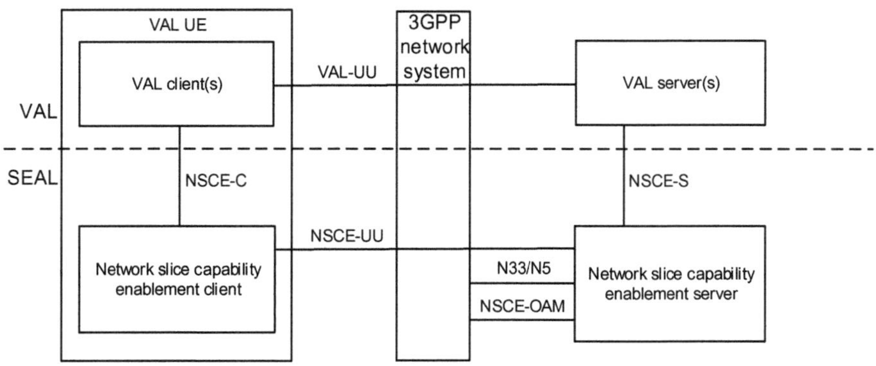

Fig. 11.5 Architecture for network slice capability enablement—reference point representation [3]

The location management is a SEAL service that offers the location-management-related capabilities to one or more vertical applications. These capabilities include location reporting configuration, location information reporting, UE information obtaining, location monitoring, and location area monitoring.

- Group management

  The group management is a SEAL service that offers group-management-related capabilities to one or more vertical applications. It enables the creation of a group to be used by one or more vertical applications within the same VAL system, while also allowing for multiple groups to be merged (temporarily or permanently) into a single group by the authorized users or VAL server, wherein all of the group members of the constituent groups are designated as members of the merged group.
- Configuration management

  The configuration management is a SEAL service that offers configuration-management-related capabilities to one or more vertical applications. It supports retrieval of VAL UE configuration data and VAL service data.
- Key management

  The key management is a SEAL service that offers key-management-related capabilities to one or more vertical applications. It supports secure distribution of security-related information and all communications in the SEAL ecosystem to be secured.
- Identity management

  The identity management is a SEAL service that offers identity-management-related capabilities to one or more vertical applications. More specifically, the procedures of user authentication and authorization for VAL services are specified.
- Network resource management

  The network resource management is a SEAL service that offers capabilities related to network resource management (e.g., unicast and multicast network resources) and monitoring to one or more vertical applications. The network resource management server sets up bearers and may need to modify the bearers for previously established VAL service communications.

11.1.3 Developing Edge Application Enablement

The edge application (EDGEAPP) allows for applications to be hosted on the edge of the 3GPP network. One of the main areas of focus is minimizing the impact to edge-based applications, to ensure they do not need major application re-development for UE use at the edge.

Edge computing is a network architecture concept that facilitates cloud computing capabilities and service environments, which are deployed close to the UE. It promises several benefits such as lower latency, higher bandwidth, reduced backhaul traffic and prospects for new services compared to the cloud environments.

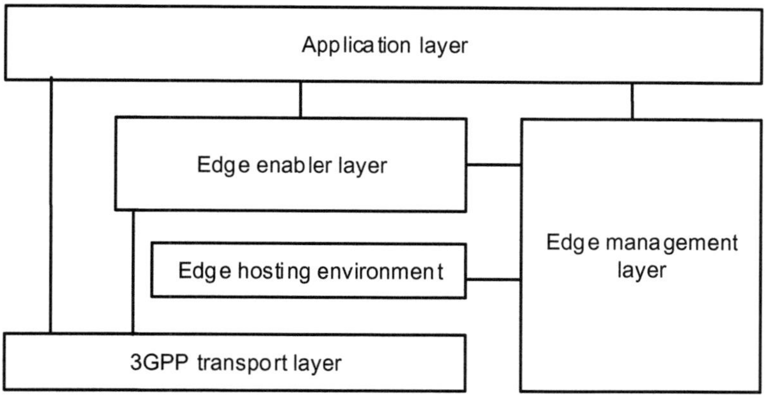

Fig. 11.6 Overview of 3GPP edge computing [5]

For edge computing, it is essential that the application clients (ACs) are able to locate and connect with the most suitable application server available on the edge data network (EDN), depending on the needs of the application. The edge enabler layer exposes APIs to support such capabilities.

The edge computing capabilities supported by 3GPP are illustrated in Fig. 11.6.

The application layer is a consumer of 3GPP-specified edge computing capabilities. The 3GPP edge computing capabilities are typically organized as follows:

- Edge enabler layer, specified in 3GPP TS 23.558 [5]
- Edge hosting environment, details of which are outside the scope of 3GPP
- 3GPP transport layer, specified in 3GPP TS 23.401 [6] and 3GPP TS 23.501 [7]
- Edge management layer, specified in 3GPP TS 28.538 [8].

11.1.3.1 Service-Based Architecture for Enabling Edge Applications

Figure 11.7 illustrates a service-based representation of the architecture for enabling edge applications.

Figure 11.8 illustrates the reference point representation of the architecture for edge enabling applications.

The EDN is a local data network. The edge application server (EAS) and the edge enabler server (EES) are contained within the EDN. The edge configuration server (ECS) provides configurations related to the EES, including details of the EDN hosting the EES. The UE contains AC(s) and the edge enabler client (EEC). The EAS, EES, and ECS all interact with the 3GPP core network. More details about the architecture for enabling edge applications are provided in 3GPP TS 23.558 [5]

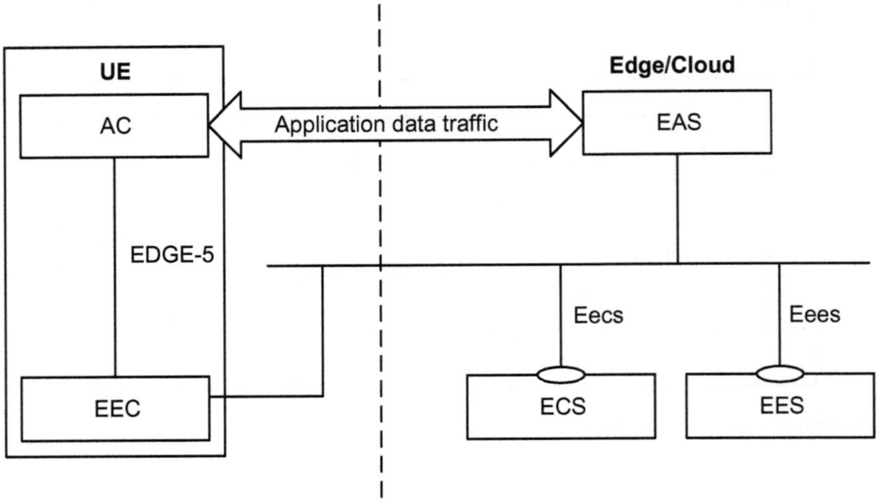

Fig. 11.7 Architecture for enabling edge applications—service-based representation [5]

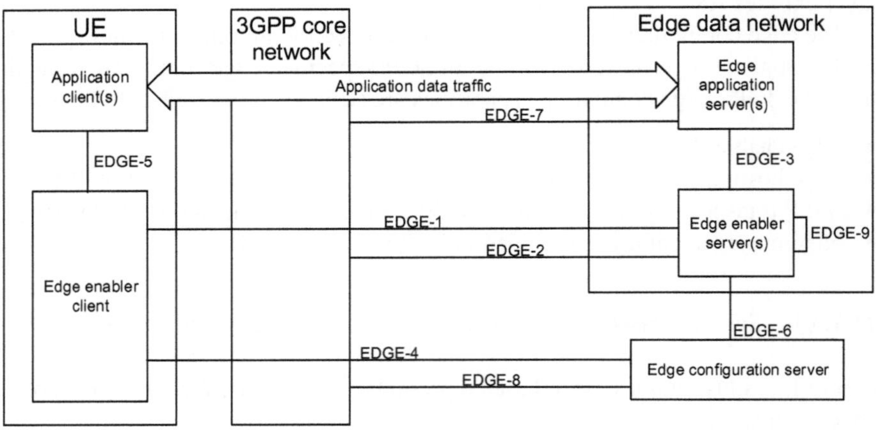

Fig. 11.8 Architecture for enabling edge applications—reference point representation [5]

11.1.3.2 Overview of EDGEAPP Features

- Service provisioning

  Service provisioning procedures supply the information required by a UE to access the edge services. The procedure takes the UE's location, service requirements, service preferences, and connectivity information into account to provide the required configuration.
- Registration

Registration allows entities (e.g., UE and application server) at the edge enabler layer to provide information about themselves to other entities at the edge enabler layer.

- EAS discovery
  EAS discovery procedures enable the UE to obtain information about suitable EASs of interest (specified as discovery filters) in the EDN.
- Capability exposure to EASs
  The edge enabler layer exposes services towards the EASs. The exposed capabilities include the services of the edge enabler layer and the re-exposed and enhanced services of the 3GPP core network. Other application layer capabilities like application enabler services and SEAL services may be exposed via the edge enabler layer as per CAPIF.
- Support for service continuity
  When a UE moves to a new location, different EASs can be more suitable for serving the UE. Such transitions can result from a non-mobility event as well, requiring support from the edge enabler layer to maintain service continuity. Support for service continuity provides several features for minimizing the application layer service interruption by replacing the source EAS connected to the AC in the UE, with a target EAS.
- Security
  The edge enabler layer supports secure communication among the enabler layer entities via EEC authentication and authorization.
- Dynamic EAS instantiation triggering
  The edge enabler layer interacts with the EAS management system to trigger instantiation of a suitable EAS depending on the application needs.

11.1.4 Overview of 3GPP SA6 Service Framework

With CAPIF to expose APIs to 3GPP functions, SEAL—as a service platform for common capabilities—to allow quick onboarding for verticals, vertical application enablers to support the creation of actual applications (e.g., V2X, Uncrewed Aerial System (UAS), and FF), and EDGEAPP to enable cloud computing capabilities and service environments, the 3GPP SA6 plays a vital role in bridging the 3GPP system and 5G verticals (Fig. 11.9).

11.1.5 Developing Enablers for Vertical Applications

The following subclauses list some enablers for vertical applications that have been completed or are under study.

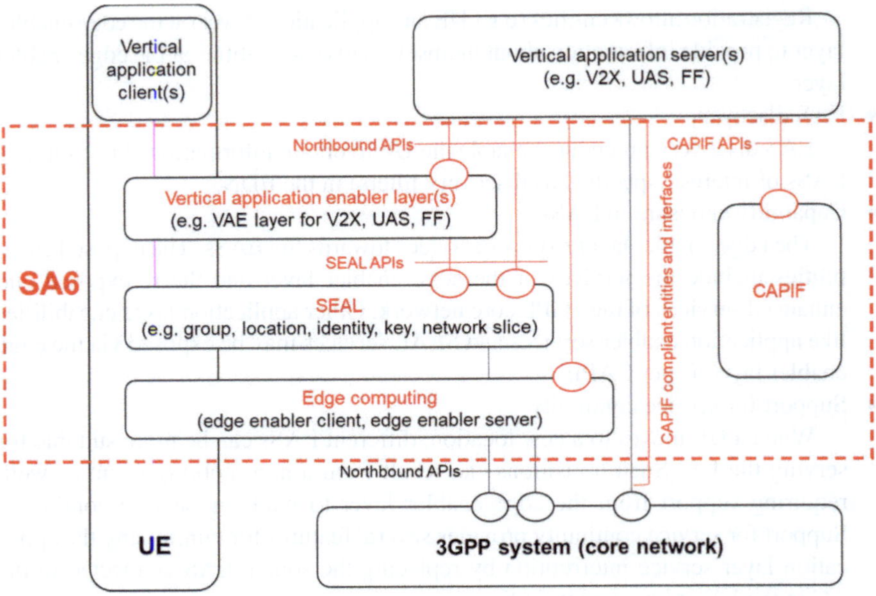

Fig. 11.9 Overview of 3GPP SA6 service framework [9]

11.1.5.1 Railway Communications

Railway communications of the future will be deeply intertwined with digital railway operations. Passengers have come to expect multimedia services during long flights and rail journeys.

This often takes the form of ground-to-train bulk transfers of multimedia data, which can be movies, TV shows, cached web pages, and other entertainment content. Such bulk transfers are completed when the train is stopped at stations or depots, and ideally proceed at a minimum rate of one gigabit. This ensures that multimedia content is highly versatile and updated on a regular basis, and that in-train networks operated by the future railway communication system can offer the multimedia content, rather than burdening the link budget between the train and the ground network, thereby minimizing the consumption of network capacity between them. Other usage cases include train-to-ground bulk transfer of CCTV archives, massive inter-carriage data transfers, and automated train control, as detailed in 3GPP TR 22.890 [10].

Railway communications services are typically mission critical services, which are the natural place to start when studying Future Railway Mobile Communication System (FRMCS).

Mission critical services can be used for public safety applications, maritime safety applications, and general commercial applications (e.g., railways). 3GPP introduces the concept of mission critical services by standardizing mission-critical push-to-talk (MCPTT) in Release 13. In Release 14, 3GPP adds additional mission critical services and enhancements to its repertoire of standardized applications, including mission critical data (MCData), mission critical video (MCVideo), enhancements to MCPTT, and a general framework for standardizing additional mission critical services. Release 15 addresses further evolutions to mission critical services, while also evaluating and studying mission critical service requirements in the railway industry.

In Release 15, 3GPP TR 23.790 [11] studies the gaps between the FRMCS requirements and existing mission critical services specifications, and provides the solutions to functional alias(es) and multi-talker feature control. In Release 16, 3GPP TR 23.796 [12] extends the scope of supported services to: (a) restricting incoming private communications; (b) group management (such as automatic group affiliation/de-affiliation by the client); (c) automatic starting of an emergency group communication after an emergency group alert; (d) support for functional aliases in private calls, private emergency calls, emergency and imminent peril group calls, and other calls; (e) support of functional aliases for file distribution, MCData control, MCVideo control, and other services; (f) use of MCPTT (e.g., list of MCPTT group members who did not acknowledge the group call request, additional commencement modes for MCPTT group calls); (g) granting application layer priorities at the request of users; and (h) support of IP connectivity.

The most recent studies on railway communications in 3GPP Release 19 include the following:

1. **Railway smart station services**. The railway station is a major touchpoint to customers including passengers. The railway community is considering the railway smart station services for the railway station operations and customers. Many railway companies plan to adopt the smart station services, and some companies have launched their own projects to provide such services, including mission critical services (MCPTT, MCVideo, and MCData) [10].
2. **Off-Network railway services**. In addition to network-oriented communications, future railway communications also include those that do not involve the network. Off-Network is a new term for the railway sector, which was introduced in the 3GPP MCX specifications. UIC will promote the use of Off-Network to replace Direct Mode. However, this term may be interpreted in slightly different ways. For example, the railway sector considers use of Off-Network communications where no network services are available, either due to network failure (e.g., coverage, capacity, etc.) or due to lack of coverage (remote areas). In 3GPP MCX specifications, Off-Network communications can be set up even when network services are available [13].

Off-Network communications, alongside voice communications, will be increasingly used for the exchange of train safety data.

Automatic train protection (ATP), automatic train operation (ATO), critical real-time video, and virtual coupling data communication are examples of applications that require Direct Mode.

11.1.5.2 UAS

A UAS consists of a UAV, remote electronic controller (such as a computer or tablet) that controls its movement, the equipment it carries, which is known as the payload (camera, radar, etc.), and a data system that connects the components (e.g., wireless connection or GPS) [14].

A typical example of this is UAS acting as an airborne wireless access node. In certain application scenarios, such as disaster monitoring, border surveillance, and emergency rescue, commercial use of UASs is ideal, due to the ease of deployment, low acquisition and maintenance costs, and strong maneuverability and hovering capabilities. The use of wireless access node mounted on a UAV, referred to as UxNB—user-defined NodeB (UAV eNodeB/gNodeB/ng-eNB)—has attracted great interest from the wireless communications community. Due to their fast deployment and broad coverage, UxNBs can be used to enhance coverage across a number of scenarios, such as emergencies and hotspot events, as well as to provide temporary coverage for mobile users [15].

Concurrent data transmission for UAVs and eMBB users is another common application of this technology. A base station may need to support simultaneous data transmission of UAVs in the air and eMBB users on the ground under limited bandwidth resources. For example, during a live broadcast, a base station will need to provide the required QoS for ground users (e.g., eMBB users), while supporting real-time uploading of pictures or videos from a UAV with an altitude of at least 100 m, at a high transmission rate and large bandwidth. This requires that interference between these two modes of communication be kept at a minimum [15].

The functional model for the UAS application layer introduced in 3GPP TS 23.255 [16] is organized into functional entities, for the purposes of outlining a functional architecture that addresses the application layer support attributes of UAS applications.

Figures 11.10 and 11.11 illustrate the simplified architectural models for the UAS application layer.

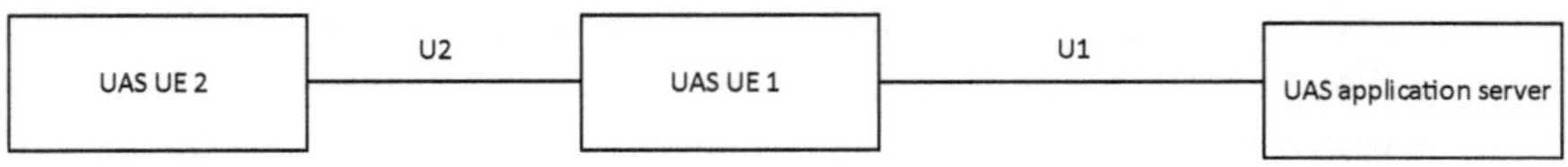

Fig. 11.10 Simplified architectural model for the UAS application layer [16]

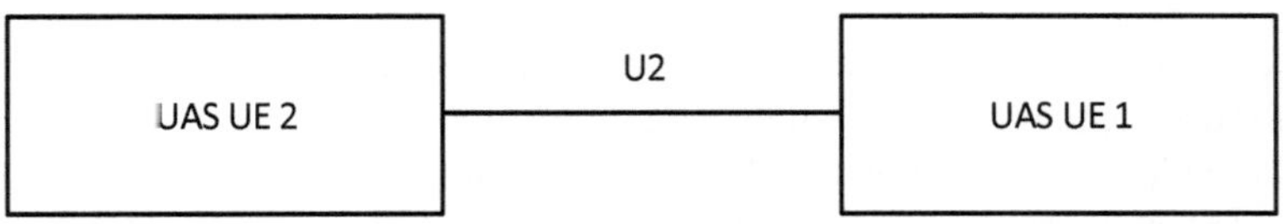

Fig. 11.11 Simplified architectural model for U2 connectivity between UAS UE 1 and UAS UE 2 at the UAS application layer [16]

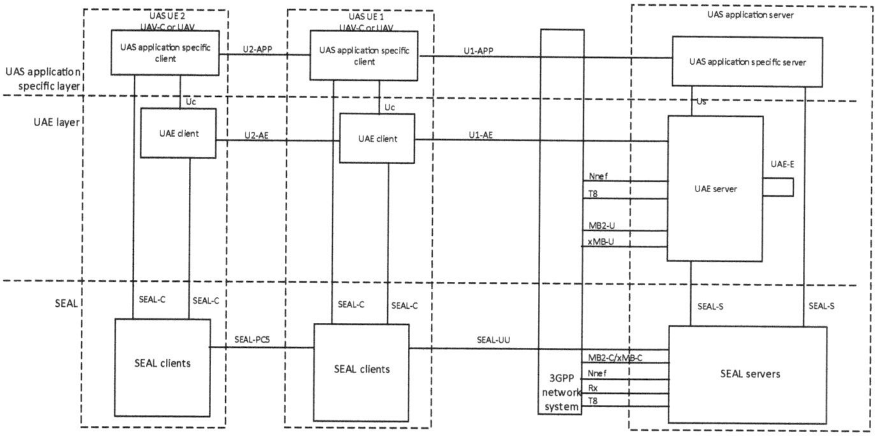

Fig. 11.12 UAS application layer functional model [16]

Figure 11.12 illustrates the detailed functional model for the UAS application layer, which enhances the simplified architectural model by specifying the functional entities at the UAS application layer.

The latest studies on the UAS application layer include the following solutions:

1. **Provisioning UAS service information**, which proposes a feasible technical solution for provisioning UAS configuration information at the UAE layer using unicast or multicast. The procedure for provisioning UAS configuration information is illustrated in Fig. 11.13.

   (1) The UAE server is triggered (e.g., after obtaining basic network connectivity) to provide UAS configuration information to the UAE client resident on the UAV/UAV-C.
   (2) The UAE server sends the UAS configuration info announcement containing the UAS configuration information to the UAE client in the UAV/UAV-C. This message can be sent via unicast or multicast.
   (3) Upon receiving the UAS configuration info announcement, the UAE client in the UAV/UAV-C stores the received UAS configuration information.
   (4) The UAE client of the UAV/UAV-C can provide an acknowledgement to the UAE server, but this is optional.

2. **Support for UAS media application session management**, which illustrates the workflow regarding how the UAE server (acting as a VAL server) requests the SEAL NRM server to fulfil the session establishment requests under the following pre-conditions: A UAV tries to establish a SIP session via the UAE server with the indication of resource requirements for the media session using SIP SDP.

   In the future, unmanned aerial system traffic management (UTM) will be used to provide a number of services in support of UAS(s), as well as to store the data required for the UAS(s) to operate. These services include but are not limited to

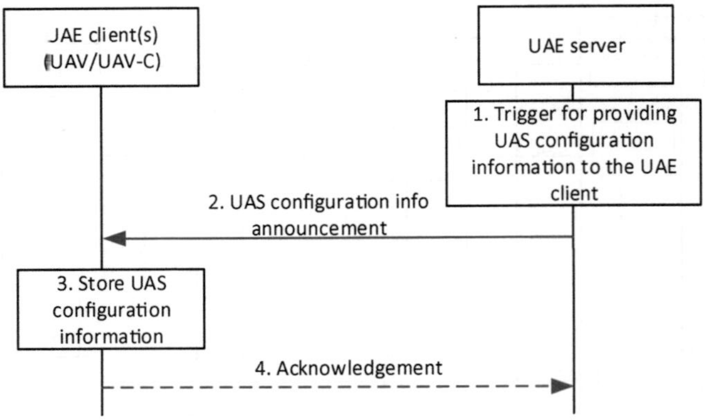

Fig. 11.13 Provisioning UAS configuration information [17]

UAS identification and tracking, authorization, enforcement, regulation of UAS operations. The UTM also makes it easy for authorized users (e.g., air traffic control, public safety agencies) to query the identity and metadata of UAVs and their UAV controllers [18].

11.1.5.3 V2X

V2X sits at the crossroads of the 5G and automotive sectors, and an area of strategy focus for national industry, in addition to a hotspot for cross-disciplinary research.

Traffic safety is one of the most common applications for V2X, as the technology is crucial to preventing traffic accidents and reducing the loss of life and property. Intersection collision warnings which require the vehicle to broadcast and receive V2X-related messages, are a prominent example of this.

Autonomous driving is another example of V2X in practice. Similar to existing methods of camera video recognition, mmWave radar, and lidar, V2X uses the interaction between information to obtain the motion status (e.g., vehicle speed, braking, lane change) of other vehicles and pedestrians, and is not easily affected by weather, obstacles, and factors like distance. Furthermore, V2X lays the groundwork for a comprehensive service system of time-shared leasing and vehicle-pedestrian-cloud collaboration, facilitating the widespread commercial deployment of autonomous driving. Current applications for autonomous driving include vehicle formation driving and remote control driving.

3GPP TS 23.286 [19] introduces the application architecture of V2X. Figure 11.14 illustrates the simplified architectural model for the V2X application layer.

V2X UE 1 communicates with the V2X application server over reference point V1. V2X UE 1 and V2X UE 2 communicate over reference point V5. V2X UE 1 can also serve as a UE-to-network relay, to enable V2X UE 2 to access the V2X application server over reference point V1.

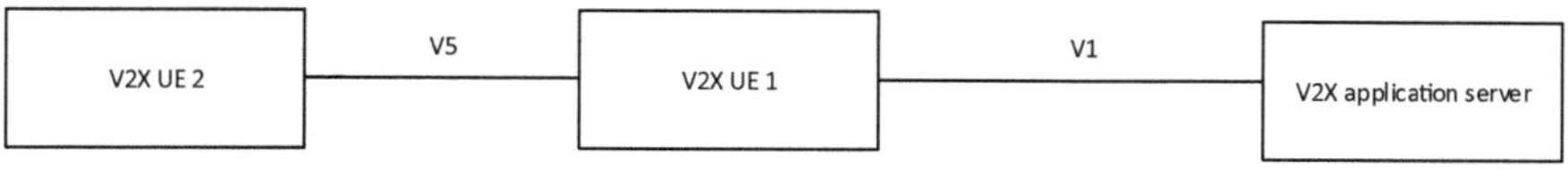

Fig. 11.14 Simplified architectural model for the V2X application layer [19]

Figure 11.15 illustrates the detailed functional model for the V2X application layer. It enhances the simplified architectural model by specifying the functional entities at the V2X application layer.

The V2X application layer functional entities for V2X UEs and the V2X application server are grouped into the V2X application specific layer and the VAE layer. The VAE layer offers VAE capabilities to the V2X application specific layer.

The latest studies on application layer support for V2X services include:

1. **Triggering of session-oriented service establishment**: This process works as follows:

   (1) The V2X application specific server triggers the VAE server to initiate a session-oriented service for the VAE client. This request can include such information as the identity of the VAE client, the identity of the V2X application specific server, the identity of the service session, the V2X service ID and the "type" or "QoS requirements" of the session.

   (2) The VAE server provides a session-oriented service trigger response to the V2X application specific server, which indicates the ability or inability of the VAE server to initiate a session-oriented service for the VAE client.

2. **Monitoring and control of QoS for eV2X communications**: The VAE server monitors events for one or multiple V2X UEs within one service or across multiple services, and may trigger a service requirement adaptation based on the actual or expected QoS change for one or more V2X services. If no QoS

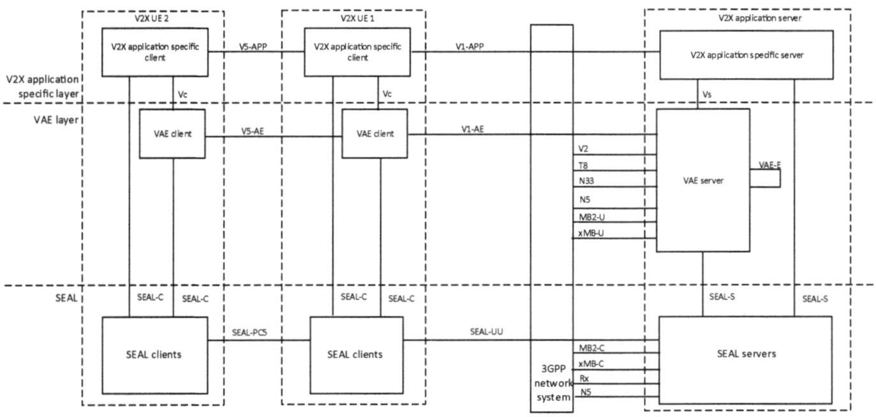

Fig. 11.15 V2X application layer functional model [19]

adaptation is required, the V2X application layer may adapt based on the adapted service requirements.

The impacts of the following high level functions on the V2X application layer are still to be studied:

1. Authorization and provisioning for V2X communications over PC5 and Uu reference points
2. V2X message transmission/reception over PC5 or Uu reference points
3. V2X application server discovery
4. QoS handling for V2X communication over PC5 and Uu reference points
5. Interworking between EPS V2X and 5G system (5GS) V2X.

11.1.5.4 Factories of the Future

"Factories of the future (FF)" refers to highly digitized manufacturing facilities that continually collect and share data through connected machines, devices, and production systems. This data can then be used by self-optimizing devices, or across the organization, to proactively address issues, improve manufacturing processes, and respond to new demands.

FF is becoming a fixture in manufacturing, and it uses connected devices to leverage advanced technologies like automation, AI, IoT, and more. These devices are capable of sensing their environments and interacting with one another. As FF continues to grow and develop, manufacturers will need to adapt the networks that connect them to ensure that they are as efficient and effective as possible. These factories will increasingly rely on sensor technology, and will feature connected attributes that make use of data to allocate work tasks.

3GPP TS 23.545 [20] introduces the application architecture of FF, and Fig. 11.16 is a service-based representation of the architecture for FF applications with SEAL and CAPIF. The FF application layer functions only use service-based interfaces for their interactions.

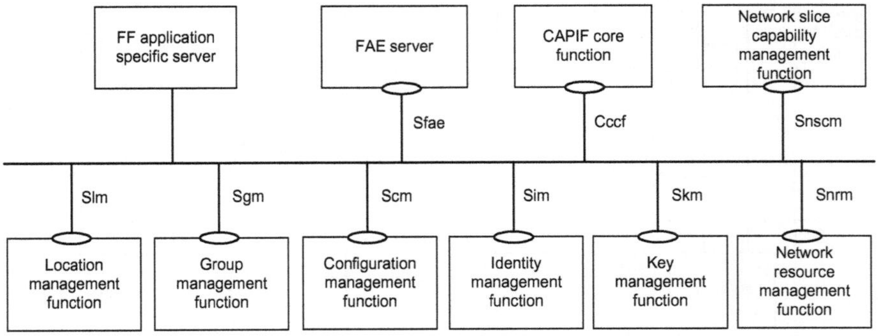

Fig. 11.16 FF application layer architecture—service-based representation [20]

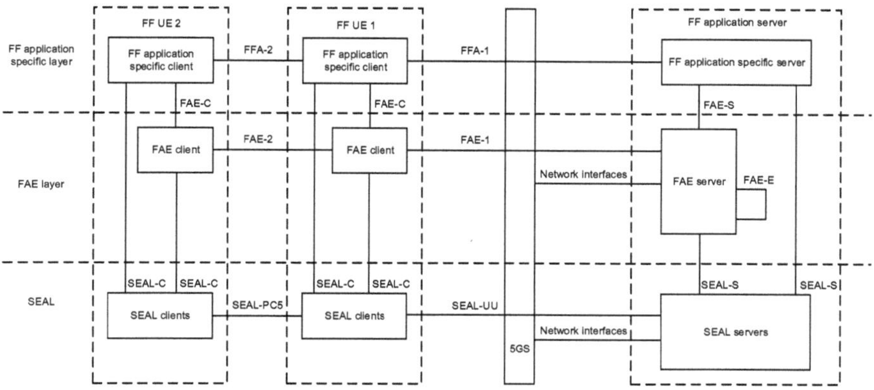

Fig. 11.17 FF application layer architecture—reference point representation [20]

The service-based interface for the FF application enabler (FAE) server is represented as Sfae. Figure 11.17 is a reference point representation of the architecture for the FF application layer.

The FF application layer functional entities for the FF UE and the FF application server are grouped into the FF application specific layer and the FAE layer. The FAE layer provides FAE capabilities to the FF application specific layer, and the FAE server is located at the FAE layer. The SEAL services utilized by the FAE layer include location management, group management, configuration management, identity management, key management, network resource management, and network slice capability management. The FF application specific layer consists of FF application specific functionalities. The FF application server consists of the FF application specific server, FAE server, and SEAL servers. The FAE server provides FF application layer support functions to the FF application specific server over the FAE-S reference point. The SEAL servers provide SEAL services to the FF application specific server over the SEAL-S reference point.

11.1.5.5 MSGin5G

Massive Internet of Things (MIoT) is one of the key sectors in the 5G market. The message service for the MIoT over the 5G System (MSGin5G) service is designed and optimized for massive IoT device communications, including thing-to-thing communications and person-to-thing communications.

SA1 has completed the work of defining the MSGin5G service requirements, as specified in 3GPP TS 22.262 [21]. SA6 investigates how to support the MSGin5G service over the 5G system. The MSGin5G service captured in 3GPP TS 23.554 [22] provides MSGin5G service messaging communications capabilities to the 5GS, in particular to MIoT.

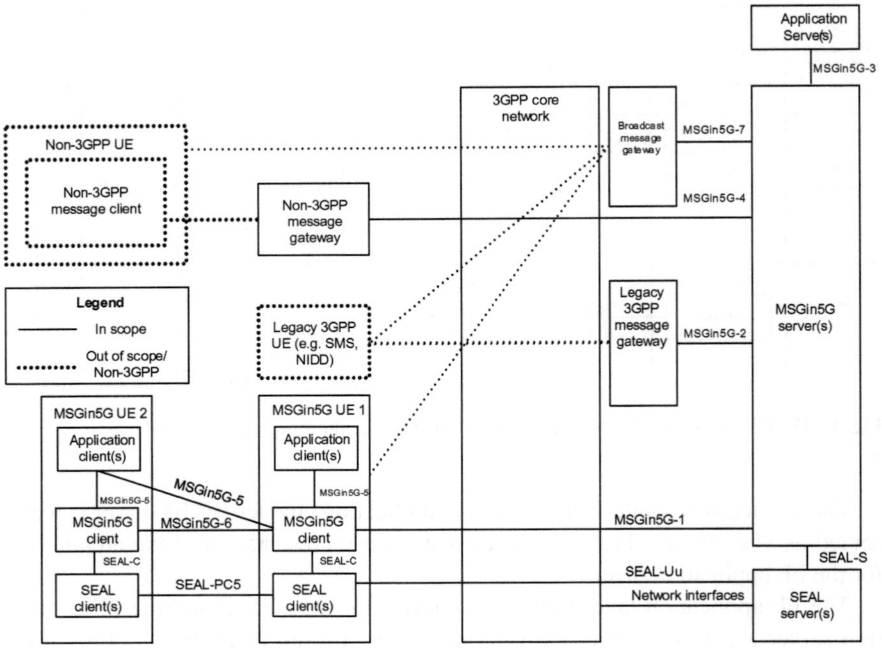

Fig. 11.18 Application architecture of the MSGin5G service [22]

The MSGin5G service includes the following message communications models:

- Point-to-point messages
- Application-to-point messages/Point-to-application messages
- Group messages
- Broadcast messages.

Figure 11.18 illustrates the application architecture of the MSGin5G service.

The MSGin5G client(s) interacts with SEAL clients over the SEAL-C reference point specified for each SEAL service. The legacy 3GPP message gateway and non-3GPP message gateway may interact with SEAL clients over the SEAL-C reference point specified for each SEAL service. The MSGin5G server(s) interacts with SEAL servers over the SEAL-S reference point specified for each SEAL service. The interaction between a SEAL client and the corresponding SEAL server is supported by the SEAL-Uu reference point specified for each SEAL service, as detailed in 3GPP TS 23.434 [2].

In 3GPP Release 18, the following technical issues are being studied:

- Procedures in scenarios with multiple MSGin5G servers, i.e., the sender and receiver of a message are served by different MSGin5G servers
- Bulk configuration and bulk registration of UEs, e.g., non-MSGin5G UEs or constrained devices, and application servers

- Support of broadcast massaging, either by invoking the cell broadcast center function (CBCF) or xMBS
- Procedures and APIs for constrained devices and relay UEs
- Service logic between certain features (including service logic between message aggregation and segment, and between store forward and device triggering).

11.1.5.6 IoT

"Personal IoT Network (PIN)" is a new paradigm based around the ever growing number of consumer IoT devices in circulation. These devices can be wearable devices that are literally worn on the body (i.e., cameras, headsets, watches, earphones, health monitors), or IoT devices, which are scattered throughout the home (e.g., smart lights, cameras, thermostats, door sensors, voice assistants, speakers, fridge, washing machines). Users create personal networks from these devices.

Consumers and small businesses no longer own just a single device (phone) or two devices (phone + computer), but are instead owners of local "networks" that are fixed in nature (e.g., the home LAN) or mobile (e.g., wearables). Devices, services, and applications within the same local "network" communicate with one another, and these local "networks" connect to the 5G network via a gateway, which ensures that the devices within the local "networks" can communicate with the UEs, services, and applications available on the 5G network.

3GPP TR 22.859 [23] documents many use cases for this technology, and the requirements in 3GPP TS 22.261 [24] have identified potential gaps in existing 3GPP specifications, resulting in the communication of the normative work initiated in "Personal IoT and Residential networks Service Requirements". The PIN also leverages requirements for user identities—the user being identified could be an individual human user, using a UE with a certain subscription, an application running on or connecting via a UE, or a device ("thing") behind a gateway UE.

SA6 studies how the 5G application layer architecture can be enhanced to support the PIN in 3GPP TR 23.700-78 [25]. Figure 11.19 illustrates the application architecture for enabling the Personal IoT Network Application (PINAPP).

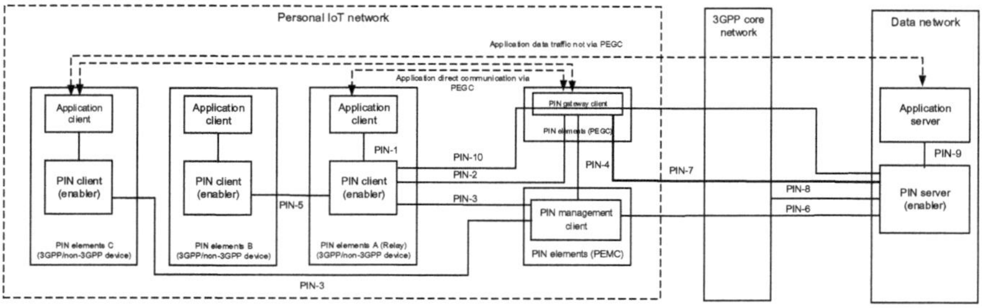

Fig. 11.19 PINAPP architecture [25]

In addition, the ongoing SA6 study analyzes the requirements to determine which PIN-related requirements have potential application level impacts, as well as to develop key issues, corresponding architecture requirements, and solution recommendations, which would allow the application layer to support PIN services over 3GPP networks. These issues include:

- Application layer architecture that supports the PIN
- Information and capabilities of PP networks to support the PIN
- Service or application discovery within a PIN, or by outside UEs via a 3GPP network
- Service or application authorization and access control
- Study on the PINAPP service switchover when path switchover occurs between PIN elements, or when the PINAPP moves between PIN elements.

11.2 5GtoB Standardization Progress in CCSA

China puts a high priority on convergence of 5G technologies for vertical industries. Multiple policies point out just how important 5G-enabled industries are, including development goals for key infrastructure. In August 2020, the Ministry of Transport of the People's Republic of China (MOT) released the *Guidelines on Promoting the Construction of New Infrastructure in the Transportation Field*. The document states "Based on 5G commercial rollout, promote the coordinated construction of transportation infrastructure and public information infrastructure, and advance the establishment of a normative infrastructure system that adapts to autonomous driving, automated port, and unmanned delivery." In June 2021, the National Development and Reform Commission (NDRC), National Energy Administration (NEA), Cyberspace Administration of China (CAC), and Ministry of Industry and Information Technology (MIIT) jointly issued *5G Application Implementation Solution in the Energy Field*, proposing: development of a batch of special technologies and products that meet the requirements of 5G applications in the energy field; development of a batch of key technical standards; and substantial improvement of basic capabilities for 5G applications to support the energy field. In July 2021, nine ministries including the MIIT, CAC, and NDRC launched the *"Sailing" Action Plan for 5G Applications (2021–2023)*, which pointed out that cross-industry standards and protocols should be converged, application standards for key industries should be developed, key standards for pivotal industries should be implemented, a 5G application standard system should be built, and 15 converged industry applications should be developed to fruition, such as 5G+ industrial Internet, 5G+ V2X, 5G+ smart grid, 5G+ smart oil and gas, and 5G+ smart mining. In September 2021, the MIIT and National Health Commission (NHC) together released a *Notice to Release 5G+ Healthcare Application Pilot Projects* to encourage all regions and departments to innovate in 5G application scenarios and accelerate decision-making in and deployment of new infrastructure such as 5G networks. With the country's support and joint efforts in industries, China's 5G technologies are on their way to maturity. Converged standards in industry segments are growing, and research on 5G converged application standards in key industries is showing positive results.

11.2.1 General Capability Standardization Progress

5G standards development gradually shifts its focus to industry requirements, furthering the convergence of 5G and vertical industries. 3GPP technical standards are also on track to gradually meet industry service requirements, promoting expansion in and evolution of the market. 3GPP Release 15 focuses on the basic architecture and capabilities of 5G networks to provide comprehensive functions. 3GPP Release 16 highlights enhancement to basic functions, introduction of new features, and expansion for vertical industries. 3GPP Release 17 expands support for vertical industries in terms of public safety, high-accuracy positioning for industrial Internet, and broadcast/multicast service standards, driving the large-scale deployment of 5G applications. China Communications Standards Association (CCSA) keeps pace with 3GPP from a macro perspective, looking at key 5G sub-fields such as networks, terminals, applications, platforms, and security. Deployment modes, key technologies, test methods, and other such detailed content have been agreed upon to meet the growing customization requirements of the industry.

11.2.1.1 Networks

- Network slicing

  Network slicing is a key feature of 5G SA networks and an enabler for 5G to serve industries, which allows operators to provide private network services for industries, achieving E2E resource isolation and Service Level Agreement (SLA) assurance. It helps meet the mobile network requirements of customers in the eMBB, URLLC, massive Internet of Things (MIoT), and V2X service scenarios as defined by 3GPP. These scenarios place different requirements on the data rate, latency, billing, and the like, all of which can be met through network slicing.

  The 5G E2E network slicing architecture [26] consists of the management domain and service domain, as illustrated in Fig. 11.20. The management domain includes the communication service management function (CSMF), network slice management function (NSMF), and network slice subnet management function (NSSMF), whereas the service domain includes the user equipment (UE), (radio) access network ((R)AN), transport network (TN), core network (CN), and data network (DN).

  CSMF is responsible for mapping between user-side functions and service requirements, and resource-side functions and services (such as network application/release and slice performance assurance). NSMF is mainly responsible for network slice management, including slice creation/update/query, lifecycle

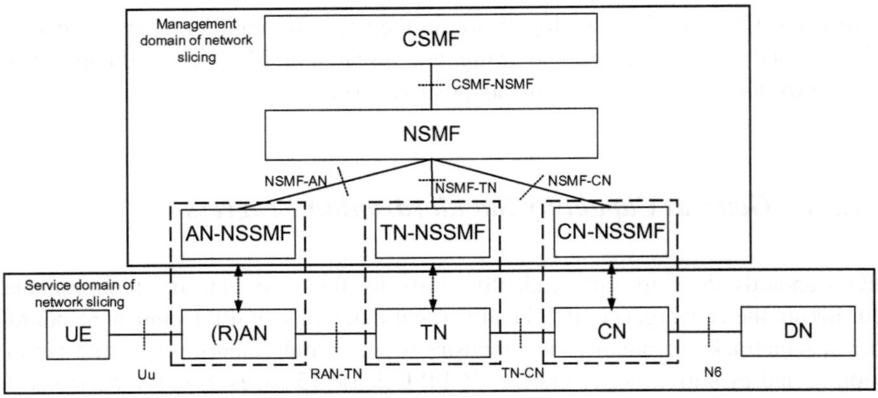

Fig. 11.20 5G E2E network slicing architecture

management, performance management, and fault monitoring. NSSMF, which consists of AN-NSSMF, TN-NSSMF, and CN-NSSMF, is mainly responsible for lifecycle management, performance management, and fault monitoring of a specific slice subnet. NSMF invokes the slice subnet lifecycle management capability through an interface, creates and manages CN slice subnet templates and instances, queries the status of subnet instances, and queries and collects resource information of slice subnet instances from NSSMF.

Each subdomain within the service domain has different network slicing requirements. A UE must support multiple slices and be able to register with the network and obtain related network slice selection assistance information (NSSAI). The AN must support CN network element (NE) selection, slice subnet data flow awareness, subnet session resource management, and slice resource sharing and isolation. The TN must provide dedicated network slice resources, soft/hard slices, and slice interconnection between the AN and CN for specific users or services. The CN must support the four standard slice types (eMBB, URLLC, MIoT, and V2X), slice configuration and configuration update, and slice roaming. The DN slice requirements are still under discussion.

- Non-public network (NPN)

  An NPN is a 5G system specially deployed for a non-public scenario. It provides an effective solution for implementation of 5G applications in industry enterprises. To meet the dedicated network resource requirements of vertical industries, 3GPP SA1 has defined the requirements of 5G NPN services in 3GPP TS 22.261. 3GPP SA2 has researched key issues and solutions of 5G NPNs. From the perspective of standard evolution, 3GPP Release 16 has standardized the architecture and processes, and 3GPP Release 17 has researched the requirements of industry-specific network deployment and application implementation. 5G continues to focus on industry application. China has been at the forefront of 5G application and deployment. In addition, China has carried out multiple

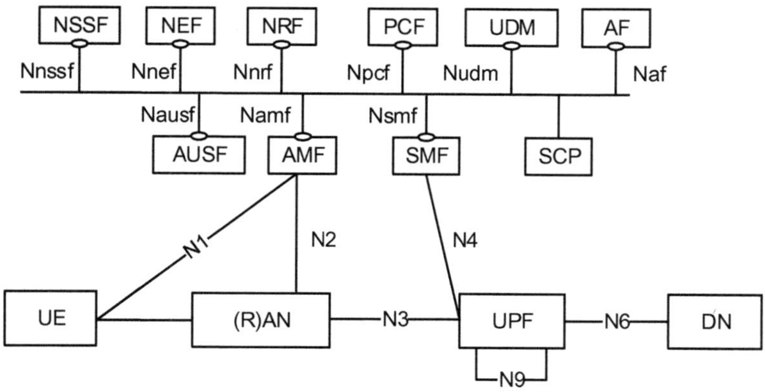

Fig. 11.21 5G system architecture

research projects and formulated related standards [27]. Figure 11.21 illustrates the 5G system architecture.

NPNs are categorized into two types: standalone non-public network (SNPN) and public network integrated non-public network (PNI-NPN). The former is operated by the NPN operator and does not depend on the network functions provided by the PLMN. The latter is deployed with support from a PLMN. Both are generated based on a 5G system architecture. SNPN access supports only 3GPP specifications and does not support interworking with 4G networks, emergency services, inter-network roaming, or inter-network handovers. The PNI-NPN is a non-public network provided by using one or more public networks (for example, one or more network slice instances). In addition, a closed access group is typically used for user access control, because a network slice cannot prevent unauthorized user access.

- Industrial proximity network

An industrial proximity network is the application of IoT technology in an enterprise intranet. To realize the digital and intelligent transformation of enterprises, the research scope of an industrial proximity network is expanding. In 2013, the ISO23247 (digital twin) standard proposed the concept of an industrial proximity network for the first time, which is recognized as the industry's earliest definition of a proximity network. An industrial proximity network implements network connection and data interworking between site devices and data collectors/controllers chiefly in industrial and transportation fields. Common technologies for industrial proximity networks include industrial connectivity technologies (such as fieldbus, industrial Ethernet, and industrial wireless connections) and low-power communications technologies (such as narrowband IoT [NB-IoT], long-range radio, and radio frequency identification [RFID]).

An industrial proximity network enables industry production and management processes by utilizing technologies such as RFID, short-distance communications, and deterministic transmission, meeting differentiated communication requirements in key scenarios such as automatic asset management, remote centralized control, and device collaboration.

Based on the principle of RF signal reflection, RFID is used for item identification, device positioning, and information query. It consumes no power and is low-cost, meaning it can be widely applied in scenarios such as asset audit and inbound/outbound statistics in industries, logistics, medical care, and transportation fields. As the current research hotspots, new system architecture, new tag technology, and anti-interference technology of RFID will help boost the communication distance and improve the item identification rate. This in turn expands RFID application scenarios.

Short-distance communications are chiefly used for small-scale interconnection between devices, featuring simple deployment and low cost. New short-distance communications technology provides microsecond-level low latency and burst interference resistance. It is designed to meet the ultimate performance requirements for low-latency and high-reliability communications for smart cars and smart manufacturing, for example.

Deterministic transmission means that the communication indicators ensuring normal service operation are predictable and fluctuate within a certain range. New deterministic transmission technology combines a high-precision clock with Ethernet data communications, and schedules and controls data communications, so that data communications have higher precision and stability. Moreover, technologies such as multipath transmission and traffic scheduling can be used to ensure the communication delay and improve reliability. The industry is actively promoting the verification and application of these technologies.

The 5G+ industrial proximity network is centered on the site gateway. Site network technologies (such as RFID and industrial Ethernet) are used to implement site device connections and communications in the southbound direction. In the northbound direction, the 5G network transmits site production and management data to the platform, which serves the industry production site. In this way, we can meet the differentiated requirements of various services. Collaboration between the industrial proximity network and 5G meets the requirements for site communications in different industries and this also improves network management and O&M capabilities. In addition, edge computing and computing awareness can improve network intelligence.

- Digital twin of an industrial proximity network

Digital twin is a comprehensive technology. It allows for a virtual twin to be mapped to a physical network and updated in real time. This technology has gradually developed into a mainstream mode of operation for IoT applications. As industrial proximity networks are widespread in smart manufacturing, flexible production lines, remote control, and asset management, problems can arise when trying to manage heterogeneous networks, and network O&M efficiency is dropping. As a result, enterprise users demand unified and convenient network management. With digital twin, technologies such as IoT perception, network service, modeling simulation, data integration, and human-machine interaction can be integrated into the virtual-physical mapping and interaction between physical entities/functions and virtual twins/functions. Ultimately, this can deliver differentiated digital management services for industry-oriented site networks, covering the entire lifecycle of network planning, construction, O&M, and optimization. Moreover, the digital twin of an industrial proximity network supports customer- and service-centric visualized, automatic, and intelligent O&M capabilities, effectively improving O&M efficiency and customer experience. Digital twin helps establish a high-value network O&M management system.

Figure 11.22 illustrates the mapping between the overall architecture and function requirements of the digital twin of an industrial proximity network [28]. The digital twin of an industrial proximity network consists of the following entities: (1) Data collection and control entity, which collects upstream perception data and executes downstream control instructions between twin objects and physical objects. (2) Core entity, which is responsible for twin generation and application expansion based on common support technologies. (3) User entity, which is responsible for man-machine interaction. (4) Cross-domain entity, which is responsible for data exchange and security assurance between entities. The physical layer represents a physical entity of the real industrial proximity network.

The data collection and control entity has a data collection sub-entity and a control sub-entity. This entity's main purpose is to collect the data that indicates characteristics of the physical network corresponding to the virtual network. It is also responsible for controlling target NE equipment in the physical network. The core entity has a model management sub-entity, an application and service management sub-entity, and a resource access and exchange sub-entity. It provides

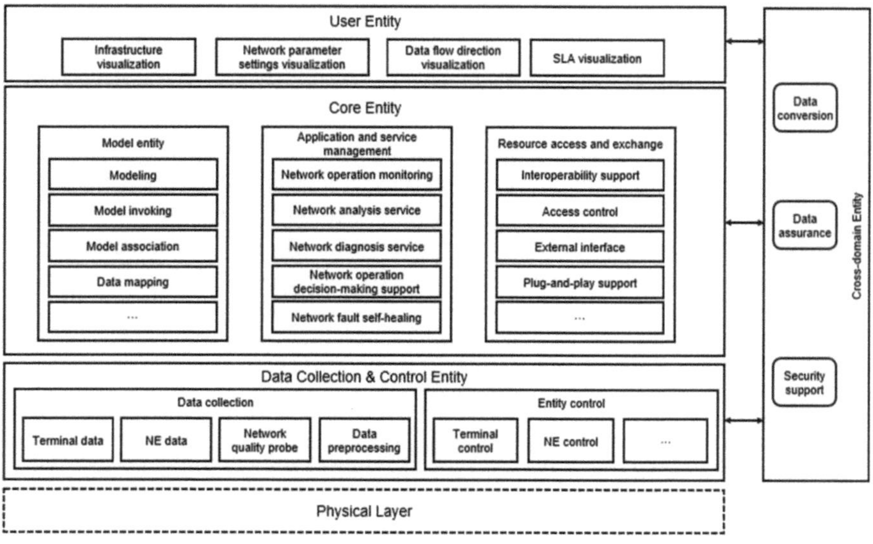

Fig. 11.22 Mapping between the overall architecture and function requirements

modeling and convergence, data integration, simulation analysis, system expansion, and other such functions. The user entity connects to the core entity, producing visualization of the network based on related models. This acts as an interactive, human-machine environment for comprehensive and intuitive management of industry twin networks. The cross-domain entity implements data exchange, interactions through instructions and intuitive interfaces, and security support.

Digital twin can be applied to industrial proximity networks in buildings, hospitals, shopping malls, industrial parks, and multiple other scenarios. Take an industrial park as an example. This is a complex environment, and heterogeneous networks typically come with many custom requirements. Enterprise users have posed higher requirements on complex network O&M, and want to converge systems into one network in addition to local data processing. Traditional operator network management systems and agent maintenance teams cannot meet these industry network requirements.

The digital twin of an industrial proximity network cannot only improve the efficiency and convenience of industry service management, but also bring value into full lifecycle management of the industrial proximity network. First, the visualization capability helps customers display network performance indicators and running status in an intuitive, real-time, and multi-dimensional manner. Second, the network planning and simulation capabilities accommodate high-level customization of costly networking solutions due to complex and diversified topologies and configurations of the industrial proximity network. Last but not least, the intelligent O&M capability solves the problem that traditional passive troubleshooting fails to quickly respond to faults, and the problem of a slow resolution process following a complaint.

11.2.1.2 Terminals

- 5G module

  Cellular wireless communication modules are basic IoT components used to connect industry terminals to the network. In addition, they are key for connecting the IoT perception and network layers, and are irreplaceable as underlying hardware. Each module maps to an IoT terminal. Given the diversity of 5G terminals key for the Internet of Everything (IoE), modules play an important role. Universal 5G modules can be applied to consumer fields and more widely across industry fields.

  Figure 11.23 illustrates the basic logical structure of a universal 5G module [29], which consists of a main chip and an RF front end. A module can include a Micro Control Unit (MCU)/access point (AP) unit, positioning unit, sensor, Subscriber Identity Module (SIM)/Universal Subscriber Identity Module (USIM), and antenna as required. Universal 5G modules support a minimum of 5G/4G dual bands. They provide access and service capabilities in 5G SA networking and access capabilities in 5G NSA networking. At least two uplink transmit antennas

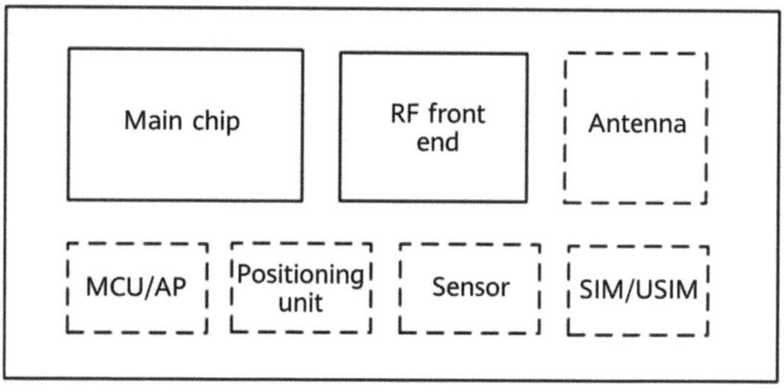

Fig. 11.23 Logical structure of a universal 5G module

are required in SA networking. In addition, modules can be classified into basic, smart, and all-in-one modules based on their computing processing capability and logical structure. A basic module is for communication only. A smart module is further equipped with a screen and a camera data port to support smart applications and provide special functions such as multimedia and artificial intelligence processing capabilities. The all-in-one module extends basic communication of the basic module with an internal antenna port, which effectively reduces the workload when developing terminals with these modules.

Complicated industry application scenarios pose higher requirements on the reliability of industry terminals, and this rings true for universal 5G modules in devices at key nodes. Faults can spiral into a series of adverse consequences. 5G modules are mainly oriented to consumer, industrial, and vehicle-mounted applications. Before we can deploy these modules, they must be tested for temperature and humidity reliability, dust and mist reliability, mechanical reliability, and electrical reliability. Only then can we ensure that they meet the requirements of the wide range of industries.

- Industry gateway

  This acts as a data egress of a production site that bridges the industrial proximity network and the cellular network. The industrial communication network, which has a complex structure, involves various industrial bus, communication, and wireless protocols. The industry gateway aggregates and converts different protocol data into standard IP data, which is transmitted to a public or private network. It is vital for integrating new operation technology (OT), information technology (IT), and communication technology (CT), given that it bears capabilities such as field-level computing and intelligence.

  As industry digital transformation forges ahead, industry networks and platforms are evolving towards a flattened, wireless, and IP-based architecture. 5G industry gateways see market potential for application innovation. With the rising demand for flexible production and device collaboration, production equipment needs to go wireless. Therefore, 5G industry gateways are used for bridging the industrial proximity network and cellular network, while meeting the latency and reliability requirements of wireless networks.

5G industry gateways are the ingress equipment that bridges 5G and industrial proximity networks, to provide sensing, edge intelligence, and protocol conversion capabilities. They serve as the infrastructure for data collection and information processing in vertical industries. On the industry gateway side, protocol adaptation enables unified access of inter-system equipment, flattening the network architecture. After integrating proximity network technologies, 5G industry gateways collect data in a unified manner to meet the requirements for industries to go wireless. Such IP-based gateways convert site production and management data of different Layer 2 protocols into standard IP data and transmit this data via 5G networks. This implements data source tracing, flexible routing, and refined network and business management, providing network scalability, maintainability, reliability, stability, and security for industry production.

- Customer-premises equipment (CPE)

  As a high-performance terminal, 5G CPE uses the industry-leading 5G technologies to convert 5G signals into WLAN signals or signals in network cables and provides high bandwidth in the uplink and downlink and low-latency data access for sensing devices (such as HD cameras and sensors) in industry scenarios. The 5G CPE connects low-latency sensing devices to 5G networks, including sensing terminals (such as cameras and industrial control equipment). In addition to expanding 5G applications, this device provides industry customers with high IP rating and reliability, and provides industrial devices and vehicles with wireless broadband services in indoor, outdoor, and onsite operation scenarios.

  The 5G CPE connects to a LAN and WAN via another non-cellular communication port and cellular communication ports/fixed network access ports, respectively. Figure 11.24 illustrates the deployment of 5G CPEs on a network [30].

  5G CPEs can be classified into multiple types. For deployment, 5G CPEs can be classified into indoor and outdoor CPEs. Indoor 5G CPEs are deployed in

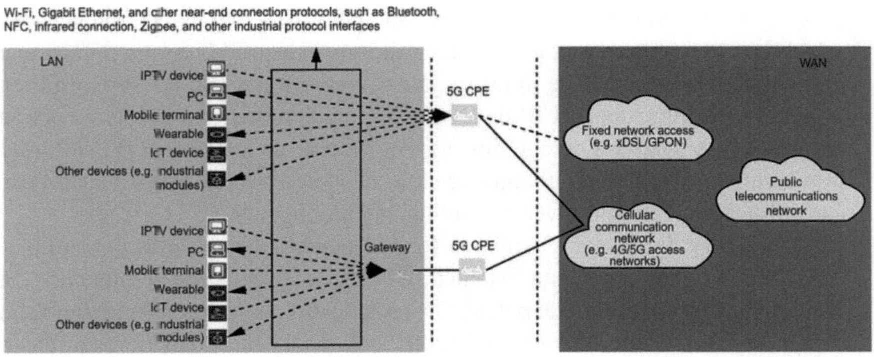

Fig. 11.24 5G CPE network diagram

enclosed spaces, while outdoor 5G CPEs are utilized in open environments. Based on functions, 5G CPEs can be categorized into access and gateway CPEs. The access 5G CPE connects one or more gateways with access to various terminals to the public network through 5G wireless technologies. The gateway 5G CPE connects to various terminals through Ethernet, WLAN, Bluetooth, Zigbee, Near Field Communication (NFC), infrared ports, industrial protocol-compliant ports, or other ports, and accesses the public network through 5G wireless technologies. Based on user groups, CPEs are classified as consumer 5G CPEs and industry 5G CPEs.

11.2.1.3 Applications

As 5G continues to develop, it will be applied to more services, including IoT services. There is a wide range of IoT services with varying characteristics and different demands. For example, low-rate collection services, such as intelligent meter reading, require many networked terminals, which are typically inexpensive, consume little power, and transmit many small data packets. On the other hand, high-speed data collection services, such as video surveillance, require a higher uplink transmission rate and traffic density in densely deployed scenarios. Not only that, there are latency-sensitive control services, such as automobile networks equipped with high-speed mobility, which require millisecond-level latency and nearly 100% reliability, while latency-insensitive control services, such as home control, require timely response.

Since IoT services are very scenario-specific, networks or terminals will need to provide different service quality and capabilities. Defect detection, for example, involves a lot of real-time data, which requires high bandwidth and low latency. HD security surveillance also involves a lot of data, so high network transmission bandwidth and network connection stability are essential. Remote control and in-park autonomous driving both require low latency and high reliability, so more attention should be paid to network latency and disconnection rate. Mobile network-based IoT services have requirements on network access success and duration, network handover success, network bandwidth and latency, service access success and duration, service transmission rate, TCP retransmission and TCP out-of-order rate, and User Datagram Protocol (UDP) packet loss rate.

The system used to monitor the service quality of 5G IoT terminals consists of a terminal service quality monitoring tool and a monitoring and analysis platform [31]. The former collects and monitors key information such as onsite device status, network coverage, and service quality. At the same time, the monitoring and analysis platform collects and stores data, delivers monitoring policy configurations, and implements functions such as data analysis, forwarding, display, and device management.

More specifically, the service quality monitoring tool collects information and reports it to the monitoring and analysis platform through the data transmission protocol. When a test command is executed, the platform sends a test request to the service quality monitoring tool, which then responds whether the request has been successfully received, establishes a connection with the dialing test server, and executes the corresponding dialing test task. Finally, the dialing test server reports the test result to the monitoring and analysis platform.

There are two ways to deploy the monitoring and analysis platform, the first of which is distributed deployment on a private network. The dialing test server and the monitoring and analysis platform are deployed on the Multi-access Edge Platform (MEP). In this case, sensitive device data and detection data, collected by the service quality monitoring tool, are distributed and transmitted within the park through the edge user plane function (UPF). The other option is centralized deployment. The test server and the monitoring and analysis platform are deployed in the operator's equipment room. Here, the data collected by the service quality monitoring tool is transmitted to the monitoring and analysis platform through the 5GC. So, the customer accesses the probe platform in multi-tenant mode.

11.2.1.4 Platforms

1. Edge Computing Platform

   Edge computing is a key link between 5G and vertical industries. An edge computing platform often works with 5G. In this context, flexible 5G user plane deployment and multiple flow-based flexible traffic steering mechanisms can be utilized to offload local traffic and localize services. This assures high bandwidth and low latency.

   Figure 11.25 illustrates the architecture of an edge computing platform system [32]. Thanks to control and user plane separation in the 5GC, the UPF can be flexibly deployed at the network edge. Also, control plane functions, such as policy control function (PCF) and session management function (SMF), can be centrally deployed. The UPF implements the user-plane functions of 5G edge computing. At the same time, the edge computing platform provides a running environment for and manages edge applications. Depending on the given scenario, the UPF and edge computing platform can be deployed separately or together. Finally, the SMF in the 5GC selects the UPF which is close to UEs, establishing a local route and distributing data.

   As illustrated in Figure 11.25, an edge computing platform system consists of the edge computing host and management functions. The former comprises the edge computing platform, virtualization infrastructure, as well as the edge computing applications and services running on the system. The virtualization infrastructure provides computing, storage, and network resources required for running edge computing apps. The edge computing platform provides the app running environment and invokes edge computing services. The MEP itself can also provide services, including load balancing, security function, bandwidth management, mobility support, local API gateway, user metering, and route control. Edge computing management functions include system-level and

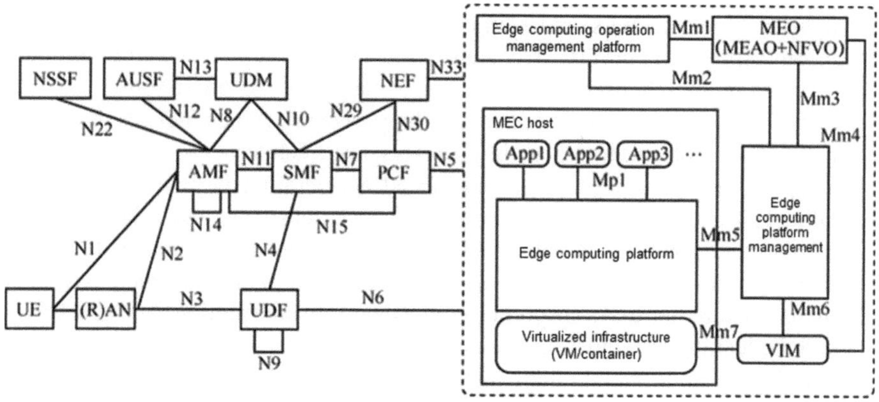

Fig. 11.25 Architecture of an edge computing platform system

host-level management. The former covers edge computing operation management and multi-access edge orchestrator (MEO), while the latter includes edge computing platform and virtualization infrastructure management.

2. Industry Message Management and Control Platform

   5G messages come in various formats that stimulate interaction between individuals and industries. Rich card, which is known as industry message, is a new service enabled by 5G, one that is complicated by difficult management and control as such messages comprise text, image, audio, and video. Such diverse types pose content security risks. Furthermore, as messaging between individuals and industries is a new service, its wide scope and high speed give rise to content security risks and require better management and control policies. This means that 5G message content security is critical to service security development. To shoulder these challenges, the intelligent management and control platform for 5G industry message content intelligently detects message elements (text, image, audio, video, and complex file) in rich cards and delivers the associated review process and policy to the Messaging as a Platform (MaaP).

11.2.1.5 Security

1. Service Security

   5G features high bandwidth, low latency, and mass connections. It promotes the integration of communication technologies into various industries, boosts the development of the world's digital economy, stimulates new social changes, and aids global economic and social growth. eMBB, URLLC, and mMTC are the three main application scenarios of 5G. Live streaming and video services dominate eMBB, which is mainly characterized by ultra-high data transmission rates and mobility assurance in wide coverage. URLLC, on the other hand, requires a connection latency of 1 ms. It also requires high reliability (99.999%), often under scenarios of high-speed mobility (500 km/h). URLLC is mainly applied to industrial Internet, V2X, and telemedicine, because of their high security requirements. And lastly, mMTC is characterized by massive IoT and IoE, and these services are mainly used in smart energy, smart city, and smart home.

5G-driven services and applications have given rise to challenges that legacy networks do not have to deal with. This is because 5G Internet access rate, latency, and performance are significantly enhanced, accelerating the arrival of big video. Ultra HD and VR/AR will be applied more extensively, and a large number of snap-and-send videos will be adopted quickly, posing challenges to content security. Moreover, 5G will be widely used in key industries such as energy, industrial Internet, and transportation, and therefore interruption to these services, and the leaking or tampering of key data are significant threats to economic and social stability, as well as national security.

Due to new applications in the industry and the discrepancies between the lab environment and onsite environment, the security of new 5G technologies and services must be assessed from multiple dimensions, including but not limited to the general security of 5G services (such as basic service information, basic user information, and service cooperation mode), key risk prevention in 5G service application scenarios (such as content security, data security, terminal security, and platform and process security in low-latency scenarios), and service security assurance capabilities (such as service platform security management, service system vulnerability management, slice security assurance, and edge computing security assurance).

2. Service Enabler Architecture Security

Requirements vary across vertical industries, but some common requirements can be extracted. Common requirements help simplify the development and deployment of new service capabilities required in vertical industries, shorten application development and deployment, and improve network utilization. 3GPP has formulated the SEAL to meet the common requirements of vertical industries. Vertical industry applications, such as V2X applications, interact with the network through SEAL.

SEAL security is closely related to the security of networks and vertical industry applications. User authentication and authorization at the VAL is one of the key requirements of SEAL security. The main requirements are as follows [33]:

(1) All users that the VAL serves must be authenticated.
(2) The VAL terminal and the VAL server should authenticate each other before the VAL server provides the user profiles for a VAL service and access to the specific service to the VAL terminal.
(3) Confidentiality, integrity, and anti-replay protection must be provided for configuration data and user profiles transmitted between authorized VAL servers and VAL terminals.
(4) Measures must be taken to detect and mitigate denial of service (DoS) attacks for VAL services to minimize the impact on the network and VAL users.

(5) A method must be provided for protecting the confidentiality of the VAL user identity and VAL signaling for VAL services. Not only that, the VAL system should take measures on the system border to protect itself from external attacks.

3. Security Capability Openness

   The commercial adoption of 5G drives up the security requirements of vertical industries. Operators take their existing advantages and expand into the security service market as network infrastructure operation units. Their data samples, smart pipes, computing resources, and performance are strong advantages. Opening up the security capabilities of operators can significantly enhance security for vertical industries while removing high costs and low benefits that can arise in scattered security construction.

   As illustrated in Fig. 11.26, operators open basic 5G security capabilities and IT cyber security capabilities to vertical industries [34]. Basic 5G security capabilities include slice authentication, Authentication and Key Management for Applications (AKMA), two-factor authentication, and analysis of UE unexpected behaviors, which are unique in 5G systems. IT cyber security capabilities are common IT network capabilities that vertical industries can leverage. These capabilities can be further divided into four sub-categories: identification, protection, detection, and response, which are applicable to building security capabilities in different phases of vertical industries. 5G network security capabilities can be customized for vertical industries through security capability subscription and unsubscription. Operators are responsible for security capability configuration and deployment, and providing security service capabilities and O&M for vertical industries. The four sub-categories of IT cyber security capabilities are denoted as follows:

   - Identification: includes vulnerability scanning, baseline check, and threat intelligence. These provide security capabilities required for system security risk identification and sorting.
   - Protection: includes web security and data security, providing security capabilities required for security protection construction.
   - Detection: includes distributed denial of service (DDoS) attack detection, anti-DDoS traffic cleaning, and malicious domain name/URL detection, providing security capabilities required for security status monitoring.
   - Response: includes attack source tracing, providing security capabilities required for quick source tracing and attack response.

11.2.2 5GtoB Standardization Progress in Key Industries

CCSA is responsible for developing 5G standards in the communications industry in China and is an important contributor to 5G+ industry convergence standards. It establishes multiple work groups specific to industrial Internet, healthcare, electric power,

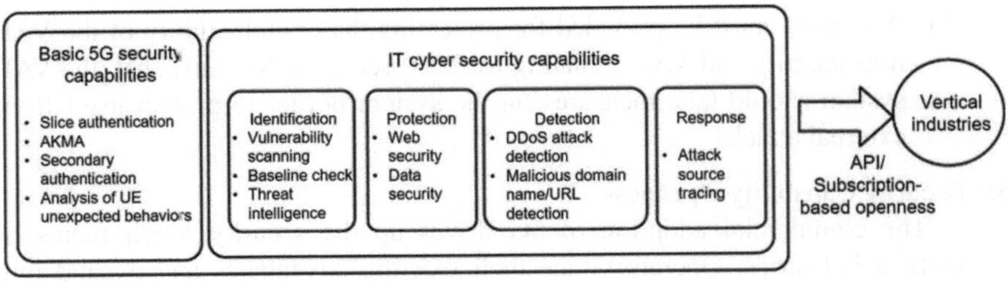

Fig. 11.26 Security capability openness system

V2X, and others, and has carried out researches on multiple industry standards. In the industrial Internet field, specific standards are being developed in an orderly manner regarding campus network deployment, data security protection, 5G application scenarios, passive optical network devices, and security test and evaluation environments. In the healthcare field, standards specific to 5G-based pre-hospital emergency care, 5G-based ultrasound diagnosis, and 5G-based telemedicine have been initiated. In the electric power field, standards specific to network slicing, network management capability, and application scenarios are the focus of the industry. In the V2X field, standards for 5G-based remote control driving, 5G V2X-oriented autonomous driving scenarios, and datasets have been formulated. Meanwhile, convergence standards have been developed for other fields such as ports, mines, steel, energy, and logistics, and this development is under the initial phase with a focus on determination of 5G application scenarios and technical requirements.

11.2.2.1 Industrial Internet

The industrial Internet is a leading industry in 5G industry applications, and 5G industrial Internet standards are also in a leading position. CCSA has carried out research on a series of standards covering networks, devices, applications, security, and services. Such standards include *Industrial Internet: 5G Network Deployment Guide of Industry Park*, *Industry Internet: The Measurement Methods of Time Sensitive Network System*, *The Framework and Technical Requirements of 5G Network Service Capability in Industrial Internet Campus Network*, *Network Technology for Industrial Internet—Test Methods for Passive Optical Network (PON)*, *Industrial Internet: Campus network—5G Application Scenarios and Technical Requirements*, *Industrial Safety Supervision Based on Edge Computing—Application Scenario and Requirements*, *Reference Architecture of Industrial Internet Security Testing and Evaluation Environment*, and *Specification on Industry Indoor Positioning Technology Based on 5G Mobile Network*. This section describes the standards related to networks and applications in detail.

1. Industrial Campus Network

   As a gathering area of industrial enterprises, industrial campuses provide a large number of infrastructure and public services for these enterprises. With the rapid development of "Internet+", industrial campuses need transformation based on next-generation information technologies to satisfy fast development requirements of enterprises and improve the service level and quality. This way, the core competitiveness and sustainable development capabilities in the information-based environment can be achieved. The industrial Internet is a key way to promote campus transformation for digitalization, networking, and intelligence. 5G technologies emerge at the right time. They help implement applications of cutting-edge information technologies in industrial campuses and meet the requirements of the industrial Internet for intelligent sensing, ubiquitous connections, real-time analysis, and precise control. As an enabler of industrial Internet, 5G creates a brand-new industrial ecosystem and promotes high-quality development of the manufacturing industry, forming an emerging business and application model that deeply integrates the next-generation ICT technologies with the advanced manufacturing industry.

With a strong demand for wireless connections, industrial campuses involve wireless access for industrial production and campus wireless public services. Deployed in campuses, 5G can give full play to its flexibility, high bandwidth, and low latency features. The industrial campus networking architecture [35] is available in three options. In option 1, the 5G network in an industrial campus shares the base station, spectrum, transport network, and core network with the public network. Multiple logical subnets are created using technologies such as network slicing and E2E QoS to provide virtual private networks with SLA assurance to meet specific industrial requirements. In option 2, the 5G network in the industrial campus adopts local deployment of the UPF (a user-plane NE of the core network), and shares the control plane of the core network with the public network. Data splitting is used to ensure that local service data can only be processed inside the campus, while the MEC platform can be deployed to dynamically adjust the service quality of the local UPF. In option 3, the 5G network in the industrial campus uses independently deployed base stations and allows for on-demand downwards movement of the core network to ensure that campus services are not affected when the public network is faulty or during network adjustment. In this way, the service control instructions and service data flows in the industrial campus can only be processed within the campus, ensuring rigid security.

5G service capabilities defined for industrial campus networks emerge, in response to the need for providing industry users with 5G network capabilities, such as network slicing, MEC, and QoS capabilities. 5G network service capabilities oriented to campus networks [36] consist of the connection capability, information capability (network status, terminal location, and network quality), resource capability (edge computing resources and network slice resources), and configuration capability (QoS scheduling and parameter configuration) of the 5G network. Standard capability service interfaces can be built to facilitate network invoking and configuration on industrial platforms, promoting in-depth convergence and collaboration between 5G and industrial Internet. As the core system for enterprises to provide 5G network service capabilities, the 5G service capability platform employs capability openness technologies as well as a series of standard application programming interfaces (APIs) and services to improve the network utilization efficiency, convenience, and control capabilities of campus network customers for industrial Internet based on network, IT, and platform capabilities.

2. Application Scenarios

The rapid development of 5G network technologies stimulates the implementation of cutting-edge information technologies in industrial campuses. 5G features high bandwidth, low latency, and high reliability and can fulfill the requirements for intelligent sensing, ubiquitous connections, real-time analysis, and precise control of the industrial Internet, accelerating intelligent industry upgrades. The industrial Internet campus network consists of the industrial production network, enterprise information network, campus public service network, and cloud infrastructure. The industrial production network is deployed inside each enterprise in the campus to interconnect various production devices, sensors, actuators, and industrial computers at the production site, collect industrial data, and perform industrial control and maintenance. The enterprise information network is deployed in the office area of each enterprise in the campus to implement department communication. The campus public service network is deployed to implement interconnection between industrial enterprises in the campus and provide basic public services for these enterprises. As an important infrastructure for information aggregation within an industrial Internet campus, the cloud infrastructure carries private and public clouds of multiple enterprises in the campus.

Based on the networks used by applications, 5G service applications in industrial campuses can be classified into campus public service applications and campus industrial production applications. Campus public service applications are carried out by campus operators and enterprises for their own operation. Such applications mainly use the campus public service network and enterprise information network, and offer basic campus management activities covering personnel, security, transportation, energy, environment, assets, and office work. They are mainly applied in 5G typical scenarios such as intelligent security protection, smart campus transportation, and energy management. Campus industrial production applications are responsible for carrying out the production and operation activities of enterprises in the campus. For such applications, the industrial production network is mainly used to cover the R&D, production, sales, and logistics activities of enterprises. They are mainly applied in such 5G typical scenarios as device information collection, remote device control, product detection, material transportation, as well as remote and onsite personnel behavior monitoring.

3. Typical Application—Video Security Monitoring Based on Industrial Internet

The development of industrial Internet has seen new security vulnerabilities emerge, posing higher requirements on industrial security monitoring, as personnel and devices in industrial campuses, workshops, and production lines must be monitored. At present, intelligent video surveillance is an important method of industrial security monitoring. Based on the monitoring subject, video security monitoring is available for personnel, environment, and device [37].

Personnel-related video security monitoring ensures personnel safety and improves production quality. Cameras are deployed to monitor the dress code and behavior of employees in real time and promptly report alarms when exceptions occur. This effectively reduces personnel non-compliance and improves security and production efficiency. Facial recognition is used in relevant scenarios, such as attendance and entry control in specific areas.

Environment-related video security monitoring helps detect exceptions quickly by using intelligent videos to monitor the production environment of a large area, effectively preventing potential risks. Example scenarios include detection of hazardous chemical storage areas and abandoned objects in important production areas.

Device-related video security monitoring uses image monitoring to conduct contact-free analysis on the running status of devices. As such, when devices without security monitoring and pre-warning are running, problems can be detected quickly and alarms can be generated before faults occur. This prevents production suspension, device damage, and casualties, and reduces losses for enterprises.

11.2.2.2 Healthcare

The trend of smart healthcare is becoming increasingly prominent nationwide. Next-generation information technologies, such as 5G, cloud computing, big data, and AI, have become powerful tools for digital transformation in the healthcare industry. However, this field faces a number of challenges, of which security is the top concern. Strict standards are fundamental to guiding the development of core links. CCSA has launched a series of researches into standards and released a number of standards that cover networks, key components, applications, and management services, including *Converged architecture and technical requirements for smart hospital private networks*, *Technical requirements for pre-hospital emergency telemedicine systems based on 5G*, *Technical requirements for medical private network SIM card*, *Technical requirements for rapid placement of hospital telemedicine systems based on 5G*, *General requirements of ultrasound system based on 5G*, *General requirements of intervention system based on 5G*, *Technical requirements for intra-hospital 5G emergency rescue system*, *Technical requirements for 5G-based intelligent epidemic control system*, *Technical requirements for rapid placement of hospital intelligent management systems based on 5G*, and *Technical requirements of remote monitoring system in rapid deployment hospital based on 5G*. This section describes the progress of the standards related to networks, applications, and key components.

1. Hospital Private Networks

   The commercialization of 5G networks has spurred the demand for private networks for smart hospitals nationwide. This is because 5G is a powerful tool for the healthcare industry in establishing private networks, as it features high bandwidth, high reliability, and low latency, and can work well with edge computing platform, SDN, cloud computing, big data, and AI technologies. According to the network status of the healthcare industry, hospitals have the following requirements on the construction of private networks for smart healthcare: (1) Currently, hospitals have built healthcare informatization systems using network technologies such as Ethernet, Wi-Fi, and 4G. Hospitals hope to cooperate with operators to make full use of their existing site locations and network transmission resources to build 5G networks, so as to smoothly integrate the existing networks and service management systems. At the same time, hospitals hope to cooperate with communications service enterprises to explore emerging healthcare applications with the help of 5G and other cutting-edge technologies. (2) Hospitals expect to have some autonomy in the O&M of private network resources. The main requirements include but are not limited to dynamic awareness of network security and performance, dynamic allocation of network resources, automatic rights- and domain-based network management, and slice template design and provisioning. (3) Data security isolation. Core service data is not transmitted out of the hospital. Hospitals expect private networks to offload, distribute, and isolate local service data. Smart hospital private networks need to integrate the existing networks and service systems in hospitals while ensuring local data security and isolation, and provide secure and easy-to-operate network resource self-management capabilities to the hospital management system.

Currently, operators in China work with equipment vendors to actively explore and build medical 5G networks with medical institutions and put them into use. As early as 2019, under the guidance of the National Health Commission of the PRC and Health Commission of Sichuan Province, the world's first medical application based on the 5G industry private network was officially put into use in the West China Second Hospital of Sichuan University. In the same year, the healthcare industry and the communications industry jointly formulated *Specification for Hospital Network Construction Based on 5G Technology*, and incorporated the specifications into the national health standard system. 3GPP has formulated the only 5G smart healthcare-oriented specification *Study on communication services for critical medical applications*, which focuses on 5G communications services in critical healthcare scenarios.

2. Typical Application—Telemedicine Systems Based on 5G

As the healthcare industry undergoes digital transformation, telemedicine (which uses email, websites, telephone, and fax) is being applied in more and more scenarios, such as checking a patient's medical background, as well as diagnosing patients and determining their treatment path. Telemedicine bridges doctors and experts, letting doctors share medical expertise with remote areas under the remote guidance of experts. This not only saves time and money for doctors and patients, but also eases unequal allocation of medical resources in China. This technology can also play an important role in responding to emergencies, but this is often hindered by poor network speed and latency. 5G overcomes these bottlenecks thanks to its high rate, large capacity, and low latency, to stimulate the development of telemedicine. 5G networks help to realize many telemedicine ideas. Compared with traditional mobile and wired networks, 5G networks are free from physical restrictions of traditional teleconsultation.

With 5G, 4K/8K medical images can be shared quickly and millisecond-level latency realizes remote ultrasound examination and remote surgery. Some typical telemedicine applications include teleconsultation, remote image quality control, and remote surgery guidance, each of which has specific functional requirements [38].

Teleconsultation must provide remote health consultation services, including online expert appointment and real-time HD video communication. Remote experts can remotely control and adjust cameras in real time. Experts at other sites can be invited in real time to join online HD consultations.

Remote image quality control can be applied to ultrasound rooms, endoscopy rooms, and operating rooms and can aggregate dynamic image data streams in real time. Real-time signal access and one-screen display of dynamic images must be supported for multiple sites. Moreover, remote experts can remotely switch between dynamic images, reshuffle them, zoom in and zoom out, and perform one-to-one audio/video interaction.

Remote surgery guidance must support HD audio/video collaboration channels between remote experts and doctors in the operating rooms. Panoramic surgery, close-up, and device images during surgery can be collected with zero signal distortion. Remote surgery guidance must allow full access of images from operating room devices including the multi-channel physiologic recorder, ultrasonic device, CT instrument, skin microscope, laparoscope, thoracoscope, and nasal endoscope while experts can remotely control close-up cameras, adjust a camera's angle, and set motion parameters. Operating room devices (surgical devices or robots) can be remotely controlled to perform remote dotting, marking, and measuring.

3. Typical Application—Ultrasound Systems Based on 5G

Ultrasonic testing is a safe, low-cost, and dynamic method of examination that is widely adopted in China. By merging 5G and ultrasound, medical institutions will be able to share ultrasound technology and scans of patients with other institutions, to help streamline patient diagnosis and treatment through collaborative means. 5G+ ultrasound is an effective way to improve primary care services in basic-level hospitals and promote equal healthcare services in urban and rural areas. Currently, China has made significant progress in 5G+ ultrasound as a number of hospitals have implemented 5G+ ultrasound, unleashing accurate ultrasound examinations in lower-level hospitals or remote areas. However, this is not without its challenges. First, the network quality is not sufficient enough for 5G+ ultrasound requirements. During 5G+ ultrasound diagnosis, a large amount of image data (in GB) will be generated for each patient. Unstable network quality may cause network fluctuations and even poor communication between doctors and patients, as well as unsatisfactory diagnosis effect. This may cause disputes between doctors and patients. Second, there is no specific performance indicator requirement for the service quality of 5G+ ultrasound networks.

Currently, the network service quality of 5G+ ultrasound is mainly measured by the traditional SLA provided by operators, which cannot accurately meet the requirements of 5G+ ultrasound scenarios. Third, the supervision system of 5G+ ultrasound lacks effective monitoring methods for network service quality and as a result, the network service quality cannot be determined and network fault points cannot be located.

4. Medical Device Component—Medical Private Network SIM Cards

   Medical communication terminals must be authenticated before accessing the 5G medical private network. 5G SIM cards are effective in preventing data tampering and are a key way of ensuring 5G network access security. Compared with individual users, the application scenarios of industry SIM cards are more complex. There are various 5G application scenarios in the healthcare industry. 5G SIM cards can be classified into the following types based on application scenarios: (1) For medical terminals and monitoring devices: They support applications such as teleconsultation, remote ultrasound, remote monitoring, and emergency rescue. The types of collected and transmitted data include physical signs of patients, images, and videos. (2) For 5G IoT and medical materials: They support applications such as smart hospital management and medical waste management. The types of collected and transmitted data include positioning information and IoT information. (3) For wearable devices: They support applications such as remote monitoring and hospital personnel management. The types of collected and transmitted data include positioning information and physical sign data. (4) For medical personnel terminals: They support applications such as mobile office, video conference, and smart hospital management. Office data is collected and transmitted. For 5G SIM cards applied to different scenarios, privacy and security protection requirements and access technology requirements vary with the terminal type, data type, and user type.

11.2.2.3 Electric Power

The electric power industry is a pillar of the national economy and society at large. 5G communications technologies have arrived in time to assist with the development of power grid services and new communications paradigms, by allowing for smart grids with enhanced sensing capabilities, interactions, and operating efficiency. The seamless flow of energy requires the seamless flow of information. CCSA has published a vast range of research content on related standards, including *Application Scenarios and Technical Requirements of 5G+ Industrial Internet for Power Grid*, *Capability Open Requirements of 5G Network Management Oriented to Vertical Industries: Power Industry*, *Security Requirements for Electric Power Industry Application via 5G Slicing*, *Technical Specification for 5G E2E Network Slice in Verticals: Power Industry*, and *Technical Specification for 5G Network Slice SLA Assurance: Power Network Slice*. These standards address the application scenarios and technical requirements related to network slicing in the electric power industry. This section provides an overview of network slice security and its corresponding application scenarios.

1. Network Slice Security

   Two major security levels have been defined for 5G smart power grids, corresponding to two major zones (production control and management information). Application scenarios can be further subdivided into four categories, based on the different SLA requirements for electric power services and geographical distribution: production control, production non-control, video services in management zone, and private LAN. Different slice isolation solutions are used for different scenarios [39]. Production control services include automatic differential protection for power distribution grids, wide-area phasor measurement units (PMUs) on power distribution grids, and telemetry, remote communication and remote control services on power distribution grids. Production non-control services encompass metering services to enable power/voltage quality monitoring and smart electric power utilization in factories, campuses, and buildings.

   Services in the production control zone are scattered across a wide range, and require full coverage. 5G networks need to provide high security isolation, low latency, high-frequency forwarding, and high-precision timing. The core services of smart power grids are in the production control zone, and therefore the highest security and reliability assurance levels are required, alongside the lowest possible service latency. All services must be physically isolated from services in other zones.

   Video services in the management zone include substation and line inspection using robots and UAVs and camera monitoring. These services fall under the category of WAN scenarios, and require 5G networks to provide high uplink bandwidth.

   Private LAN services enable electric power services in LAN scenarios, such as intelligent campus and smart substation. They feature limited coverage in specific areas. 5G networks are required to provide such capabilities as high uplink bandwidth and local data processing.

   The communication pipeline network for a 5G smart power grid consists of RAN, transport network, core network, and MEC edge nodes. Each part must meet the requirements of security zones. In addition, data security for sensitive power services must be ensured throughout the entire communication process. On air interfaces, QoS or resource blocks (RBs) are provided for services based on slices. On transport networks, Flexible Ethernet (FlexE) is employed for slicing, to achieve hard isolation of production control and management information services. Virtual local area network (VLAN) or virtual extensible LAN (VXLAN) is utilized on core networks to provide slices for services. Hardware firewalls or virtual firewalls are deployed at physical or virtual network borders to control access. Physical isolation of slices is achieved through physical deployment, to ensure that each slice has relatively independent physical resources. In addition to the basic terminal authentication for access to the 5G network, autonomous and controllable secondary authentication and slice authentication are provided for electric power terminals. The terminals themselves are capable of power grid security encryption. These various capabilities interact to meet the stringent security requirements of the power grid industry.

2. Application Scenarios

   Smart power grids feature a range of different service scenarios and a wealth of different sensors. Each type of service has its own unique characteristics. High bandwidth, low latency, high reliability, and massive connectivity are all necessary attributes of 5G. With the rapid development of power grid services and emergence of new communications needs, next-generation communications technologies, represented by 5G, are required to empower tomorrow's sensitive, interactive, and efficient smart power grids. The seamless flow of energy requires the seamless flow of information.

   Power grid operations, from production to consumption, consist of five phases in total: power generation, transmission, transformation, distribution, and consumption [40]. In the power generation phase, common 5G application scenarios include dynamic power generation monitoring, intelligent power generation control, intelligent power plant inspection, and comprehensive micro-grid applications (including PV monitoring, small wind power generation, and solar thermal power generation). In the power transmission phase, common 5G application scenarios include online monitoring and video surveillance of transmission line status, and UAV-assisted inspection of transmission lines. In the power transformation phase, common 5G application scenarios include comprehensive substation monitoring and substation inspection robots. In the power distribution phase, common 5G application scenarios include intelligent distributed feeder automation, telemetry, remote communication, and remote control services on power distribution grids, intelligent power distribution rooms, synchronous phasor measurement on power distribution grids, and precise load controls. In the power consumption phase, common 5G application scenarios include advanced metering, industrial enterprise and campus power usage, smart power consumption in homes, charging pile controls, and smart lamp poles.

11.2.2.4 V2X

The evolution of V2X services is highly dependent on the development of communicative and collaborative intelligence. V2X services place higher requirements on performance indicators such as communication latency, connection reliability, and communication throughput, while also requiring that multiple performance indicators are supported at the same time. 5G-based V2X communication technology provides effective guidance for the evolution of V2X services. CCSA has conducted research into related standards, such as *5G Enables Remote Driving: Technical Requirements for 5G Communication System*, *Autonomous Driving Scenario Analysis and Data Set for 5G V2X*, and *Research on 5G based Internet of Vehicles Communication Technology*. This section details the remote driving standards.

As vehicles become more intelligent and networked, vehicle remote control via 5G wireless communications can be implemented in the event of traffic accidents or challenging driving conditions. This capability enhances driving safety and comfort, and also improves traffic accident handling efficiency. Now that 5G systems have been deployed commercially on a large scale, a number of experiments have been initiated for remote driving. Remote driving poses differing communication feature and performance indicator requirements on the 5G communication system, depending on factors like the vehicle control level, road conditions, and distribution modes for sensing, decision-making and control-related features. 5G-based remote driving refers to the act of a people or machine remotely driving vehicles, once such vehicles have been interconnected with road side units (RSUs), cloud control platforms, and remote control cockpits [41].

The communications architecture for 5G-based remote driving consists of four subsystems, which are central subsystem, MEC platform, RSU, and vehicle-mounted (passenger transport/freight transport vehicle) subsystems. Figure 11.27 illustrates the communications architecture for 5G-based remote driving. The vehicle-mounted subsystem senses the surrounding environment and vehicle status, and sends this information to the central subsystem. The MEC platform allows the cloud control platform and remote control cockpit to perform remote operations. The RSU sends the remote control execution monitoring information to the MEC platform and central subsystem, which then receives, stores, processes, and distributes global data, senses information globally, and controls the global service policies.

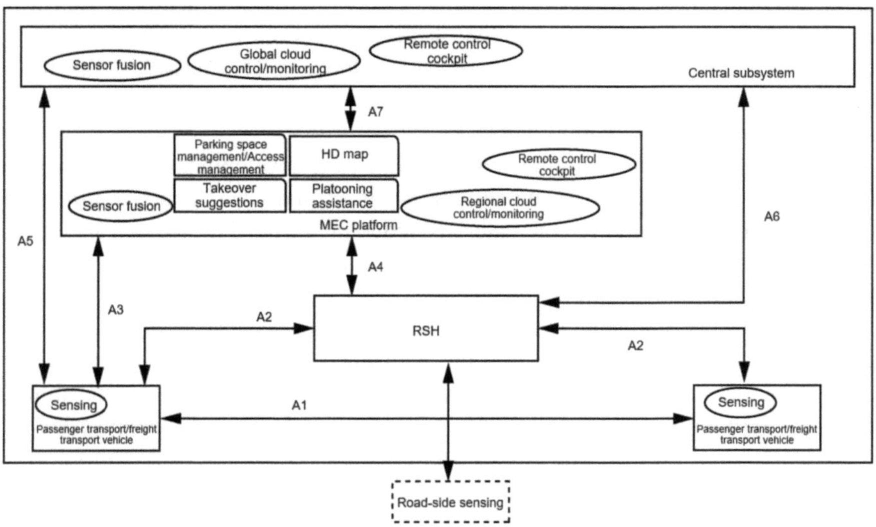

Fig. 11.27 Communications architecture for 5G-based remote driving

11.2.2.5 Mining

Smart mining has revolutionized the mining industry in recent years, but unified standards for smart mining-related 5G+ industrial Internet have yet to be formulated. CCSA has conducted research into this with the publication of *Application Scenarios and Technical Requirements of 5G+ Industrial Internet for Mine*.

The bedrock of smart mining is the deployment of intelligent and automated mining equipment. By utilizing high-speed, large-capacity, and bidirectional integrated digital communication networks as the pipes, and intelligent design and production management software systems as the platform, smart mining is able to ensure real-time and dynamic monitoring and control of objects and production. It makes mining safer and more efficient than ever before, producing tangible economic benefits [42]. Based on mineral deposits, conditions, and operations with mines, smart mining application scenarios can be classified into five common scenarios, which fall under the following categories: surface mining and underground mining. Surface mining is the process of extracting minerals by removing overlying soil and rock, or extracting at the exposed surface. It involves two types of operations: remote control of surface mining equipment, and operation of overground unmanned mining vehicles. In underground mining, specialized devices are used to drill a series of entries or shafts. This involves three types of operations: underground unmanned mining, underground AI-assisted HD detection, and underground device information collection.

Remote control of surface mining equipment is implemented by installing a remote control system and monitoring and sensing devices on the equipment. The following equipment can be remotely controlled: drillers, power shovels, hydraulic shovels, excavators, dump trucks, road scrapers, and bulldozers. The massive connectivity and low-latency communications inherent to 5G make remote control of surface mining equipment remarkably seamless.

Overground unmanned mining vehicles require the development of an unmanned driving system and an assisted autonomous driving system. Unmanned mining vehicles are used to transport mineral materials, reducing the need for tedious and unsafe labor. Unmanned mining vehicles in mines require the ultra-low latency and high bandwidth characteristics of 5G connections.

In underground unmanned mining scenarios, remote control systems and video surveillance devices are installed on the unmanned mining equipment. Operating instructions are delivered to the mining equipment to implement remote mining, reducing the number of personnel who need to go underground, which makes the work much safer. URLLC and eMBB supported by 5G are involved in this service scenario. High uplink and downlink rates are a must.

In underground AI-assisted HD detection service scenarios, HD video surveillance, made possible by low latency and high bandwidth of 5G, is used to detect the operating status of equipment in key areas and status of personnel. This helps identify any abnormalities in underground equipment, or any improper personnel behavior (violation of rules regarding commanding, operating procedures and workplace conduct), so that personnel can be alerted to any risks well in advance. This solution facilitates seamless collaboration between personnel and mobile equipment, and effortless monitoring of all operations. Underground AI-assisted HD video services require high uplink and downlink rates of 5G and the on-demand addition of edge computing.

Underground device information collection services mainly require the massive connectivity and high bandwidth features in 5G. They collect the information and operating status related to underground fixed equipment to sense the location, status and security of each piece of mobile equipment. This collected data is then sent to the data processing platform.

References

1. 3GPP TS 23.222: Common API Framework for 3GPP Northbound APIs
2. 3GPP TS 23.434: Service Enabler Architecture Layer for Verticals (SEAL); Functional architecture and information flows
3. 3GPP TR 23.700-99: Study on Network Slice Capability Exposure for Application Layer Enablement
4. 3GPP TS 23.435: Procedures for Network Slice Capability Exposure for Application Layer Enablement Service
5. 3GPP TS 28.558: Architecture for enabling Edge Applications
6. 3GPP TS 23.401: General Packet Radio Service (GPRS) enhancements for Evolved Universal Terrestrial Radio Access Network (E-UTRAN) access
7. 3GPP TS 23.501: System architecture for the 5G System (5GS)
8. 3GPP TS 28.538: Management and orchestration; Edge Computing Management
9. https://www.3gpp.org/common-api-framework-capif
10. 3GPP TR 22.890: Study on supporting railway smart station services
11. 3GPP TR 23.790: Study on application architecture for the Future Railway Mobile Communication System (FRMCS); Stage 2
12. 3GPP TR 23.796: Study on application architecture for the Future Railway Mobile Communication System (FRMCS) Phase 2
13. https://www.3gpp.org/ftp/tsg_sa/TSG_SA/TSGS_88E_Electronic/Docs/
14. https://www.antarcticglaciers.org/glacial-geology/uncrewed-aerial-vehicles-uavs/
15. https://www.5gamericas.org/wp-content/uploads/2019/11/5G-Services-Innovation-FINAL-1.pdf
16. 3GPP TS 23.255: Application layer support for Uncrewed Aerial System (UAS); Functional architecture and information flows
17. https://www.3gpp.org/ftp/tsg_sa/WG6_MissionCritical/TSGS6_041-e/inbox/Drafts/CC_Jan26_Closing_Call/S6-210318_Rev1_was156-FS_UASAPP-Solution-to-KI%237.doc
18. https://www.etsi.org/deliver/etsi_ts/122100_122199/122125/16.03.00_60/ts_122125v160300p.pdf
19. 3GPP TS 23.286: Application layer support for Vehicle-to-Everything (V2X) services; Functional architecture and information flows
20. 3GPP TS 23.545: Application layer support for Factories of the Future (FF)
21. 3GPP TS 22.262: Message service within the 5G System (5GS); Stage 1
22. 3GPP TS 23.554: Application architecture for MSGin5G Service; Stage 2
23. 3GPP TR 22.859: Study on Personal Internet of Things (PIoT) networks
24. 3GPP TS 22.261: Service requirements for the 5G system

25. 3GPP TR 23.700-78: Study on Application layer support for Personal IoT and Residential Networks
26. CCSA (2021) General technical requirement of 5G E2E network slice: YD/T 3973-2021
27. CCSA (2021) Technical requirement for 5G non-public network
28. CCSA (2021) Digital twin of proximity network part 3: general technical requirements
29. CCSA (2021) Technical requirements of 5G superior universal module (5G S-Module)-phase I: YD/T 3988-2021
30. CCSA (2021) Technical requirements for 5G customer premises equipment (CPE)
31. CCSA (2021) Study on key technologies of QoS monitoring for typical scenarios of vertical industries on 5G IoT terminal side (in Chinese)
32. CCSA (2021) Technical requirement for edge computing platform of 5GC network (in Chinese)
33. CCSA (2021) Technique requirement for service enabler architecture layer for verticals security
34. CCSA (2021) General requirements of network operator security capability exposure for 5G vertical industry (in Chinese)
35. CCSA (2021) Industrial Internet: 5G network deployment guide of industry park
36. CCSA (2021) Industrial Internet: campus network—5G application scenarios and technical requirements
37. CCSA (2021) Industrial safety supervision based on edge computing—application scenario and requirements: T/CCSA 333-2021
38. CCSA (2021) Technical requirements for rapid placement of hospital telemedicine systems based on 5G
39. CCSA (2021) Security requirements for electric power industry application via 5G slicing
40. CCSA (2021) Application scenarios and technical requirements of 5G+ industrial Internet for power grid
41. CCSA (2021) 5G enables remote driving: technical requirements for 5G communication system
42. CCSA (2021) Application scenarios and technical requirements of 5G+ industrial Internet for mine

Chapter 12
6G Vision, Performance Metrics, and Innovative ToB Enabling Technologies

Mobile communications has never ceased to evolve, and has undergone radical transformation from first generation (1G), featuring analog communication, to fifth generation (5G), realizing the connectivity of everything. Mobile communications has profoundly changed the way people live, becoming an engine for accelerating the digitalization and informatization of the social economy. 5G is already on the fast track to large-scale commercialization. It will provide high-speed, low-latency, and large-capacity connections for Internet of Things (IoT) and usher in a brand-new era of Internet of Everything (IoE), penetrating various industries (such as manufacturing, transportation, and agriculture) and stimulating innovation and development in these industries.

The sixth generation (6G) is expected to be commercialized around 2030, according to the industry trend that three generations are in progress simultaneously, which are in the commercialization, deployment, and research stages individually. To this end, the mobile communication industry will focus on envisioning, requirement analysis, key technology research, and proof of concepts from 2018 to 2025, and will aim to realize standards formulation, industrialization, and initial commercialization between 2026 and 2030. Countries around the world have competed to lay out and carry out relevant works intensively, as 6G has gradually become a new target of research.

Global 6G research is still in the phase of proposing the vision and requirements, and global research institutions are still plotting technical paths. Finland took the lead in launching a 6G flagship project in May 2018, while the EU released a 6G innovation plan that started at the beginning of 2020. The U.S. began their own research into new services using the terahertz (THz) spectrum in the range of 95 GHz to 3 THz in 2018. Meanwhile, China has launched its own research and development, and has drawn up blueprints for a number of technological fields, such as 6G network architectures, air interface technologies, and networking technologies. The IMT-2030 (6G) Promotion Group has been set up to study the 6G vision, requirements, key technologies, and the feasibility of developing a unified global standard. The Working Party 5D (WP 5D) of ITU Radiocommunication Sector (ITU-R) has planned to develop a research

P. Sun, *A Guidebook for 5GtoB and 6G Vision for Deep Convergence*,
Management for Professionals, https://doi.org/10.1007/978-981-99-4024-0_12

report on "IMT vision for 2030 and beyond" from 2021 to 2023, determine 6G spectrum requirements at the end of 2023, and complete 6G spectrum allocation by the end of 2027. However, no consensus has been reached yet on key technologies and network architectures, as 6G research is still in the exploration stage.

1G to 5G mobile communications have undergone varying degrees of capacity expansion and technological transformation in terms of business models, service objects, network architectures, and bearer resources. 6G mobile networks are expected to break through in mobile network architectures, boost their capabilities to support industry applications, and strengthen their ubiquitousness, intelligence, reliability, broadband, and eco-friendliness. Mobile networks in the future will enhance their key capabilities in unleashing industry applications and meet differentiated performance requirements in diverse application scenarios, frequency bands, and radio access technologies (RATs).

12.1 6G Vision

6G will satisfy higher-level demands generated with the rapid penetration of 5G applications, continuous breakthroughs in key technologies, and deep integration of big data and ICT. It will fully support digitalization on the basis of 5G and achieve ubiquitous intelligence and empower everything alongside technologies such as AI.

6G will derive a digital twin from the physical world from 2030. Humans and things in the physical world will be able to transmit information and intelligence via the digital twin. The digital twin will be used to accurately reflect and predict the state of the physical world, and maintain the physical world in advance. This gives people greater freedom, makes life better and more convenient, and increases the efficiency of manufacturing, and enhances the governance of society. In addition, the digital twin will merge with the real world in depth and push 6G to open up a brand-new era that intelligently connects humans, machines, and things.

6G will create new applications, such as holographic communication, digital twin application, smart interaction, communicate-sense integration, and super transportation, in production, life, and society. From a production perspective, 6G will better support ToB application and technology evolution, and deliver higher performance than 5G. For example, 6G-oriented holographic communication will bring an immersive holographic connection experience, which requires extremely high Tbit/s data transmission. As for our daily lives, digital twin technology will primarily face the challenge of delivering low E2E latency, which is expected to be less than 1 ms. To deliver high-precision and guaranteed services (for example, remote surgery, cloud power line communication (PLC), and smart transmission), data packet transmission needs to be switched from in-time transmission to on-time transmission. As for society, super transportation calls for super-high-speed real-time communication and high-precision positioning, while mixed reality (MR) requires high throughput, low latency, consistent experience anytime and anywhere, reliable connection, and a long battery life.

6G will revolutionize applications. IMT-2030 (6G) Promotion Group has proposed eight potential applications, including not only immersive cloud XR, holographic communication, sensory interconnection, and intelligent interaction, but also communication for sensing, inclusive intelligence, digital twins, and seamless global coverage, in *White Paper on 6G Vision and Candidate Technologies* [1]. Among these applications, communication for sensing can provide more service capabilities, such as imaging, environment reconstruction, and precise positioning; inclusive intelligence enables each device to act as an intelligent agent (also called an agent), allowing high-speed data transmission, collaboration, and learning between them. With these applications, 6G will signify the newest digital wave of innovation in industries, becoming the engine for productivity growth in 2030 and decades from then on, and more importantly will boost the development of the social economy based on vertical applications.

6G will realize full connectivity and integration. Full connectivity refers to the interconnection of not only humans but also machines and things, forming ultra-dense intelligent connections. Integration refers to the convergence of ground and satellite communications, realizing full-coverage information transmission over ground, sea, air, and space. Deep integration with cloud computing, AI, big data, and other technologies will help implement all-domain, all-spectrum, and universal intelligent wireless communication. The ultimate goal of 6G is to achieve the intelligent connectivity of everything, one where network signals will reach any rural area, opening up expert telemedicine care to patients and distance education to children. Not only that, but network signals will also cover the sky and sea, so that people will no longer be isolated from society. Full-coverage networks will help people predict weather and quickly respond to natural disasters as satellite positioning systems, telecom satellite systems, earth image satellite systems, and 6G terrestrial networks work as one [2].

12.2 6G Network Performance Metrics

6G networks will be more systematic and loaded with scenario-specific capabilities than 5G networks, particularly for ToB application scenarios. This requires 6G networks to improve key performance as much as possible and introduce new performance metrics for new businesses, such as high-precision positioning, deterministic QoS provisioning, AI, security guarantee, computing, and sensing.

6G networks are expected to fulfill the following performance requirements:

- Peak rate: at Tbit/s level
- User-perceived rate: 10–100 Gbit/s
- Spectral efficiency: 2–3 times higher than that for 5G
- Network energy efficiency: 10–100 times higher than that for 5G
- Mobility: greater than 1000 km/h
- Air interface latency: less than 0.1 ms
- Connection density: 10^8–10^9 connections per km^2 in 3D coverage scenarios

- Reliability: greater than 99.99999%
- Traffic density: 0.1–10 Gbit/s/m^2 in 3D coverage scenarios

6G networks will introduce new performance metrics related to ultra-high positioning accuracy (at submeter-level outdoors and centimeter-level indoors), ultra-low delay variation, sensing, AI, computing, security, and so on.

Not only do 6G networks need to fulfill these performance requirements, they also need to deliver high performance for full-domain 3D coverage, simplified network distribution, ubiquitous endogenous intelligence, native trustworthiness, and the autonomous operation of digital twins, to better cater for new application scenarios and satisfy refined and differentiated service demands at both individual and industrial levels [3, 4].

6G networks need to be more reliable, deterministic, and intelligent to complete diverse tasks in fundamentally changed communication scenarios, which might intertwine multiple communication modes, such as multipoint-to-multipoint, human-to-machine, and machine-to-machine. It is worth noting that 6G is neither an extension of 5G's three major application scenarios, technical specifications, or key technologies, nor a simple combination of 5G with satellite, AI, or THz communication. 6G has its own radically new features, breakthrough technical specifications, new application scenarios, and revolutionary key technologies, including communicate-sense-compute integration, semantic communication, and endogenous intelligence.

12.3 Innovative ToB Enabling Technologies

Key technologies for wireless communications of the next generation need to accommodate diversified scenarios, frequency bands, and RATs, enabling 6G networks to profoundly change the way people live, empower various industries, and achieve mutual development.

Key technologies oriented to 6G have attracted attention from both academia and industry, in terms of air interface, network architecture, networking, and security. These technologies involve ultra-large-scale antenna technology, THz communication, visible light communication, intelligent reflecting surface (IRS), integrated sensing and communication (ISAC), intelligent network, semantic communication, and satellite communication. The following sections will present three types of technologies that can be directly used in industries.

12.3.1 6G-oriented Semantic Communication Technologies

Future mobile communications needs to consider not only accurate data transmission but also semantic information transmission, motivated by various emerging intelligent services, such as smart vehicle-to-everything (V2X), smart healthcare, and holographic projection. Mobile communications will undergo fundamental changes, with a stronger emphasis on intelligent and effective control of digitalized information flows, to meet the growing demand for intelligent communication and personalized services in more vertical industry scenarios [5, 6].

Shannon, in 1948, abstracted the classical system model of information transmission in his book *A Mathematical Theory of Communication* and proposed the concept of information entropy to solve the theoretical problem of reliable point-to-point communication, laying the theoretical foundation of communication systems. However, Shannon's original information theory can only solve the problem of reliable information transmission. Later, Shannon and Weaver further pointed out that communications should tackle three levels of problems [7]:

- Level 1 (technical problem, also called syntactic problem): How accurately can the symbols of communication be transmitted?
- Level 2 (semantic problem): How precisely do the transmitted symbols convey the desired meaning?
- Level 3 (effectiveness problem): How effectively does the received meaning affect conduct in the desired way?

Shannon's classical information theory, a syntactic information theory, leaves aside the impacts of semantic and effectiveness problems on communication, focuses on reliable information transmission, and expects that minimal resources are used to implement reliable transmission of data bits at the maximum rate. However, a large number of bits transmitted in a mobile communication system does not mean a large amount of semantic information. "Semantic" is widely used in computer vision and AI fields, for example, semantic segmentation, semantic analysis, semantic understanding, and even semantic computing, with semantic representation being unable to be manipulated but possible in the future thanks to AI and computing technologies. Semantic communication has gradually become a major research subject in the mobile communication field, with the increasing intelligence level and cognitive ability of devices integrated with communication and computing [1, 7], and will be one of the potential key technologies in future communication networks.

In contrast to syntactic communication, semantic communication focuses on exchanging information rather than transferring data, and aims to accurately convey meanings between agents [8–12]. To streamline information exchange in the future, semantic communication performs joint coding based on source information characteristics and channel characteristics, accurately conveying meanings and boosting communication efficiency and user experience [8, 9]. Unlike syntactic communication, semantic communication is not concerned with the accurate transmission of data or communication symbols, but rather it looks to ensure that semantic information restored by the receiver matches that provided by the transmitter [5], improving the efficiency of collaboration between communication nodes by speeding up information exchange and understanding. As semantic communication loosens the error-free requirements of information transmission, it is expected to break through the transmission bottleneck of communication systems which are based on the classical information theory, fundamentally address the cross-system, cross-protocol, cross-network, human–machine compatibility and interconnection issues experienced by traditional data communication, to realize the transparent and intelligent connectivity of everything [11].

12.3.2 6G-oriented Enhanced Air Interface Technologies

Air interface technologies will be enhanced for vertical industries, breaking the capacity bottleneck that holds back innovation and rapid growth, and improving safety and reliability.

In terms of expanding capacity by improving spectral efficiency, large-scale antenna technology is a standout candidate. Actually, multi-antenna technology together with dense networking has been widely used to improve spectral efficiency from 3G to 5G [13, 14]. It has increased the transmit/receive antennas of a base station from 1T1R to 64T64R, the number of parallel data streams from 1 to 16, and the system spectral efficiency from 0.5 bit/s/Hz to 100 bit/s/Hz, in contrast to 2G. The design of future base station antennas must overcome the challenges of larger and heavier antennas, more antenna elements, more RF channels, and more complex feeding networks. To handle these challenges and improve spectral efficiency, cell-free cooperative massive multiple-input multiple-output (MIMO) is introduced to enable all access points to jointly serve all users using the same time–frequency resources [15, 16].

In terms of improving safety and reliability, visible light communication is an important candidate. It works on a network built on existing lighting nodes to avoid high-power RF signals from affecting the services of vertical industries, for better safety. It also prevents interference to/from existing microwave systems due to the huge gap between visible light and microwave frequency bands, for better reliability.

12.3.3 6G-oriented Enhanced Networking Technologies

6G will satisfy higher requirements on computing resources, latency, differentiated services, and network autonomy, with the continuous development of AI, IoT, cloud computing, and big data technologies as well as the transformation and upgrade to intelligent and digital networks.

To meet the demands of vertical industries for expanding computing resources and reducing latency, cloud-edge-network-device collaboration will integrate communication, sensing, and computing to connect devices and connect users and devices, which is becoming ever more important. This collaborative architecture will see computing and storage resources be distributed in clouds, edges, and devices, and communication resources be allocated by the network that connects clouds, edges, and devices. This architecture offers a complete set of network services and a convenient development framework for vertical industries, improving efficiency and reducing management costs.

To deliver customized solutions for services of 6G industry applications, flexible networking and configuration of 6G private networks are introduced, as traditional network capacity expansion methods cannot accommodate the ever-increasing number of differentiated services. More specifically, multiple customized, virtualized, dedicated, and isolated logical networks are constructed on a universal physical infrastructure platform, with different logical networks fulfilling different service needs.

To realize 6G network autonomy, digital twin technology provides new concepts [17]. A digital twin network is able to interact in real time with a physical network. This kind of system can analyze, diagnose, simulate, and control the physical network through the digital twin network, data, and models. Building digital twin networks can help 6G networks implement autonomy, including self-optimization, self-evolution, and self-growth, and even more interact and merge with digital twin networks through the endogenous intelligence plane and data plane to achieve high autonomy throughout their lifecycle. 6G networks can select proper architectures, build, orchestrate, and adjust various models, and generate, update, and implement digital twins and digital planners to achieve the required level of autonomy. In this process, digital twin networks strongly rely on high-performance capabilities related to intelligence and data.

12.4 Summary and Prospects

Mobile communications, from 1G to 5G, have undergone capability expansion and technological transformation in terms of business models, service objects, network architectures, and bearer resources. The next phase of evolution will see 6G break through in network architectures, enhance their ubiquitousness, intelligence, trustworthiness, broadband, and eco-friendliness, boosting key capabilities for industry applications. 6G is not simply an upgrade of 5G's three major application scenarios, technical specifications, or key technologies. It will have its own radically new features, revolutionary key technologies, breakthrough technical specifications, and new application scenarios. Moreover, 6G will evolve from connecting humans and things to connecting intelligence, and will promote the digitalization of vertical industries. 6G will act as a huge distributed neural network to integrate communication, sensing, and computing, and converge the physical, biological, and digital worlds. It will usher in a brand-new era of everything having connection, intelligence, and perception [18].

References

1. China Academy of Information and Communications Technology (CAICT), IMT-2030 (6G) Promotion Group (2021) White paper on 6G vision and candidate technologies
2. Liu GY, Huang YH, Li N et al (2020) Vision, requirements and network architecture of 6G mobile network beyond 2030. China Commun 17(9):92–104
3. You XH (2020) 5G application innovation and 6G technology evolution (in Chinese). Broadcasting Realm 6:9–11
4. Zhao YJ, Yu GH, Xu HQ (2019) 6G mobile communication networks: vision, challenges, and key technologies (in Chinese). Sci Sin Inform 49(8):963–987
5. Zhang P, Xu XD, Han SJ et al (2020) Entropy reduced mobile networks empowering industrial applications. J Beijing Univ Posts Telecommun 43(6):1–9
6. Zhang P, Niu K, Tian H et al (2019) Technology prospect of 6G mobile communications. J Commun 40(1):141–148
7. Weaver W (1953) Recent contributions to the mathematical theory of communication. ETC Rev Gen Semant (10):261–281
8. Zhang P, Xu WJ, Gao H et al (2022) Toward wisdom-evolutionary and primitive-concise 6G: a new paradigm of semantic communication networks. Engineering 8:60–73
9. Zhang YC, Zhang P, Wei JB et al (2021) Semantic communication for intelligent devices: architectures and a paradigm. Sci Sin Inform. https://doi.org/10.1360/SSI-2020-0379
10. Niu K, Dai JS, Zhang P et al (2021) 6G-oriented semantic communications (in Chinese). Mobile Commun 45(4):85–90
11. Shi GM, Li YY, Xie XM (2018) Semantic communications: outcome of the intelligence era. Pattern Recognit Artif Intell 31(1):91–99
12. Shi GM, Xiao Y, Li YY et al (2021) Semantic communication networking for the intelligence of everything. Chin J Internet of Things 5(2):26–36
13. Wei KJ, Zhao Y, Xu XY (2020) 6G vision and analysis of potential key technologies (in Chinese). Mobile Commun 44(6):17–21
14. You XH, Pan ZW, Gao XQ et al (2014) 5G mobile communication development trend and key technologies (in Chinese). Sci Sin 44(5):551–563
15. Wang DM (2021) 6G-oriented wireless transmission technology for cell-free massive MIMO (in Chinese). Mobile Commun 45(4):10–15
16. Tang WK, Dai JY, Chen MZ et al (2020) MIMO transmission through reconfigurable intelligent surface: system design, analysis, and implementation. IEEE J Sel Areas Commun 38(11):2683–2699
17. China Mobile Research Institute (2022) White paper on 6G wireless network autonomy based on digital twin networks (2022) (in Chinese)
18. Huawei Technologies Co., Ltd. (2022) 6G: the next horizon

Chapter 13
Semantic Communication Technologies

13.1 From Syntactic Communication to Semantic Communication

The flow of information can be divided into three stages: generation, transmission, and processing. For information transmission, Shannon abstracted a classical system model in 1948, with the following key theorems: lossless source coding, noisy channel coding, and lossy source coding. These theorems establish the theoretical foundations for source compression coding, channel coding, and lossy compression coding in mobile communication systems. Source coding is used for data compression to ensure that the content transmitted on the network is interpretable after decoding, while channel coding is used to reduce the error rate of the transmitted content, according to the information theory proposed by Shannon on the assumption of discrete memoryless lossy transmission for point-to-point communication. It is this basic theory that has seen the mobile communication industry evolve from 1 to 5G, which has experienced an upgrade around every 10 years.

The next generation aims to create a smart society where everything has connection, intelligence, and perception, thanks to the expansion of the global information infrastructure. In a smart society, all things will be connected and all data will be shared online; intelligence will spread all across the network, with everything being connected intelligently through big data and AI; perception will allow smart devices to sense everything in the physical world and transform it into digital information for sharing. Future communications will support multiple communication modes, such as point-to-multipoint, multipoint-to-multipoint, human-to-machine, and machine-to-machine, in contrast to traditional communications, which build on the Shannon capacity and focus on reliable transmission for human-to-human and point-to-point communication. To handle information of not only humans but also machines in a future smart society, information digitalization requires a more efficient information coding mechanism. If we continue to use the existing data coding method, the amount of data transmitted in communication will exceed its limit and the network will encounter unprecedented challenges. Therefore, 6G mobile communications

P. Sun, *A Guidebook for 5GtoB and 6G Vision for Deep Convergence*,
Management for Professionals, https://doi.org/10.1007/978-981-99-4024-0_13

needs to consider the essence of communication from the perspective of the information theory and make breakthroughs in the basic theory and transformations in technologies to enable intelligent interconnection and industry digitalization in the future.

Semantic communication was first proposed by information theory scholars Weaver and Shannon on the basis of the information theory. Semantic communication aims to convey meanings and its essence is to transmit information expressed by semantic symbols [1]. Traditional syntactic communication serves humans and transfers signals following the principle of ensuring waveform or bit fidelity, where it is humans who finally extract and interpret information in signals. In contrast, semantic communication, which is an emergent technology in the intelligent era, serves not only humans but also things and machines. It aims to establish a universal semantic knowledge base that can be understood by humans, things, and machines, and accurately convey meanings between intelligent nodes [2–6]. The basic theory of semantic communication, a communication means of higher organisms, is likely to surpass the classical information theory, implementing transmission with a higher degree of information compression.

Semantic representation will be possible thanks to AI and computing technologies. Beyond that, semantic sensing, recognition, analysis, understanding, and inference will represent key 6G network capabilities, with the rollout and development of new technologies and architectures, such as brain-computer interaction, brain-like computing, integrated sensing and communication (ISAC), and endogenous intelligence. This will motivate a paradigm shift for the network architecture from data communication to semantic communication [7]. Semantic communication intends to break through the limitations of classical information theory in research category, research level, and research dimension, and expand the field of information research from only syntactic information to semantic information. As an enabler of vertical industries, semantic communication has greater application potential than traditional communications and is considered a key candidate technology for 6G.

Semantic communication boasts a number of features. First, the transmitter can understand the purpose and environment of communication in advance, so that it can reduce transmission of redundant data. This helps relieve the wireless resource shortage faced by mobile communication systems [3]. Second, background knowledge in semantic communication helps improve the effectiveness and accuracy of a communication system in an undesirable communication environment with limited bandwidth, low signal-to-noise ratio (SNR), or high bit error rate [8]. Context information, individual communication purposes, and such background knowledge accumulated by communication parties help intelligently correct and restore intended information based on received incomplete information in a poor communication environment. Third, semantic communication can conceal information better [3]. Semantic communication is based on the context of the transmitter and receiver, in which case the same sentence may convey a significant meaning for those who understand the background but nothing for those who do not. Fourth, traditional communications assume that transmitted data has the same value for different receivers in a network, but this assumption is no longer applicable to new technologies, such as

IoT and cloud service. Therefore, the method of maximizing spectral efficiency by considering only radio resources, such as access time and spectrum, cannot meet the requirements of communication networks for automation, intelligence, and service diversification. By contrast, semantic communication considers the meanings of information and the requirements of users for information. It will fundamentally implement cross-system, cross-protocol, cross-network, human–machine communication, and finally achieve transparent and intelligent connectivity of everything [6].

13.2 Semantic Communication System Framework

Traditional syntactic communication based on Shannon's information theory ensures that information after being decoded by the receiver is strictly consistent with information before being coded by the transmitter, that is, to ensure error-free transmission at bit level. By contrast, semantic communication fully utilizes technologies such as AI to extract semantic information from the source, and performs joint encoding and decoding at the transmit and receive ends. Its purpose is not to ensure that the sequence after being decoded strictly matches the sequence before being encoded. Rather, it only ensures that the semantic information restored by the receiver matches that delivered by the transmitter [4].

As the error-free requirements of information transmission are loosened, semantic communication is expected to break through the transmission bottleneck of syntactic communication and provide a new solution for evolution towards 6G mobile networks [2]. To cater for potential scenarios, mobile networks are expected to connect humans and machines with varying degrees of intelligence (either natural or artificial), forming a closed loop of humans, machines, things, and intelligence. In this loop, various objects no longer merely transmit data bits between them, but also ensure the effective transmission of meanings, which cannot be overlooked anymore [7–9]. Mobile networks are designed for carrying out variable complex intelligent tasks, and are expected to use their intelligence together with semantic and on-purpose communication to realize accurate control and efficient communication between humans, machines, things, and intelligent agents.

Figure 13.1 illustrates the architecture of the 6G-oriented semantic communication system.

- At the transmit end, the information generated by the source is first sent to the semantic analysis and encoding module that is built on a semantic base (Seb), semantic features are extracted, compressed, and encoded there to form a semantic representation sequence according to the Seb and the background knowledge base at the transmit end. The semantic representation sequence is then sent to the channel encoder to generate an encoded sequence that can be transmitted on the radio channel. The Seb-based semantic analysis and encoding module selects a deep learning model with a specific structure based on the source redundancy

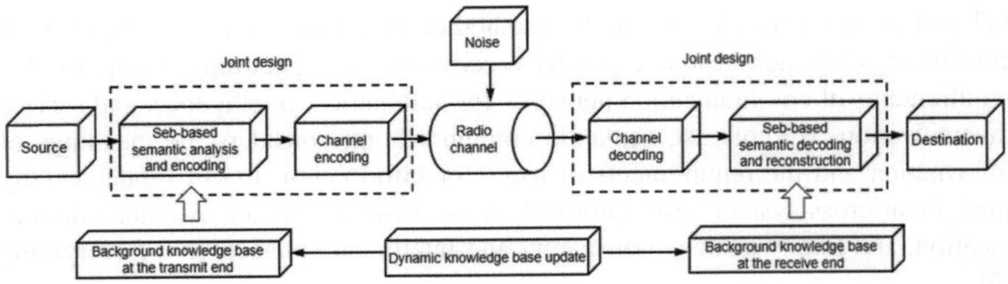

Fig. 13.1 The 6G-oriented semantic communication system architecture [2]

feature. For example, it uses a recurrent neural network (RNN) model for a text source, uses a convolutional neural network (CNN) model for an image source, or uses a graph convolutional network (GCN) model for a graph source. If the source is of multiple modes or heterogeneous ingredients, semantic synthesis needs to be performed during semantic extraction and encoding.

- At the receive end, the channel output signal is first sent to the channel decoder to generate a decoded sequence. This sequence is then processed by the Seb-based semantic decoding and reconstruction module to produce a semantic representation sequence according to the deep learning model and the background knowledge base at the receive end, and the restored information is finally sent to the destination.

In the semantic communication system, background knowledge bases are required for processes such as semantic analysis, encoding, decoding, and reconstruction, cloud knowledge bases are shared between the receive end and the transmit end, and scenario-specific prior information is fed to neural networks in a data-driven manner [2–4].

Future technological breakthroughs in semantic communication are expected to support multimodal data, which typically includes text data [10–13] and image data [1, 13] according to multimodal data research in academia. The following uses text data and image data as examples to briefly introduce recent advancements.

In terms of text data, its meaning is a major concern during transmission, whereas its structure (including auxiliary verbs, conjunctions, and punctuation marks) is related to a smooth and grammatical expression. Therefore, text data contains semantic redundancy as well as statistical redundancy. Semantic extraction and association modeling for text data can be performed using the Transformer or RNN model [10–13]. The feature vectors produced by the model based on context can be used by the decoder for text reconstruction. Figure 13.2 illustrates an example of semantic coding and transmission for text data. Semantic coding can well resist channel transmission errors according to reference [14], whereby the meaning of reconstructed text is semantically equivalent to the original meaning, though the form may differ from the original. Its anti-noise capability and its fidelity in consecutive segments are significantly better than those of traditional source coding and channel coding.

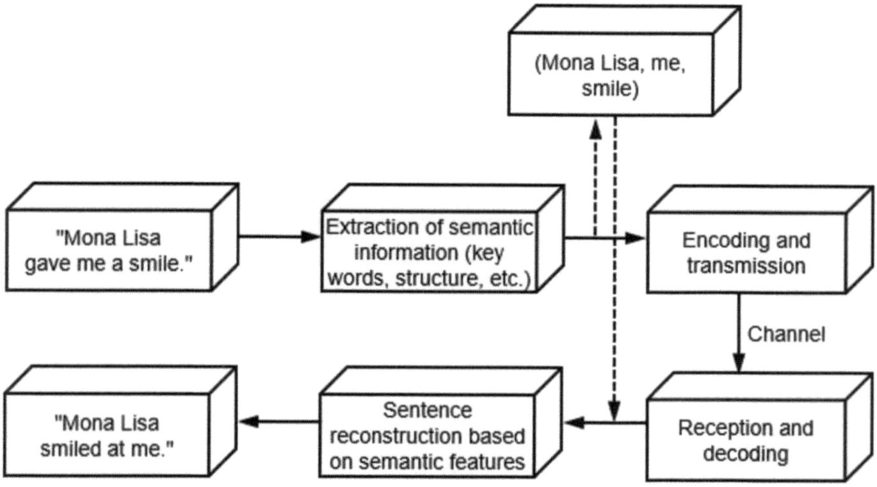

Fig. 13.2 Example of semantic coding and transmission for text data

In terms of image data, a CNN integrates source compression coding, error-control channel coding, signal modulation, and other relevant modules. Along with this, a semantic channel iterative receiver compatible with digital communication is used for data restoration. It boosts the capability of fault tolerance in mobile communication, and better balances efficiency and reliability. Figure 13.3 provides an example of semantic coding and transmission for image data. The semantic coding rate is only 1/5 of the H.264 coding rate, under almost the same learned perceptual image patch similarity (LPIPS), according to reference [2]. Therefore, semantic coding can significantly reduce transmission bandwidth overheads, thereby boosting spectral efficiency, compared with the classical H.264 + LDPC image coding scheme.

13.3 Application Prospect of Semantic Communication Technologies

Communication between agents is an urgent issue as agents such as robots, intelligent assistants, and intelligent terminals will be developed rapidly, distributed everywhere, and used in different scenarios where intelligent systems are under local or remote control. Communication between agents aims to accurately convey the meaning of symbols transmitted for a specific task, whereas traditional syntactic communication focuses on error-free transmission at bit level. Semantic communication is expected to become the most effective communication mode for agents in the future because it can help agents understand context more accurately when performing interactive services, which are the main type of services between agents. Moreover, semantic inference, interpretation, and other technologies can assist communication

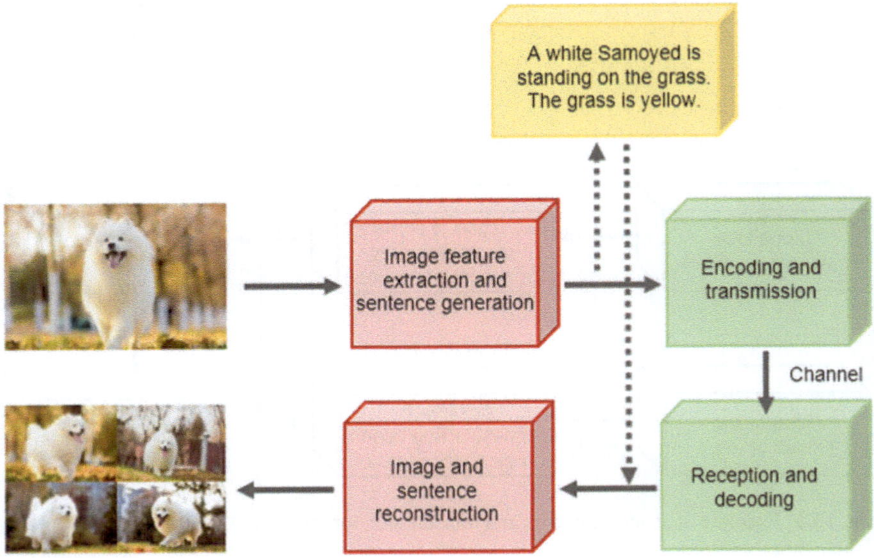

Fig. 13.3 Example of semantic coding and transmission for image data

between agents and humans. Semantic communication will stimulate the generation of language between agents as they shift over from signal exchange to intent exchange. This language will enable agents to understand each other and cooperate efficiently, and will help solve more complex, open, time-varying, and robust communication issues [3].

Semantic communication will also benefit intelligent human–machine communication, in addition to communication between agents. Human–machine intelligent connection will have tremendous complex applications and services, with the emergence of applications such as the Tactile Internet and human–machine interaction network [6]. It focuses on a variety of sensors as well as virtual and real interaction devices, and relies heavily on AI-based sensing, learning, and adaptation to identify and perceive users' emotions, personalities, and others. However, such human–machine collaboration and improved performance are hindered by the current human–machine interaction mode oriented to signal transmission. To address this issue, AI-based semantic communication can be used. It can overcome difficulties in intent understanding by machines and natural interaction between humans and machines. Its goal is to provide natural and smooth interaction for on-purpose communication and collaboration between communication parties based on the knowledge bases accumulated through continuous learning. To this end, it standardizes the semantic representation framework, self-learning knowledge bases, and semantic encoding and decoding modes. Human–machine interaction based on semantic communication is a natural, human-like mode, enabling the direct understanding of information. It will build a bridge between humans and machines, greatly improving the efficiency of human–machine collaboration [1].

13.4 On-Purpose Networks Based on Semantic Communication

13.4.1 Introduction to On-Purpose Networks

On the basis of Shannon's information theory, Weaver proposed in 1953 that communications should not only focus on the syntactic problem but also on the semantic problem and effectiveness problem [1]. The syntactic problem is a major concern of Shannon's information theory, limited by the special needs of communication engineering at that time. However, the semantic problem has become more widespread in recent years. The semantic problem focuses on how the transmitted symbols precisely convey the desired meaning, to address the issue of transferring information under reliable machine inference. Beyond that, the effectiveness problem focuses on intent or purpose, that is, how the received meaning effectively affects conduct in the desired way [6]. On-purpose communication will be realized as a result of developing technologies such as mobile communication technologies, AI, and natural language processing, and with the continuous enhancement of intelligence and cognitive capabilities of machines. Smart machines can interact with each other, understand each other's intent, and communication networks can be built flexibly and quickly. This lays a foundation for on-purpose networks in the future [6].

On-purpose networks will gradually enter the communication research field and will drive the rapid growth of a new industry ecosystem, including intelligent vehicle networks, telemedicine, and smart manufacturing. Smart-machine-oriented networks in the ecosystem will become a new engine for the development of future mobile communications and will drive new growth of future mobile communications. The global UAV market reached USD80.9 billion in 2020 [15], while the vehicle-to-everything (V2X) market will reach USD162.9 billion by 2022 [16]. The education robot market will reach USD84.1 billion by 2023 [14] and the Industry 4.0 market will reach USD214 billion by 2023 [17]. About 125 billion smart machines will appear on communication networks by 2030 [18]. However, current on-purpose networks driven by user intents [19–26] cannot support smart, immersive, and all-domain service applications of future mobile communications [27–29]. Machines in such networks are short of native intelligence and unable to cope with high-speed changes of network topologies and environments. This means that intent decomposition, adaptation, intelligent decision-making, and interaction between machines are all accomplished by users and O&M personnel. To improve the native intelligence of machines and realize on-purpose communication, on-purpose networks driven by machine intents need to be constructed, which will be a key engine for the digitalization of network infrastructure.

In future on-purpose networks, machines will be equipped with native intelligence, decision-making capability, and the ability to think, freeing up O&M personnel and users so that they can focus on exploratory and creative tasks. For example, in a V2X network driven by machine intents, smart machines such as roadside base stations, sensors, vehicles, and mobile edge computing servers can autonomously perceive the entire road status, interact on demand, and make decisions. When a vehicle is about to turn right, it autonomously interacts with the right nodes based on its intent to acquire the road status in the blind spot. By utilizing the perceived real-time information and the road prediction information released by roadside base stations, the vehicle determines the location of the optimal communication target promptly through further intent analysis and model processing. Then, the vehicle transmits the information to the communication target on the right side in a directional manner. It can be seen that the existing network nodes can become more intelligent based on machine intents and therefore can find the optimized transmission solution based on available information. In short, on-purpose networks can automatically perceive the network situation, form the overall network objective for a specific scenario based on historical experience and prior information, and implement self-configuration and self-organization through interaction between network elements.

13.4.2 Architecture of On-Purpose Networks

The future on-purpose network architecture is similar to the current cloud-edge-device structure [30].

The cloud layer can be composed of clustered servers with powerful computing and storage capabilities. It plays a central role in overall planning and control and provides a powerful guarantee for self-optimization, self-configuration, and self-healing to achieve optimal system performance. Cloud servers can aggregate and process various types of information from edge servers in different areas. For example, they can acquire the running status and topology structures of smart machines within the coverage of edge servers in different areas, and reconstruct communication links and networks based on scenarios and requirements.

The edge layer mainly consists of base stations and multi-access edge computing (MEC) servers, acting as a link between the cloud layer and the device layer. It can draw a unified plan for smart machines within its coverage, accelerate the synchronization between them, and reduce communication overheads based on prior information and demands. It can also perform unified convergence and computing for the intents of smart machines, realizing autonomy within its coverage. Moreover, the edge layer can use digital twin technology to build the digital twins of smart machines in specific areas, mapping massive smart machines from the physical space to the digital space, and manage the whole lifecycles of smart machines.

The device layer (also called the terminal layer) comprises a large number of various smart machines, which can perceive and obtain multimodal data in surrounding environments, and interact with other machines in real time by using ISAC technology. Then, they can analyze and interpret perceived information based on prior information and understand the custom intents of specific users. According to the intents, smart machines can extract customized features from perceived information to obtain customized semantic information. Furthermore, they can collaboratively converge and process perceived information through ubiquitous computing or distributed computing, compensating for insufficient computing capability. This way, the terminal layer realizes autonomous perception, cognition, and decision-making.

13.4.3 Key Technologies for On-Purpose Networks

Innovative technologies, such as AI and semantic communication, stimulate the emergence of on-purpose networks. Essential technologies, including instant sensing by agents and the proper allocation of computing and communication resources, enable on-purpose networks to achieve the goal of on-demand and on-purpose communication. Beyond that, digital twin technology provides real-time monitoring, computing, and controlling for on-purpose networks to manage and control the whole lifecycles of agents and facilitate decision making. This section will present the key technologies for on-purpose networks, including computing resource sharing, digital twin technology at the edge layer, sensing-assisted on-demand fast networking, and intelligent management and control.

13.4.3.1 Computing Resource Sharing for Reliable Prior Information Provisioning

Computing resource sharing is based on distributed computing technology in the cloud-edge-device architecture [27]. In this mode, smart machines can optimally use multi-layer computing resources anytime and anywhere based on intelligent and automatic scheduling [27]. Particularly in an on-purpose network, smart machines can share computing resources among themselves and use the computing resources of servers, including cloud servers, as required to complete tasks such as customized semantic feature extraction, intent analysis, interpretation, and decision making. This way, smart machines can utilize a massive amount of perceived information, predict the tracks of communication targets in real time through online training, provide prior information for on-purpose communication, and implement fast communication and self-decision-making. Therefore, how to coordinate heterogeneous computing resources, interconnect multi-layer computing nodes, and optimize the scheduling of computing resources is a major challenge to on-purpose communication.

13.4.3.2 Edge Digital Twin for Real-Time Digital Image Building

Digital twin technology can build digital models for existing or future physical entities. It can perceive, diagnose, and predict the status of physical entities and further control their behavior in real time through actual testing, simulation, and data analysis [31]. It enables agents to iteratively update digital models through transfer learning between models, and further guide, adjust, and improve decision-making within the lifecycle of physical entities [31]. Digital twin technology has become an important tool for promoting the growth of the digital economy. It has established a universally adaptable theoretical and technological system, and found wide applications in the aeronautic industry, astronautic industry, health care, urban management, construction industry, electric power industry, and so on [32].

Digital twin technology can be used at the edge layer of an on-purpose network to build real-time images of physical entities within the coverage, forming a virtual twin network, and to analyze situation changes based on images and promptly make decisions, by utilizing the computing power of edge servers. Specifically, servers at the edge layer can construct a network topology that closely couples the digital space and the physical space, based on perceived information uploaded by the terminal layer. This process depends on technologies such as surveying, mapping, scanning, geometric modeling, network modeling, and IoT to represent physical objects as digital models that can be recognized by computers and networks [32]. In this process, smart machines at the terminal layer can optimize their behavior based on the network topology of the virtual world. Based on a large amount of information provided by the terminal layer, the regional control center at the edge layer can use digital twin technology to effectively analyze, diagnose, simulate, and control physical edge networks, and to further manage the specific content of smart machines and edge networks, enabling real-time interplay between the physical space and the digital space. Moreover, edge servers can continuously analyze and predict the intents of smart machines through the virtual twin network and massive information provided by the edge layer, and finally form an on-purpose network and implement on-purpose communication.

Using digital twin technology to build a virtual world for an on-purpose network requires massive data, including the location, speed, energy consumption, storage capability, and computing capability of agents at the terminal layer, the temperature, humidity, and illumination of the surrounding environment, as well as the network node density, road information, channel status, user behavior characteristics, and more. Therefore, how to collect multi-dimensional data on a large scale and perform accurate and efficient analysis and modeling on the obtained data are the two key issues to be solved regarding the application of digital twin technology to on-purpose networks.

13.4.3.3 Sensing-Assisted On-Demand Fast Networking

Precise sensing of node mobility and network topology changes in an on-purpose network is critical to building a virtual twin network as well as adjusting and optimizing decisions based on the virtual twin network for on-purpose communication. For example, in a smart V2X network, smart machines can quickly discover available neighbor nodes in a very small number of timeslots based on perceived information, and can therefore provide accurate network topology information for quick networking and decision-making when the smart machines intend to communicate. Not only that, network-event-driven on-demand sensing can be used to deliver greater flexibility and intelligence to intent- and meaning-driven networks. This sensing policy enables smart machines to predict communication status based on the network environment and their communication intents, and at the data link layer, allocate timeslots and frequency domain resources to communication links as required in advance for fast network access.

Rich perceived information is a strong guarantee for on-purpose communication. In an on-purpose network, smart machines can perceive a lot of prior information about environments using their sensors such as mmWave radars, laser radars, and cameras, and perform on-demand sensing based on their intents to reduce data collection and significantly lower information processing overheads. Moreover, smart machines can effectively converge heterogeneous perceived information. For example, smart vehicles in a V2X network can comprehensively process information captured by cameras and mmWave radars, to accurately acquire the position, speed, and shape of a communication target and precisely locate and recognize the target. In short, perceived information enables the network to achieve visualized communication, enhance on-demand autonomy, and implement efficient and accurate on-demand networking.

13.4.3.4 Intelligent Management and Control

Intelligent management and control are necessary for on-purpose network construction promoted by the native intelligence of machines. Intelligent management and control enable computing and intelligent modules to fully play their roles in on-purpose networks, and will be more and more important with the evolution of networks. In terms of network structure, future on-purpose networks will become intelligent and autonomous with the ever increasing environment sensing and decision-making capabilities of nodes, and the coupling between terminals and humans in the network environment will be weakened [27]. In terms of information transmission, on-purpose networks require a large amount of interactive information for instant and efficient data convergence, self-decision-making, and intelligent transmission, and therefore have high demands for both uplink and downlink bandwidths. In terms of network evaluation, on-purpose networks will further focus on ultra-reliable low-latency communication (URLLC), massive access, strong indoor coverage, and flexible scalability while improving spectral efficiency and peak traffic volume. Therefore, in on-purpose networks, agents will exploit the distributed network architecture to collect, clean, and process data, so that they can build models, analyze network data, and sense, predict, manage, and control network status in real time. Intelligent nodes at the edge layer will use digital twin technology to monitor, model, and deduce the network environment, implementing intelligent network management, control, autonomy, and optimization.

13.5 Summary and Prospects

Traditional syntactic communication targets accurate data transmission, which cannot meet the requirements of intelligent and collaborative communication between humans and machines, between machines, and between intelligent nodes in the future. Semantic communication aims to accurately convey meanings between communication parties, significantly improving communication efficiency, especially for communication between smart devices. Semantic communication will give rise to new development opportunities for mobile communications based on Shannon's information theory in the intelligent era with a view to evolving to 6G. As an efficient communication manner, it will transform the interaction of intelligent nodes into on-purpose communication in the future. Moreover, it will promote joint innovation in various industries, exploration of new opportunities, models, and businesses, and construction of next-generation digital information communication network infrastructure, production infrastructure, and social infrastructure to help industries grow.

References

1. Rapoport A (1954) Recent contributions to the mathematical theory of communication (vol 10, pg 261, 1953). Etc Rev Gen Semant 11(3):240
2. Zhang YC, Zhang P, Wei JB et al (2021) Semantic communication for intelligent devices: architectures and a paradigm. Sci Sin Inform. https://doi.org/10.1360/SSI-2020-0379
3. Shi GM, Xiao Y, Li YY et al (2021) Semantic communication networking for the intelligence of everything. Chin J Internet of Things 5(2):26–36
4. Zhang P, Niu K, Tian H et al (2019) Technology prospect of 6G mobile communications. J Commun 40(1):141–148
5. Niu K, Dai JS, Zhang P et al (2021) 6G-oriented semantic communications (in Chinese). Mobile Commun 45(4):85–90
6. Zhang P, Xu WJ, Gao H et al (2022) Toward wisdom-evolutionary and primitive-concise 6G: a new paradigm of semantic communication networks. Engineering 8:60–73
7. Shi GM, Xiao Y, Li YY et al (2021) From semantic communication to semantic-aware networking: model, architecture, and open problems. IEEE Commun Mag 59(8):44–50
8. Kountouris M, Pappas N (2021) Semantics-empowered communication for networked intelligent systems. IEEE Commun Mag 59(6):96–102
9. Calvanese Strinati E, Barbarossa S (2021) 6G networks: beyond Shannon towards semantic and goal-oriented communications. Comput Netw 190:107930
10. Xie HQ, Qin ZJ (2021) A lite distributed semantic communication system for Internet of Things. IEEE J Sel Areas Commun 39(1):142–153
11. Lu K, Li RP, Chen XF et al (2021) Reinforcement learning-powered semantic communication via semantic similarity. https://doi.org/10.48550/arXiv.2108.12121
12. Lu K, Zhou QY, Li RP et al (2021) Rethinking modern communication from semantic coding to semantic communication. IEEE. https://doi.org/10.1109/MWC.013.2100642
13. Xie H, Qin Z, Li GY (2021) Task-oriented semantic communications for multimodal data. https://doi.org/10.48550/arXiv.2108.07357
14. Smart Learning Institute of Beijing Normal University (2019) White paper on global education robot development in 2019 (in Chinese).

15. QYResearch (2021) Global and China AI manned UAV market status and development trend in 2021–2027 (in Chinese)
16. Huajing Industry Research Institute (2020) Report on the supply, demand, and prospect of China's V2X Industry in 2021–2026 (in Chinese)
17. BCC Publishing (2018) Global markets for 5G technologies
18. Sharma SK, Wang XB (2020) Toward massive machine type communications in ultra-dense cellular IoT networks: current issues and machine learning-assisted solutions. IEEE Commun Surv Tutor 22(1):426–471
19. Zhou YC, Yan S, Peng MG (2020) Intent-driven 6G radio access network. Chin J Internet of Things 4(1):72–79
20. Liu TZ, Mou XQ, Chen XZ (2020) Research on the architecture of internet of things based on intention driven. Mod Inf Technol 4(10):176–179, 182.
21. Zhang JM, Yang CG, Pang L et al (2019) Intent driven Internet of things. Chin J Internet of Things 3(3):5–10
22. Wang JY, Zhou C, Zhang L et al (2021) Knowledge-defined intent-based network autonomy. Telecommun Sci 37(9):1–13
23. Mehmood K, Kralevska K, Palma D (2021) Intent-driven autonomous network and service management in future networks: a structured literature review. https://doi.org/10.48550/arXiv.2108.04560
24. Yang H, Zhan KX, Bao BW et al (2021) Automatic guarantee scheme for intent-driven network slicing and reconfiguration. J Netw Comput Appl 190:103163
25. Gomes PH, Buhrgard M, Harmatos J et al (2021) Intent-driven closed loops for autonomous networks. J ICT Stand:257–290
26. Banerjee A, Mwanje SS, Carle G (2021) Contradiction management in Intent-driven cognitive autonomous RAN. IEEE. https://doi.org/10.23919/IFIPNetworking55013.2022.9829768
27. China Academy of Information and Communications Technology (CAICT), IMT-2030 (6G) Promotion Group (2021) White paper on 6G vision and candidate technologies (in Chinese)
28. Ouyang Y, Wang LL, Yang AD et al (2021) The next decade of telecommunications artificial intelligence. https://doi.org/10.48550/arXiv.2101.09163
29. Careglio D, Spadaro S, Cabellos A et al (2021) Results and achievements of the ALLIANCE project: new network solutions for 5G and beyond. Appl Sci 11(19):9130
30. Yan GQ, Mao SJ (2020) Cloud-edge-end simulation system architecture. In: 2020 2nd International Conference on Advances in Computer Technology, Information Science and Communications (CTISC), 2020.
31. CAICT, IMT-2030 (6G) Promotion Group (2021) White paper on 6G network architecture vision and key technology outlook (in Chinese)
32. China Electronics Standardization Institute (2020) White paper on the digital twin (in Chinese)

Chapter 14
Cell-Free Cooperative Massive MIMO

Cell-free cooperative massive multiple-input multiple-output (MIMO) adopts a scalable cell-free radio access network (CF-RAN) architecture, which is completely different from the traditional cellular network architecture.

Cell-free cooperative massive MIMO for 6G is classified into four types, namely, basic, reconfigurable intelligent surface (RIS)-assisted, network-assisted full duplex (NAFD), and mmWave cell-free cooperative massive MIMO, as illustrated in Fig. 14.1a–d and described in Sects. 14.1, 14.2, 14.3, and 14.4, respectively.

Cell-free cooperative massive MIMO addresses scalability and complexity issues by improving spectral efficiency in low frequency bands. For example, it can address the issues by adding expandable spatial units, moving computing capabilities from access points (APs) to these units, and using a central processing unit (CPU) to manage and coordinate these units.

Cell-free cooperative massive MIMO can overcome the challenges related to large bandwidth, near-field optics, and easy blocking by studying the traffic volume in high frequency bands. In addition, it can handle coverage challenges in complex environments, which cannot be effectively solved in traditional cellular systems, but can be solved through the RIS technology.

14.1 Basic Cell-Free Cooperative Massive MIMO

Basic cell-free cooperative massive MIMO is a general term covering virtual MIMO, network-assisted MIMO, distributed MIMO, (coherent) coordinated multipoint transmission, and distributed antenna system (DAS). It utilizes geographically dispersed base station antennas to jointly provide unified high-quality services for all user equipment (UEs) in the network based on advanced backhaul technology.

P. Sun, *A Guidebook for 5GtoB and 6G Vision for Deep Convergence*,
Management for Professionals, https://doi.org/10.1007/978-981-99-4024-0_14

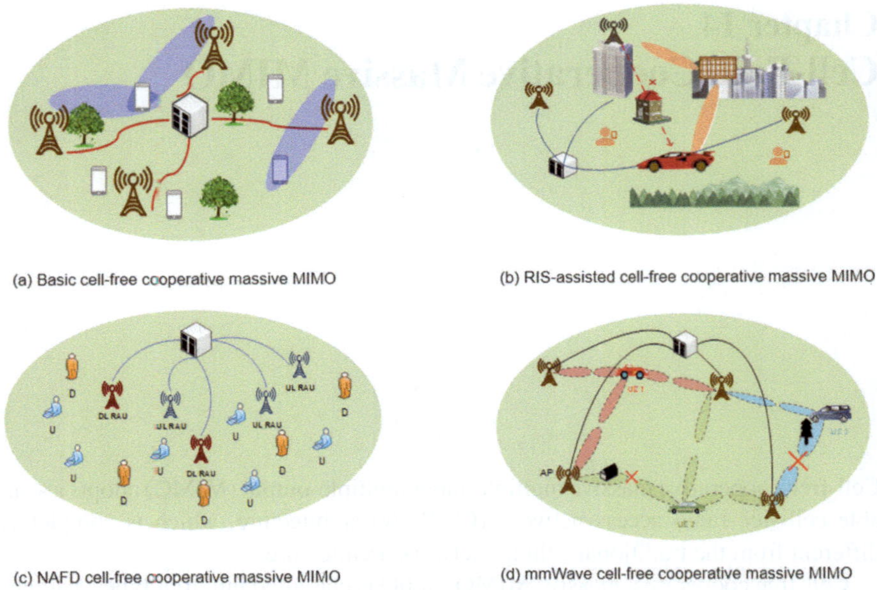

(a) Basic cell-free cooperative massive MIMO

(b) RIS-assisted cell-free cooperative massive MIMO

(c) NAFD cell-free cooperative massive MIMO

(d) mmWave cell-free cooperative massive MIMO

Fig. 14.1 Cell-free cooperative massive MIMO for 6G

In a basic cell-free cooperative massive MIMO system, as illustrated in Fig. 14.1a, a lot of APs are distributed in a large area and allocated the same time–frequency resources to serve many UEs. As this basic system integrates distributed MIMO and massive MIMO, it has the advantages of both of them. It provides a higher coverage probability by deploying APs closer to UEs. It performs most processing on APs by employing conjugate beamforming and matched filtering, which are commonly used in uplink and downlink transmission and can be conducted in a distributed manner due to low calculation complexity.

In this basic system, however, information exchange between APs and the CPU is limited by payload, power control coefficients are updated slowly, and channel state information (CSI) is not shared promptly between APs or at the CPU for downlink data precoding and uplink data detection.

This basic system is prominent in its operation mechanism, adopting low-complexity signal processing techniques such as conjugate beamforming to enable a large number of single-antenna APs to serve a large number of UEs simultaneously. Therefore, it can exploit favorable propagation (FP) and channel hardening that are widely studied, globally optimal power control algorithms with high computing efficiency, and simple pilot allocation solutions. In short, it can be considered an effective and scalable form of network-assisted MIMO and DAS, just as cellular massive MIMO is considered an effective and scalable form of conventional multi-user MIMO.

14.2 RIS-Assisted Cell-Free Cooperative Massive MIMO

In actual communication environments (for example, in urban areas), buildings, vehicles, and vegetation cause inevitable blocking, fading, scattering, and many coverage holes. Therefore, delivering wide area coverage with low cost and low power consumption is one of the fundamental capabilities of a communication system. RIS is foreseen to be a promising intelligent radio technology for future 6G communications because it can enhance coverage and expand capacity while saving energy.

RIS, in terms of physical structure, is made from artificial sub-wavelength two-dimensional material and usually composed of metal, medium, and adjustable components. Its application in the mobile communication field benefits from the development of new electromagnetic material science. Typically, it can intelligently reconstruct the radio propagation environment between transmitters and receivers. Its core features include quasi-passive characteristic, filled aperture, software programmability, wide frequency response, and low thermal noise [1].

RIS can be deeply integrated with cell-free cooperative massive MIMO, thanks to its strengths in low cost, low power consumption, and easy deployment. Figure 14.1b illustrates an RIS-assisted system, where an RIS and distributed base stations are deployed in a large area to jointly serve all UEs, implementing user-centric transmission design, effectively mitigating interference through efficient cooperation between base stations, and expanding network capacity. The RIS is controlled by certain base stations through the corresponding smart controller, all base stations are connected to the CPU through optical cables or wireless backhaul, and the CPU is responsible for control and planning. The RIS can reconstruct the phases and amplitudes of spatial signals and add multipath channels to raise the multiplexing gain and further improve system performance and quality of service (QoS).

RIS-assisted cell-free cooperative massive MIMO mainly solves the problem of communication blocking in coverage holes, for example, in shadow areas of high-rise buildings. In this case, RISs can be properly deployed between base stations and coverage holes to establish effective connections between them and provide coverage for UEs in these coverage holes. Similarly, in indoor scenarios, where a large number of communication services occur, received signals are poor as indoor walls and furniture cause blocking, refraction, and scattering. In this case, RISs can be deployed on the glass surface of a building to effectively receive signals from base stations and transfer the signals indoors, enhancing indoor communication quality. In short, RIS technology helps effectively handle coverage holes in actual communication scenarios and enhances the performance of communication systems.

14.3 NAFD Cell-Free Cooperative Massive MIMO

Duplex is another major concern of mobile communication standards. 5G adopts flexible duplex, while 6G may use co-frequency co-time full duplex (CCFD) as this duplex mode matures over time. Despite their different applications, flexible duplex and CCFD inevitably experience cross-link interference (CLI) during networking.

NAFD is a cell-free duplex mode [2] that can be used with cell-free cooperative massive MIMO to reduce CLI, which may exist in flexible duplex, full duplex, or hybrid duplex mode. Figure 14.1c illustrates NAFD cell-free cooperative massive MIMO, where multiple remote antenna units (RAUs) with a large number of antennas are densely distributed in an area and connected to a CPU responsible for baseband processing. Each RAU may perform uplink reception or downlink transmission and different RAUs may have varying choices in each timeslot. This way, uplink reception and downlink transmission can be performed in the same time–frequency resource block. In this process, no self-interference is produced inside an RAU in the NAFD system, compared with a conventional full-duplex system, lowering transceiver complexity, reducing system latency, and improving energy efficiency and spectral efficiency.

NAFD, together with cell-free cooperative massive MIMO, allows uplink and downlink resources to be efficiently and flexibly utilized by eliminating interference between uplink and downlink and supporting dynamic TDD, flexible FDD, and full duplex. When CCFD RAUs are used, each can be regarded as two RAUs, one for uplink reception and the other for downlink transmission. Interference can be eliminated in the digital field. One reason for this is that reference signals can be used to estimate downlink and uplink channels between RAUs at very low overheads due to the quasi-static characteristic of links. The other reason is that the CPU can acquire precoded downlink signals of all UEs in advance, as uplink and downlink baseband signals are all processed on the CPU in cell-free cooperative massive MIMO. Speaking of UEs, existing half-duplex hardware, rather than CCFD transceivers, is used to implement in-band full duplex in cell-free cooperative massive MIMO because of its lower complexity. As each UE works in flexible duplex mode, interference may occur between uplink and downlink but can be mitigated through joint scheduling on the CPU or interference cancellation on UEs.

NAFD has different characteristics and advantages from traditional duplex modes. First, NAFD can provide services with lower latency than traditional TDD, and can support asymmetric services without reducing spectral efficiency in contrast to traditional FDD. Second, NAFD allows RAUs to work in half duplex or CCFD mode and use joint processing to reduce CLI, compared with flexible duplex used in 5G. Third, NAFD is compatible with flexible TDD used in 5G, and is able to reduce CLI in scenarios where some RAUs send data while others receive data if all these RAUs work in half duplex mode but have different timeslot structures. Theoretically, the relationship between NAFD and CCFD is similar to that between distributed MIMO and centralized MIMO. NAFD can achieve better performance than CCFD due to a higher RAU density, just as distributed MIMO can obtain the additional power gain and macro diversity gain in contrast to centralized MIMO.

14.4 mmWave Cell-Free Cooperative Massive MIMO

Sub-6 GHz spectrum is widely used in mobile communications after undergoing substantial development. Its strengths are in long propagation distance, wide coverage, and strong signal penetration, but in crowded hotspot areas, it cannot effectively cope with network congestion, to provide higher transmission rates, or meet diverse QoS requirements. Against this backdrop, mmWave has gradually become a key technology for next-generation mobile communications.

mmWave cell-free cooperative massive MIMO [3], as illustrated in Fig. 14.1d, shows great potential. First, it leverages the short wavelength of mmWave. It adopts a small-sized antenna panel with a huge number of antennas rather than using large-sized co-sited antennas in order to lower hardware complexity, thanks to advances in complementary metal–oxide–semiconductor (CMOS) device miniaturization technologies. Second, it overcomes the disadvantage that mmWave is not suitable for long-distance propagation due to easy blocking and large channel fading. It selects the nearest APs among densely deployed APs to serve UEs, which in turn reduces latency and improves system robustness while boosting spectral efficiency. All of these highlight the importance of mmWave cell-free cooperative massive MIMO for 6G in delivering low latency and high peak rates, spectral efficiency, and reliability.

mmWave cell-free communication faces more challenges than traditional sub-6 GHz communication. The most typical challenge is that it calls for a different propagation mechanism, with large antenna arrays used on both sides of base stations to offset path losses [4–7]. Another challenge is that the complexity of calculation for beam management continuously increases, as more and more antennas are used and beams are expected to be more concentrated in certain directions. AI is expected to improve beam management performance, stimulated by the successful application of AI technologies in the communications field. This means that learning how to reasonably use AI and build models has become a challenge.

A promising solution is to use different machine learning techniques to learn about the surrounding environment. Machine-learning-based beam management can be classified into two categories [8]: CSI-based and CSI-independent. CSI-based intelligent beam management is widely applied to beam selection and tracking [9–11]. Its main algorithms include deep learning represented by classifiers and multi-armed bandit (MAB). The former formulates the mmWave beam selection problem as a multiclass classification problem to learn optimal beam pairs. The latter formulates the beam management problem as an environment-sensing contextual MAB problem and solves the problem by learning historical information. The preceding algorithms present fast solving speeds and show the nearly optimal performance. Another algorithm finds a beam selection policy while considering other factors. For example, it

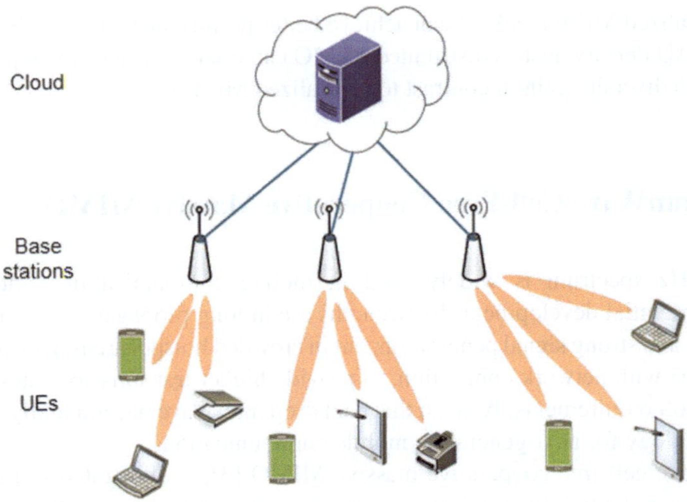

Fig. 14.2 mmWave cloud-based multi-base-station multi-UE communication scenario

uses a distributed algorithm named multi-agent reinforcement learning (MARL) to find a beam selection policy while considering UE scheduling and is very effective in the mmWave cloud-based multi-base-station multi-UE communication scenario illustrated in Fig. 14.2. Such policies can provide stable and good communication services with low system overheads when channels between two communication parties change rapidly [12].

CSI-independent intelligent beam management implicitly acquires channel information by using an intelligent algorithm and searches for optimal beams by using a tool such as a neural network. Its main algorithms are classified into three types [10], which are based on direct training, numerical search, and positioning. These technologies are prominent in avoiding system overheads for CSI acquisition.

- Beam management based on direct training uses deep reinforcement learning (DRL) to model action spaces as adjacent areas of beams indicated in codebooks, so that it can search for beams within a relatively small range. Moreover, it extracts environmental characteristics from feedback information, such as received signal strength (RSS) obtained through beam search, so that it can find a beam management policy with good performance.
- Beam management based on numerical search generally uses a randomization method to obtain a beam pair and then uses a numerical search algorithm such as Rosenbrock or Tabu for adjustment to approach the optimal beam pair.
- Beam management based on positioning formulates the optimal mmWave beam selection problem as a multiclass classification problem and trains the network based on information such as the locations of neighbor vehicles. Outdoor location information is acquired mainly based on a positioning system such as the GPS, and indoor location information is acquired mainly based on technologies like Wi-Fi.

14.5 Summary and Prospects

Cell-free cooperative massive MIMO is an important user-centric enabler for 6G. It coordinates a large number of distributed small base stations to eliminate inter-UE interference, boost system spectral efficiency, reduce communication latency, and enhance system scalability and reliability. It will stimulate further technological transformation and framework innovation to enable various industries.

References

1. Alexandropoulos GC, Shlezinger N, Del Hougne P (2021) Reconfigurable intelligent surfaces for rich scattering wireless communications: recent experiments, challenges, and opportunities. IEEE Commun Mag 59(6):28–34
2. Wang MH (2020) A research on performance theory and scheduling technique of network-assisted full-duplex for cell-free massive MIMO. Southeast University
3. Femenias G, Riera-Palou F (2019) Cell-free millimeter-wave massive MIMO systems with limited fronthaul capacity. IEEE Access 7:44596–44612
4. Gao XY, Dai LL, Han SF et al (2016) Energy-efficient hybrid analog and digital precoding for MmWave MIMO systems with large antenna arrays. IEEE J Sel Areas Commun 34(4):998–1009
5. Zhang Y, Huo YM, Wang DM et al (2020) Channel estimation and hybrid precoding for distributed phased arrays based MIMO wireless communications. IEEE Trans Veh Technol 69(11):12921–12937
6. Zhang Y, Wang DM, Huo YM et al (2020) Hybrid beamforming design for mmWave OFDM distributed antenna systems. Sci China Inf Sci 63(9):1–12
7. Zhan JL, Dong XD (2021) Interference cancellation aided hybrid beamforming for mmWave multi-user massive MIMO systems. IEEE Trans Veh Technol 70(3):2322–2336
8. Elhalawany BM, Hashima S, Hatano K, et al (2022) Leveraging machine learning for millimeter wave beamforming in beyond 5G networks. IEEE Syst J 16(2):1739–1790
9. Klautau A, González-Prelcic N, Heath RW (2019) LIDAR data for deep learning-based mmWave beam-selection. IEEE Wirel Commun Lett 8(3):909–912
10. Wang YY, Klautau A, Ribero M et al (2019) MmWave vehicular beam selection with situational awareness using machine learning. IEEE Access 7:87479–87493
11. Cheng M, Wang JB, Wang JY et al (2019) A fast beam searching scheme in mmWave communications for high-speed trains. In: Proceedings of ICC 2019—2019 IEEE International Conference on Communications. IEEE Press, Piscataway, pp 1–6
12. Xu CM, Liu SH, Zhang C et al (2020) Joint user scheduling and beam selection in mmWave networks based on multi-agent reinforcement learning. In: Proceedings of 2020 IEEE 11th Sensor Array and Multichannel Signal Processing Workshop. IEEE Press, Piscataway, pp 1–5

Chapter 15
Sensing-Assisted Fast Network Access

Massive smart devices in the future call for communication technologies that can satisfy both latency and reliability demands when a large number of devices access a network simultaneously. However, the conventional Long Term Evolution (LTE) uses a grant-based access solution, which schedules devices in a complicated manner and with a large delay. Not only that, but the conventional cellular network architecture limits the system throughput in massive access scenarios, making it difficult to reliably meet the service requirements of each user. Fortunately, massive devices provide communication systems with a lot of information, such as the occasional access characteristics and rich channel characteristics. Implementing fast network access for massive devices based on auxiliary information and scalable system architecture will become a key research topic in 5G and future networks.

15.1 Environment Sensing

15.1.1 Fingerprint-Based Positioning

Higher positioning accuracy, even at centimeter level, is required in 6G application scenarios such as in industrial IoT [1, 2]. However, traditional positioning methods can provide only relatively low accuracy due to obstacles in signal transmission paths, as they utilize information acquired by base stations during signal reception to estimate user equipment (UE) location.

A more suitable method is fingerprint-based positioning, which is able to resist multipath effect in rich scattering environments (for example, in indoor environments) of a cell-free cooperative massive multiple-input multiple-output (MIMO) system. Fingerprint-based positioning is conducted in two basic steps. First, it extracts channel characteristics as fingerprints from signals sent by reference points in a known geographical area, and establishes a one-to-one correspondence between channel characteristics and geographical locations of the reference points. Second,

P. Sun, *A Guidebook for 5GtoB and 6G Vision for Deep Convergence*,
Management for Professionals, https://doi.org/10.1007/978-981-99-4024-0_15

it extracts channel characteristics from signals sent by a specific UE, compares them with fingerprints in the fingerprint database, and estimates the UE location according to the similarity between the fingerprints and the channel characteristics as well as the weight values of the reference points' coordinates.

Fingerprint-based positioning converts positioning into a pattern recognition issue, reducing communication overhead and improving UE positioning accuracy in rich scattering environments. Fingerprint-based positioning can be used not only in two-dimensional (2D) but also in three-dimensional (3D) space. By building 3D channel models, 3D positioning can utilize spatial information more effectively and locate UEs more accurately than 2D positioning [3].

15.1.2 Channel Knowledge Map

A channel knowledge map refers to a database with transmitter/receiver locations. It is another means of realizing environment sensing. This database includes useful location-related channel information. From the database, UE channel information can be obtained instantly once UE locations are available and can be used for system optimization. If UE locations are obtained in advance (for example, using a GPS), we can draw a channel knowledge map by obtaining channel information through online or offline measurement using ground apparatuses or through offline calculation based on radar tracking data and then establishing a mapping from geographical locations to channel information using a machine learning technology [4].

The channel knowledge map can help acquire channel information based on location information, learn parameters in channel models, reconstruct channel matrices, and optimize parameters. It can also help directly find the mapping from location information to optimized parameters. This method does not require channel estimation based on reference signals, reducing overhead, but requires necessary location information be obtained in advance.

15.1.3 Combination of Fingerprint-Based Positioning and Channel Knowledge Map

Fingerprint-based positioning is used to obtain UE locations based on the mapping between UE channel information and reference point channel information, through offline fingerprint database establishment and online fingerprint matching. Channel knowledge map is used to obtain channel information at a specific location through offline calculation based on transmitter/receiver location information and channel-related information collected using ground apparatuses or radars.

Location information in a channel knowledge map is generally obtained using a GPS, but in an environment with rich scattering or many obstacles, GPS-based

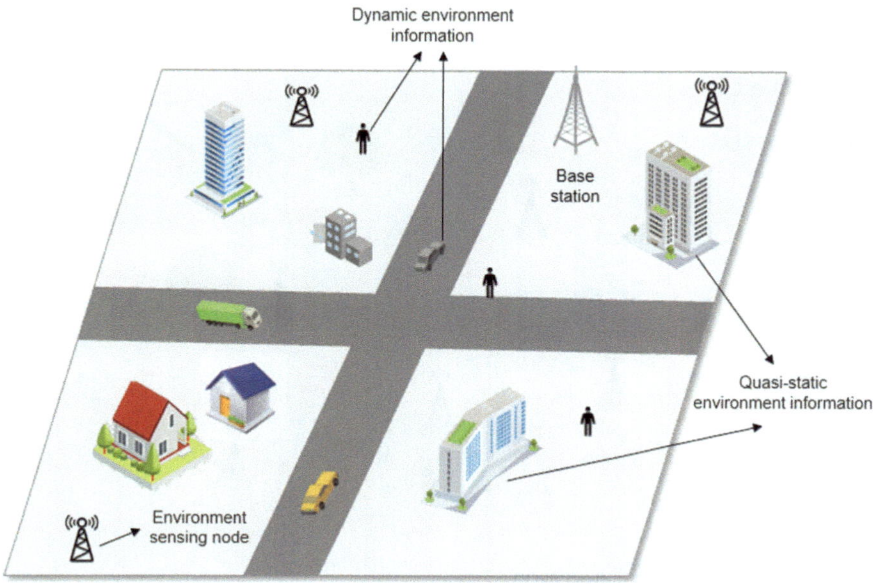

Fig. 15.1 Environment sensing scenario

positioning results are inaccurate and are not updated in real time. In this case, fingerprint-based positioning can provide relatively accurate positioning. Moreover, reference point information in fingerprint-based positioning can be obtained by fully utilizing many ground apparatuses employed for establishing the channel knowledge map, and therefore a fingerprint database with rich information can be set up without other reference points. The two methods mutually promote and complement each other to effectively implement environment sensing. Figure 15.1 illustrates an environment sensing scenario.

15.2 Active UE Detection

15.2.1 Covariance-Based Active UE Detection

In Massive Machine-Type Communications (mMTC), massive UEs conflict with limited allocable resources, which is an important factor that restricts access. For example, in orthogonal multiple access (OMA), assigning an orthogonal preamble to each UE is not practical. Fortunately, in most cases, only a small number of UEs are active, alleviating access conflicts, and a grant-free access mode can be used, allocating orthogonal or non-orthogonal preambles with a finite length to UEs in advance [5] for signal transmission. However, it is difficult to perform low-complexity and high-accuracy estimation on channels between UEs and base stations [6] when only

Fig. 15.2 Active UE detection scenario

a few UEs are active and non-zero channel gains are low. Therefore, it is necessary to detect active UEs before channel estimation. Figure 15.2 illustrates an active UE detection scenario.

Active UE detection generally uses covariance-based algorithms (such as gradient descent) and compressed sensing algorithms (such as approximate message passing). Covariance-based algorithms are less complex and more accurate, with neither limitation on the relationship between the number of active UEs and the number of preambles, nor demand for complex mathematical approximation and calculation, whereas compressed sensing algorithms require relatively long preambles, not applicable to massive access scenarios. Covariance-based algorithms use covariance matrix as a sufficient statistical indicator for active UE detection. A covariance matrix, consisting of received signal vectors, contains channel information and UE activity information. The larger the covariance value, the more likely the UE is active.

In a centralized network, active UEs may be overlooked or incorrectly detected. In such a network, all antennas are deployed on a single base station. The closer a UE is to the base station, the stronger the signal received by the base station. Nevertheless, UEs in the same location cannot be distinguished based on the signal strengths received by different antennas, as all these strengths are the same. Once multiple UEs are assigned the same preamble, there will inevitably be strong interference between UE signals on the base station side, leading to missed and incorrect detection of active UEs. In a cellular network, each base station is responsible for detecting active UEs in its own coverage, but the signals of UEs at coverage edges also experience relatively strong interference, degrading the detection performance.

However, in a network with cell-free cooperative massive MIMO deployed, active UEs can be detected using covariance matrices on different access points (APs). In such a network, APs distributed in a wide area receive signals from UEs and calculate covariance matrices based on received signals. Covariance matrices, containing UE activity information and channel information, vary with not only UE activity distributions but also APs. Therefore, covariance matrices on different APs can be selected for active UE detection, demonstrating the advantages of active UE detection in cell-free cooperative massive MIMO.

15.2.2 Scalable Active UE Detection Based on Intelligent Solutions

Covariance-based active UE detection is applicable to massive access scenarios as it does not need long preambles, but it is not applicable to latency-sensitive scenarios such as massive URLLC (mURLLC) scenarios because its iterative calculation needs time [7]. Owing to improved hardware computing and the popularization of deep learning, many active UE detection approaches based on intelligent solutions have emerged. One of them is data-based deep learning. It builds models based on application scenarios and applies them to solve real-world problems through simple transfer learning. However, this approach is limited to only active UE detection at fixed locations, considering that it generally requires numerous samples with labels to be collected in advance and multi-layer neural network models to be built in dynamic scenarios with massive UEs. Another approach is to use an E2E neural network architecture that integrates preamble design and active UE detection, and accelerates network convergence through covariance calculation or compressed sensing [8]. However, acquiring channel matrices might prove difficult in view of the fact that the acquisition depends on real channel training. Therefore, it is better to combine the two approaches to adapt to dynamic scenarios with massive UEs, achieving scalable active UE detection.

Scalable active UE detection can be better supported in a cell-free cooperative massive MIMO system than in a centralized or cellular system. First, as this cell-free system provides better coverage and significantly improves spectral efficiency, it can work with distributed computing methods such as mobile edge computing to lower the requirement on CPU performance and implement scalable detection through distributed training [9]. Second, it only needs to train the parameters of models by introducing an intelligent algorithm named Federated Learning (FL), to significantly lower the requirement on CPU performance and enhance detection reliability. Third, it can provide real-time guarantee for active UE detection through offline training and online deployment based on deep learning, and reduce the computational complexity of covariance-based algorithms and compressed sensing algorithms using the simple mapping models obtained through offline training based on deep learning.

15.3 UE Clustering and Dynamic Association

15.3.1 UE Clustering

In massive access scenarios, a huge number of UEs compete for limited resources to concurrently access a network, causing network overload and collision, which further leads to long latency and a low access success rate due to data retransmission [10]. In this case, adaptation between limited resources and multiple UEs is very important for improving the resource utilization based on resource competition and resource multiplexing.

UE clustering is such a technology. It classifies UEs into clusters by using *K*-means and Federated Reinforcement Learning (FRL) algorithms according to similar features such as service requirements or geographical locations, to reduce simultaneous contention-based access and coordinate interference during contention-based access. It enables intra- and inter-cluster resource allocation through intra-cluster resource contention and inter-cluster resource multiplexing, improving the resource utilization and access success rate. It allows high-priority UEs to be allocated more dedicated resources to reduce interference from low-priority UEs and implement high-reliability low-latency access [5].

15.3.2 Dynamic Association

In a cell-free cooperative massive MIMO system, a large number of APs are arbitrarily distributed in the coverage area and connected to a CPU. With coordination and computational assistance of the CPU, the APs jointly provide services for a huge number of UEs using the same time–frequency resources. In this case, the APs need to calculate the information of all UEs, and send it to the CPU via fronthaul links. It seems impractical for all the APs to serve all the UEs, as the calculations of APs will be very complex, fronthaul links will become overloaded, and the cell-free cooperative massive MIMO system will not be scalable [11].

To address all these issues, dynamic association can be used as a substitute to realize the scalability [12]. It classifies APs into sets in a user-centric manner and associates each UE with a set of APs, based on intelligent optimization algorithms or factors such as large-scale information. This way, each UE is served by certain APs, and each AP serves only a limited number of UEs, preventing AP calculation complexity or fronthaul link load from increasing with the number of UEs.

15.3.3 Integration of UE Clustering and Dynamic Association

A cell-free cooperative massive MIMO system may experience network overload and collision when a large number of UEs concurrently access the network, and may suffer very complex calculation and fronthaul link overload when APs provide services to these UEs. To cater for such scenarios, UE clustering and dynamic association are integrated for access management. By classifying UEs or APs according to requirements, this integrated method not only coordinates collision and improves access efficiency, but also reduces calculation complexity, increases transmission efficiency, and boosts scalability. This way, it can accommodate massive access of UEs with various QoS requirements.

15.4 Non-orthogonal Multiple Access

Conventional wireless communication systems mostly use OMA schemes to provide connectivity for multiple UEs and avoid multiple access interference. In OMA, UEs are allocated radio resources that are orthogonal in the time, frequency, or code domain. Ideally, there is no interference between UEs, and receiver complexity is low. OMA is adopted in 4G, as it maintains a low receiver cost while providing a high system throughput. One drawback of OMA is that the system cannot allocate orthogonal resources to each UE due to limited radio resources when a large number of UEs have access requirements [13].

Non-orthogonal multiple access (NOMA) emerges as a promising technology in response to the explosive growth of mobile data traffic stimulated by the rapid development of mobile communication technologies and continuous evolution of technological specifications. This technology can meet future requirements for high spectral efficiency, low transmission latency, high data rate, and massive connections.

NOMA allows multiple UEs to share time–frequency resources in the same spatial layer through code domain or power domain multiplexing. The core concept of NOMA is to employ superimposed coding (SC) at the transmitter and successive interference cancellation (SIC) at the receiver to implement correct demodulation. SIC uses a level-by-level interference cancellation policy. Specifically, it selects a UE after decision-making based on received signals, restores the signal amplitude of this UE, subtracts the multiple access interference generated by this UE, and then selects another UE from the remaining UEs and repeats the preceding operations until all multiple access interference is eliminated [14]. In the power domain, NOMA allocates different signal power to different UEs at the transmitter to obtain the maximum system performance gain and differentiate UEs, and uses SIC to determine the sequence of UEs at the receiver for eliminating interference based on different signal power levels of UEs. This way, it implements multiple access in the power domain.

NOMA based on power domain multiplexing has obvious performance advantages over traditional OMA in massive UE scenarios. Though the required receiver is more complicated, it can improve spectral efficiency and network throughput and significantly reduce transmission latency and signaling overhead. 6G is expected to require new access schemes, for example, a scheme that can dynamically switch between OMA and NOMA [14].

References

1. Alves H, Jo GD, Shin J et al (2021) Beyond 5G URLLC evolution: new service modes and practical considerations. https://doi.org/10.48550/arXiv.2106.11825
2. Zhou CY, Liu JY, Sheng M et al (2021) Exploiting fingerprint correlation for fingerprint-based indoor localization: a deep learning based approach. IEEE Trans Veh Technol 70(6):5762–5774
3. Liao CJ, Xu K, Xia XC et al (2020) AOA-assisted fingerprint localization for cell-free massive MIMO system based on 3D multipath channel model. In: Proceedings of 2020 IEEE 6th International Conference on Computer and Communications. IEEE Press, Piscataway, pp 602–607
4. Wu D, Zeng Y, Jin S et al (2021) Environment-aware and training-free beam alignment for mmWave massive MIMO via channel knowledge map. In: 2021 IEEE International Conference on Communications Workshops (ICC Workshops), pp 1–7
5. Weerasinghe TN, Balapuwaduge IAM, Li FY (2020) Priority-based initial access for URLLC traffic in massive IoT networks: schemes and performance analysis. Comput Netw 178:107360
6. Shao XD, Chen XM, Qiang YY et al (2021) Feature-aided adaptive-tuning deep learning for massive device detection. IEEE J Sel Areas Commun 39(7):1899–1914
7. She CY, Sun CJ, Gu ZY et al (2021) A tutorial on ultra reliable and low-latency communications in 6G: integrating domain knowledge into deep learning. Proc IEEE 109(3):204–246
8. Cui Y, Li SC, Zhang WQ (2021) Jointly sparse signal recovery and support recovery via deep learning with applications in MIMO-based grant-free random access. IEEE J Sel Areas Commun 39(3):788–803
9. Ke ML, Gao Z, Wu YP et al (2021) Massive access in cell-free massive MIMO-based Internet of Things: cloud computing and edge computing paradigms. IEEE J Sel Areas Commun 39(3):756–772
10. Yang HL, Xiong ZH, Zhao J et al (2021) Deep reinforcement learning based massive access management for ultra-reliable low-latency communications. IEEE Trans Wireless Commun 20(5):2977–2990
11. Björnson E, Sanguinetti L (2020) Scalable cell-free massive MIMO systems. IEEE Trans Commun 68(7):4247–4261
12. Chen SF, Zhang JY, Björnson E et al (2021) Structured massive access for scalable cell-free massive MIMO systems. IEEE J Sel Areas Commun 39(4):1086–1100
13. Tominaga EN, Alves H, Souza RD et al (2021) Non-orthogonal multiple access and network slicing: scalable coexistence of eMBB and URLLC. In: Proceedings of 2021 IEEE 93rd Vehicular Technology Conference. IEEE Press, Piscataway, pp 1–6
14. Liu YW, Yi WQ, Ding ZG et al (2021) Application of NOMA in 6G networks: future vision and research opportunities for next generation multiple access. https://doi.org/10.48550/arXiv.2103.02334

Chapter 16
Cloud-Edge-Network-Device Synergy, and Convergence of Communication, Sensing, and Computing

16.1 Cloud-Edge-Network-Device Synergy

16.1.1 Cloud Computing

6G poses higher requirements on computing resources and latency. Cloud computing, as a type of distributed computing, uses a system consisting of high-performance servers to process and analyze data and then return results to users. With this technology, massive amount of data can be processed within seconds, providing powerful network services. Cloud computing stimulates innovation in software development and deployment, and has emerged as a key infrastructure for a variety of applications. It not only provides basic support for emerging fields such as big data, IoT, and AI, but also promotes the convergence of manufacturing and the Internet. Despite this, cloud computing has a number of drawbacks, such as poor real-time performance, high energy consumption, limited bandwidth, and unsatisfactory data privacy and security protection. For this reason, edge computing is emerging.

16.1.2 Edge Computing

In scenarios where the cloud computing-based centralized solution is not suitable, edge computing can provide superior services. Various edge-based technologies, such as mobile edge and fog computing, are now available.

As channel capacity and data rates increase, transferring big data from IoT devices to the cloud is usually inefficient or even infeasible, for example, when bandwidth is insufficient. The cloud cannot offer the low latency required by location-sensing and time-sensitive applications, such as real-time manufacturing, patient monitoring, UAV cluster, autonomous driving, and cognitive services. What's more, storing data in the cloud is not feasible for some applications due to privacy issues and concerns.

P. Sun, *A Guidebook for 5GtoB and 6G Vision for Deep Convergence*,
Management for Professionals, https://doi.org/10.1007/978-981-99-4024-0_16

To address issues concerning latency, bandwidth, sensitive privacy, and geographical distribution, computing must be performed closer to connected devices. Edge devices, which are more closer to devices, are used instead to provide computing and storage capabilities.

Edge computing, performed at the network edge, has natural advantages in resolving latency issues. It can process and analyze data more quickly, reduce network traffic, shorten service time, and improve transmission efficiency. Edge computing connects IoT devices to the cloud, expanding the connectivity of these devices and enhancing data storage, processing, and management capabilities [1, 2]. Since IoT devices are close to users, the latency of edge computing is usually lower than that of other computing platforms. Therefore, edge computing does not depend on a highly centralized platform to provide services. Compared with cloud computing, edge computing uses smaller data centers. However, edge computing is not suitable for scenarios with high computing and storage requirements. That's why cloud-edge synergy is introduced.

16.1.3 Cloud-Edge Synergy

Cloud-edge synergy and the like are introduced to compensate for the drawbacks of traditional centralized cloud services. The cloud provides powerful computing capabilities, rich resources, and round-the-clock availability. Meanwhile, the edge supports computing and storage over the radio access network, while devices obtain a large amount of data through sensing technologies. To realize cloud-edge synergy, smart devices must be deployed in an edge environment so that the edge can process and respond to mission-critical data in real time, providing short-distance data transmission and analysis.

Local task deployment can greatly reduce the dependency on and consumption of transmission resources and significantly improve the local response speed. However, global optimization is difficult to achieve since edge devices can process only partial data. Therefore, the cloud computing platform is still required in practice to collect data for further evaluation, processing, and analysis to realize information convergence and governance. This not only meets data privacy and security requirements but also leverages fast iteration and update of cloud services. Distributed computing and resource scheduling are used to manage edge computing and storage resources and cloud computing resources to develop a cloud-edge synergy platform, which is of a logically centralized but physically dispersed architecture.

16.1.4 Federated Learning-Based Cloud-Edge-Network-Device Synergy

Data from intelligent services on a future network will be collected by multi-source sensing nodes, and the data will be of various structures, functions, and semantics. Obtained data must be processed in real time, to reduce communication overhead (the amount of data to be transmitted) and increase information perception efficiency (the value of information included in a data unit). The computing power of open-source wireless NEs can be flexibly deployed, which is ideal for diversified IoT services. Unfortunately, computing power is limited, highly fragmented, and random. To resolve data silos and data security issues in communication, federated learning is introduced. It enables joint modeling while ensuring data privacy and security, so that data in cloud computing centers and on edge devices are continuously encrypted and exchanged [3].

Federated learning is essentially a distributed machine learning technology or framework. Its purpose is to realize joint modeling and improve AI models on the basis of data privacy and security protection as well as legal compliance [4]. In traditional machine learning, data required for model training is usually collected by a data center, where a model is trained and prediction is performed. In federated learning, the model training is sample-based and distributed, which means that all data is distributed to different machines. Each machine downloads a model from a server, trains the model by using local data, and returns the parameters to be updated to the server. The server aggregates the parameters returned by all the machines, updates the model, and feeds back an updated model to each machine. In a network that features cloud-edge-network-device synergy, the edge, network, and device sides download a global model from the cloud, use local raw data to quickly infer the model, and return updated model parameters to the cloud server to update the global model. Figure 16.1 illustrates a cloud-edge-network-device model based on federated learning. Inference space can be reduced on the edge, network, and device sides by means of model segmentation and tailoring, to implement fast inference. This distributed network structure efficiently and reliably ensures data privacy, implements converged and open data management, and effectively allocates network resources.

16.2 Convergence of Communication, Sensing, and Computing

In the future communication systems, communication, sensing, and computing will be converged. When information is transmitted through channels, the physical characteristics of the surrounding environment are sensed by proactively identifying and analyzing channel characteristics. In this way, the communication and sensing functions are mutually enhanced. For example, signals received by base stations can be used to sense the surrounding environment and design communication links,

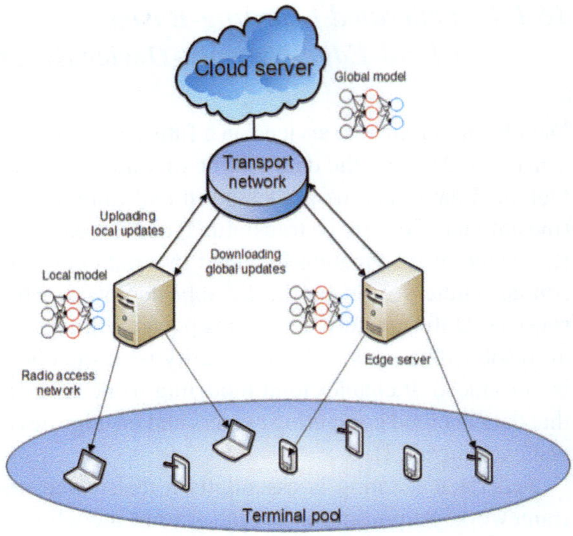

Fig. 16.1 A cloud-edge-network-device model based on federated learning

so that obstacles can be better avoided and the communication performance is improved. From 1 to 4G, communication and sensing function independently. In 4G, for example, the communication system is merely responsible for transmission, whereas the radar system is responsible only for measuring speed, sensing images, and the like. The separation of these systems results in wasted wireless spectrum and hardware resources, and independent functions also lead to a relatively long information processing delay. With the advent of the 5G and 6G eras, millimeter wave (mmWave) is now being used and terahertz (THz) bands are to be used for communication. In the future, the spectrum for communication will overlap with the spectrum for sensing, meaning new technologies need to be studied for spectrum convergence. Under the circumstances, integrated communication and sensing can be used to implement joint scheduling of communication and sensing resources. Moreover, the digital and physical worlds will be integrated on a future network, not only providing transmission channels but also perceiving everything through sensors and machine learning so that all things will be intelligent. Through the use of data centers (which act as the brain) and machine learning, the communication network will be optimized and managed to realize self-generation, self-governance (for example, channel adaptation), and self-evolution. Data is required when the network needs to carry native AI. Integrated communication and sensing provides basic data for AI services to form a network with the convergence of communication, sensing, and computing [5].

This convergence mainly depends on integrated wave communication and sensing. Specifically, radio communication and radar functions are performed simultaneously by using the same RF transceivers and spectrum resources. Unlike the separation of communication and sensing functions, this convergence reduces interference and enhances both functions. To improve the performance of integrated communication and sensing, intelligent computing needs to be introduced for efficient interference management. With intelligent computing, the network can learn the channel features of received signals, and then use these features to infer integrated communication and sensing information. To sum up, integrated waveform design and interference management can improve communication and sensing efficiency and reduce deployment costs in an effective way. Signal waveforms required for optimal communication and sensing are generally different. As a result, utilizing distributed computing resources to coordinate beams between different nodes and implementing multi-target multi-point perception are critical to the overall sensing precision and communication performance of the convergence of communication, sensing, and computing in the future [6].

16.3 Synergy and Convergence

16.3.1 Collaborative Data Processing on the Converged Network

A converged network, a network featuring the convergence of communication, sensing, and computing, works with the cloud-edge-network-device architecture. Such a network consists of the central network cloud, distributed NEs, and UEs. The central network cloud and distributed NEs collaborate to obtain, transmit, and calculate UE information. The central network cloud comprises high-performance computing servers and large-capacity storage devices, and provides three global functions, which are network information awareness, model training, and network performance optimization. Distributed NEs comprise devices for integrated communication and sensing as well as edge computing devices. They can detect their physical and electromagnetic environments during communication. In addition, they can manage and control the computing for UEs based on AI algorithms, and provide intelligent and dynamic allocation and adjustment of computing power based on scenario and service awareness as well as requirements and performance. The convergence of communication, sensing, and computing is implemented based on the close coupling relationship among the three. On one hand, communication is needed to expand the breadth and depth of sensing. On the other hand, distributed computing power is required to integrate data to reduce data dimensions and explore the deeper meaning of data, implementing user intent perception and optimizing network performance. The perceived multi-dimensional data in turn promotes algorithm performance, while the enhancement of the communication capability further improves the computing capability. In this way, algorithms can be optimized both locally and globally. Convergence works in various scenarios. When there is a new service, sensed data is summarized layer by layer with communication and sensing support, and the network gradually learns the features of the new service and performs optimization. If a fault occurs, the central network cloud identifies the abnormal area through data mining, analyzes the possible causes based on historical data, replans distributed NEs, isolates the faulty area, and triggers an active/standby switchover, without being noticed by users. When a task is complete, the distributed NEs actively reduce irrelevant energy consumption, coordinate with each other to complete tasks in hotspot areas, and share the load of other NEs.

16.3.2 Multi-dimensional Resource Allocation

When a converged network works with the cloud-edge-network-device architecture, computing and storage resources are distributed on the cloud, edge, and device sides. The network connecting the cloud, edge, and devices provides communication resources and realizes convergence through synergy. The stable development of the 6G industrial Internet creates increasingly important connections between devices and those between devices and users. By building a cloud-edge-network-device architecture, enterprises and developers will acquire a complete set of network services and a convenient development framework. On one hand, the convergence of cloud-edge technologies can greatly improve efficiency, reduce management costs, and bring more flexibility to enterprise operations [7, 8], yet on the other hand, it is more difficult to coordinate and integrate multiple technologies (involving cloud, edge, transmission, storage, AI, and big data) as opposed to using a single technology. More enterprises would find it challenging to control costs and to balance the input and output.

References

1. Tan K, Bremner D, Kernec JL et al (2020) Federated machine learning in vehicular networks: a summary of recent applications. In: Proceedings of 2020 International Conference on UK-China Emerging Technologies (UCET). IEEE Press, Piscataway, pp 1–4
2. Lin FPC, Brinton CG, Michelusi N (2020) Federated learning with communication delay in edge networks. In: Proceedings of GLOBECOM 2020—2020 IEEE Global Communications Conference. IEEE Press, Piscataway, pp 1–6
3. Ye DD, Yu R, Pan M et al (2020) Federated learning in vehicular edge computing: a selective model aggregation approach. IEEE Access 8:23920–23935
4. Liu Y, Yuan XL, Xiong ZH et al (2020) Federated learning for 6G communications: challenges, methods, and future directions. China Commun 17(9):105–118
5. Ma D, Lan GH, Hassan M et al (2020) Sensing, computing, and communications for energy harvesting IoTs: a survey. IEEE Commun Surv Tutor 22(2):1222–1250
6. Cui YH, Liu F, Jing XJ et al (2021) Integrating sensing and communications for ubiquitous IoT: applications, trends, and challenges. IEEE Netw 35(5):158–167
7. Khan LU, Majeed U, Hong CS (2020) Federated learning for cellular networks: joint user association and resource allocation. In: Proceedings of 2020 21st Asia-Pacific Network Operations and Management Symposium (APNOMS). IEEE Press, Piscataway, pp 405–408
8. Yan M, Chen BL, Feng G et al (2020) Federated cooperation and augmentation for power allocation in decentralized wireless networks. IEEE Access 8:48088–48100

Chapter 17
Intelligent Network Slicing

17.1 Overview

Operator networks are geared towards providing differentiated services for industrial customers with the support of 6G, AI, IoT, cloud computing, and big data technologies, the transformation and upgrade to intelligent and digital networks, as well as the introduction of diverse network applications. Different 6G applications in industries pose different service requirements. For example, wide-area services are needed in scenarios such as mobile office, media assurance, financial services, and smart inspection, whereas wireless cloud connections are required by cost-sensitive small- and medium-sized enterprises (SMEs). Smart policing, smart campus, industrial manufacturing, transportation and logistics, and stadiums have relatively high requirements on data isolation and service latency. Nuclear power, wind power, and industrial manufacturing as well as closed geographical areas, such as mines, oil fields, and wharfs pose extremely high requirements on data localization and independent O&M, management, and control. Traditional methods of expanding network capacity cannot meet wide-ranging 6G service demands, and a single network cannot accommodate multiple customized services in an ecosystem. If an operator uses the conventional "one network" deployment to deliver differentiated services, investment will be high and network efficiency will drop considerably. Therefore, 6G private networks need to be flexibly deployed and configured to provide customized solutions in different scenarios. Specifically, multiple customized, virtualized, dedicated, and isolated logical networks need to be constructed on a universal physical infrastructure platform, with different logical networks fulfilling different service needs.

A network slice is a logical network that provides specific network capabilities and has specific network characteristics. It consists of a group of network function instances and the computing, cache, and network resources for running these instances. The logical network is able to provide custom network services. Each network slice includes at least a wireless network sub-slice, a transport network sub-slice, and a core network sub-slice. There are many specific network slicing methods.

P. Sun, *A Guidebook for 5GtoB and 6G Vision for Deep Convergence*,
Management for Professionals, https://doi.org/10.1007/978-981-99-4024-0_17

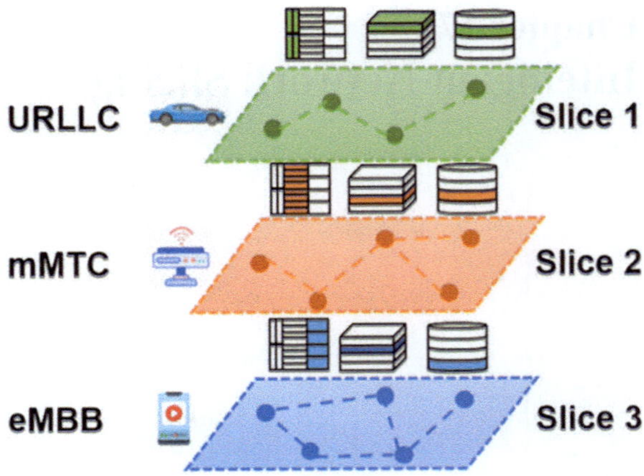

Fig. 17.1 Custom services provided by 6G network slices

The most basic method is to slice networks based on service scenarios: enhanced Mobile Broadband (eMBB), Massive Machine-Type Communications (mMTC), and ultra-reliable low-latency communication (URLLC). Figure 17.1 illustrates the custom services provided by 6G network slices. The required technologies, especially wireless network technologies, vary by scenario. Further classification can be achieved in the three scenarios based on the service level, RAT, or enterprise.

17.2 Convergence Advantages and Research Challenges

The integration of network slicing technologies into the cell-free cooperative massive MIMO system architecture has a number of strengths. First, network slicing can implement resource sharing, reduce costs, and improve efficiency. On a physical infrastructure, multiple network slices work at the same time by using a unified architecture, which significantly reduces network construction costs and improves physical infrastructure utilization. Second, since each slice is logically isolated and has its own lifecycle, creating and removing a network slice does not affect other slices. Secure and reliable communication can be ensured. Not only that, network slices can be customized for services based on the native architecture of cloud computing. Service resources can be customized according to scenario-specific requirements. Cloud-based monitoring can monitor the resource utilization of network slices in real time, provide highly reliable and flexible scaling, and adapt to the tidal effect of communication networks. Furthermore, since service requirements determine the network slice performance, end-to-end (E2E) differentiated services can be provided through resource division over a core network, a transport network, and an access network, fully meeting diversified 6G service requirements.

As one of the most promising technologies in 6G, network slicing can cope with all kinds of service demands and provide differentiated services for users. Despite its strengths, research has shown that network slicing in 6G also faces some challenges, which are mainly reflected in the following aspects.

1. Radio Access Network (RAN) Virtualization
   A core challenge of infrastructure virtualization lies in RAN virtualization. Allocating different spectrum blocks to virtual base station instances (slices) in advance is a solution of RAN virtualization, which can directly realize radio resource isolation, but also leads to a low usage of radio resources. RAN virtualization based on spectrum sharing is regarded as a substitute. It is dynamic and fine-grained. However, ensuring isolation of radio resources is a challenge to this solution. This can be solved by adjusting a software-defined RAN (SD-RAN) controller.
   In RAN virtualization, implementing the Robot as a Service (RaaS) model is a major challenge. Besides sharing radio resources and physical infrastructure, the RaaS model is expected to support individual slices and services through customized virtual control functions (such as scheduling and mobility management) and ensure isolation between different slices (virtual RAN instances) at the same time.
2. Service Composition with Fine-grained Network Functions
   How difficult it is to combine available network functions to provide services is directly determined by the granularity of these functions. If the granularity is excessively large, differentiated network services cannot be provided for users and the resource utilization decreases. If the granularity is excessively small, flexibility increases while the difficulty in network slice management and orchestration also increases. Coarse-grained functions can be combined easily because only a few interfaces need to be defined for their connection. However, this reduces the adaptability of slices and the flexibility of service requirement satisfaction. Fine-grained functions are more popular because they do not have these shortcomings, but we lack a scalable and interoperable way to combine fine-grained functions of different vendors for services. The simple method of defining new standard interfaces for all new functions will fail in scaling as the number of functions increases and the granularity of functions is refined.
3. E2E Slice Orchestration and Management
   A major challenge in network slicing is how to transform service descriptions to specific infrastructures and network functions. A solution to this is developing a domain-specific description language that supports a comprehensive expression of service characteristics, key performance indicators (KPIs), network element functions, and requirements while retaining a simple and intuitive syntax. Moreover, the language must be flexible and scalable to accommodate potential network elements and to improve applicability in a multi-vendor environment. An ideal feature is the ability to combine simple rules and expressions into complex rules and expressions, introducing abstraction layers into the expressions of service requirements.
   To enable the overall orchestration of different slices to meet the service requirements of each slice and effectively utilize underlying resources, network

slicing requires a complex E2E orchestration and management plane. Such a plane should not be limited to simple slice generation, that is, slice mapping to network components and static resource allocation. Instead, the plane should be adaptive, ensuring that the performance and flexible requirements of deployed services are met. To achieve this, it must make decisions based on the current state of each slice and the predicted state and requirements in the near future to effectively manage the overall resources.

Many specific solutions have been proposed after preceding issues have been studied in the context of cloud computing and data centers. While basic principles are available, mechanisms for 6G network slicing must be adapted and extended. Not only cloud resources (memory, cache, and networks) but also radio resources must be taken into account and their correlation, mutual impacts, and influence on the overall service quality need to be considered. How to meet the requirements of different services and effectively manage underlying network resources through 6G network slicing is similar to QoS provisioning on the network.

17.3 Joint Resource Optimization Based on Network Slicing

To adapt to the explosive growth of data traffic on 6G networks, some network nodes will be assigned computing and storage capabilities to enhance network performance. Evolution in software-defined networking (SDN), network functions virtualization (NFV), and multi-access edge computing (MEC) means that 6G networks will support more and more mobile users, provide outstanding network performance, and satisfy a wider range of service requirements. Despite this, these technologies will bring additional challenges to network deployment and resource allocation.

To improve the efficiency of network deployment and resource allocation, the three types of resources (computing, cache, and communication resources) of a mobile communication system may be scheduled together and jointly optimized. When the total number of these resources is limited, resources can be flexibly allocated for certain service requirements based on network slicing technology, ensuring the efficient utilization of network resources as well as the user's QoS. In addition, global resource allocation as well as flexible and real-time performance tracking can be performed by expanding the temporal scale and spatial scale. The joint optimization and scheduling of these three types of resources on a mobile network have become a major research topic, in terms of both theory and technology. That said, most existing wireless network resource allocation algorithms focus on improving a certain indicator of a network system, such as spectral efficiency, energy efficiency, and delay. Therefore, resources in multiple types need to be comprehensively enhanced to optimize the performance of the entire network system.

Due to the diversity and real-time requirements of IoT services, network service types change on a dynamic basis. As a result, traditional resource allocation algorithms struggle to cope with complex application scenarios, due to high computational complexity and slow convergence when solving complicated resource allocation issues. These algorithms can hardly be applied to next-generation mobile networks and IoT, putting a spotlight on research on efficient resource allocation algorithms.

Different policies for allocating computing, cache, and communication resources are required in different network scenarios. The ultimate goal is to maximize the total utility for mobile virtual network operators (MVNOs). Therefore, it is necessary to trade off the utility of these resources to construct a unified utility function. The unified utility function can be constructed in different forms, and a suitable one must have strict mathematical characteristics and intuitive physical meaning. Currently, most studies focus on resource allocation issues in core network slicing, but research on RAN slicing is highly valuable, since user access greatly affects the performance of the entire network. RAN slicing mainly involves the allocation of power and physical resource blocks. After MEC is introduced, the processing delay is calculated based on computation offloading and cache placement, and the optimal solution is obtained by limiting the delay and energy. The primary goal when using RAN slicing is to maximize the overall system utility, which is related to multiple factors, such as spectral efficiency, delay, and reliability. Therefore, the ultimate goal of network slicing optimization is to maximize system performance when resources are limited.

17.4 AI-Enabled Network Slicing

In most cases, computational complexity is high when a conventional method is used to solve an optimization problem of network slicing. Several technical barriers hinder optimal network slicing. First, spectral efficiency needs to be ensured when RAN spectrum resources are scarce. Second, there are strict requirements in Service Level Agreements (SLAs) for slice users. And finally, the actual slicing requirements depend on the service request mode of mobile users. In a dynamic and complex network environment, it is not easy to obtain prior knowledge for diversified data requests, which complicates optimization solutions. Reinforcement learning-based algorithms can solve these issues. Currently, reinforcement learning mainly applies to autonomous driving, answering questions, gaming, and smart grids. It is also used for dynamic routing and traffic distribution in a communication network, improving the efficiency of resource allocation. Reinforcement learning generally uses three approaches: dynamic programming, Monte Carlo, and temporal difference. Traditional reinforcement learning algorithms, however, need to consider the cumulative returns of policies during problem solving. As a result, the calculation workload is huge and it is difficult to solve complex decision-making problems. Deep reinforcement learning, which combines deep learning and reinforcement learning, is a breakthrough point for complex decision-making. Deep reinforcement learning uses deep neural networks with complicated architectures to estimate value functions, benefits, and policies in reinforcement learning, promoting the development and application of reinforcement learning. Furthermore, the state and action spaces of reinforcement learning are too large in a network slicing environment that is complex and dynamic and has multi-dimensional target constraints and resource allocation. In this case, computational complexity is high and deep learning must be combined with reinforcement learning for improved problem-solving efficiency.

There are several algorithms for deep reinforcement learning, and each has its own advantages. Currently, policy-based and value-function-based learning are mainly used for network slicing. Policy-based deep reinforcement learning can directly search for the optimal policy, which provides higher sampling efficiency when there are a large number of parameters. For example, the multi-agent deep deterministic policy gradient (MADDPG [1])-based algorithm features centralized training and distributed execution. It can solve issues related to continuous action spaces, and can better process issues on a multi-objective collaborative communication network. A major problem that hinders traditional reinforcement learning algorithms is that each agent is continuously learning and improving its policy. Therefore, from the perspective of each agent, the environment is unstable, which does not meet the convergence conditions of traditional reinforcement learning. To some extent, it is impossible to adapt to the dynamic and unstable environment simply by changing the policy of the agent. Due to the unstable environment, the essential skills of the deep Q-network (DQN [2]), such as experience replay, cannot be directly used. Basic policy gradient algorithms make the variances of random variables even higher due to the increase in the number of agents. The MADDPG algorithm has the following features: First, the optimal policy obtained through learning can provide the optimal action by using only local information. Second, the dynamic model of the environment and special communication requirements do not need to be known. And finally, the algorithm can be used not only in cooperation environment, but also in competitive environment.

For multi-dimensional customized network slicing, the MADDPG algorithm converts the multi-objective joint optimization problem into the multi-agent deep reinforcement learning problem, and completes the one-to-one mapping between elements in communication models and elements in AI models. In the algorithm training phase, each agent contains an actor network and a critic network. The actor network is responsible for updating the policy, which uses local information to provide the best action and continuously optimizes action selection based on the feedback of the critic network. The critic network is responsible for updating Q values (also called action values), that is, using the policies of other agents for global learning and updating Q values to evaluate the actions selected by the actor network. Through the interaction between the actor network and environment, the samples generated during the interaction are stored in the experience pool, and then the data in the experience pool is sampled in small batches and transferred to the actor and critic networks for training. After a large amount of training during experience episodes, convergence can be achieved. Therefore, the MADDPG algorithm can effectively solve the multi-objective joint optimization problem of network slicing and implement custom services for thousands of industries.

References

1. Lowe R, Wu YI, Tamar A, et al (2017) Multi-agent actor-critic for mixed cooperative-competitive environments. Adv Neural Inf Process Syst 30
2. Mnih V, Kavukcuoglu K, Silver D, et al (2013) Playing atari with deep reinforcement learning. arXiv preprint arXiv:1312.5602

Chapter 18
Future-Oriented Satellite-Ground Integrated Network

18.1 Brief Introduction

Future 6G will integrate satellite communications and ground networks into services, systems, and the like, constructing a satellite-ground integrated communication system that achieves seamless global coverage. Figure 18.1 illustrates the communication architecture of a satellite-ground integrated network.

The satellite-ground integrated network is a stereo-multilayer and heterogeneous dynamic complex network that covers multiple spatial platforms. A scientific and reasonable architecture design is the basis for efficient management and control and heterogeneous integration of the entire network. Currently, the architecture of the integrated network can be classified into the following three types: space-satellite-ground integrated network, space-based network, and space & ground network. (a) The space-satellite-ground integrated network is simply structured because satellites are not networked with each other. Multiple satellite gateways on the ground implement most functions and satellites generally use the transparent forwarding mode. Typical systems include the International Telecommunications Satellite (Intelsat), Mobile User Objective System (MUOS), and global satellite system. (b) In the space-based network, satellites interwork with each other to form a satellite network and implement functions such as collecting, processing, and transmitting information. Satellites generally use a processing and forwarding mode or a flexible forwarding mode to deliver strong robustness, while satellite gateways take only a supplementary role. The space-based network is often used for military purposes and typical systems include the Iridium's satellite network and the Advanced Extremely High Frequency (AEHF) system of the United States Armed Forces. (c) The space & ground network is in between. Specifically, it interconnects satellite and ground networks into an integrated spatial information network. In this way, seamless global coverage is realized, and the entire system can be managed and controlled without deploying satellite gateways around the world, reducing system maintenance costs and technical complexity to some extent. Typical systems include the Transformational Satellite Communications System (TSAT) of the United States Armed Force

P. Sun, *A Guidebook for 5GtoB and 6G Vision for Deep Convergence*,
Management for Professionals, https://doi.org/10.1007/978-981-99-4024-0_18

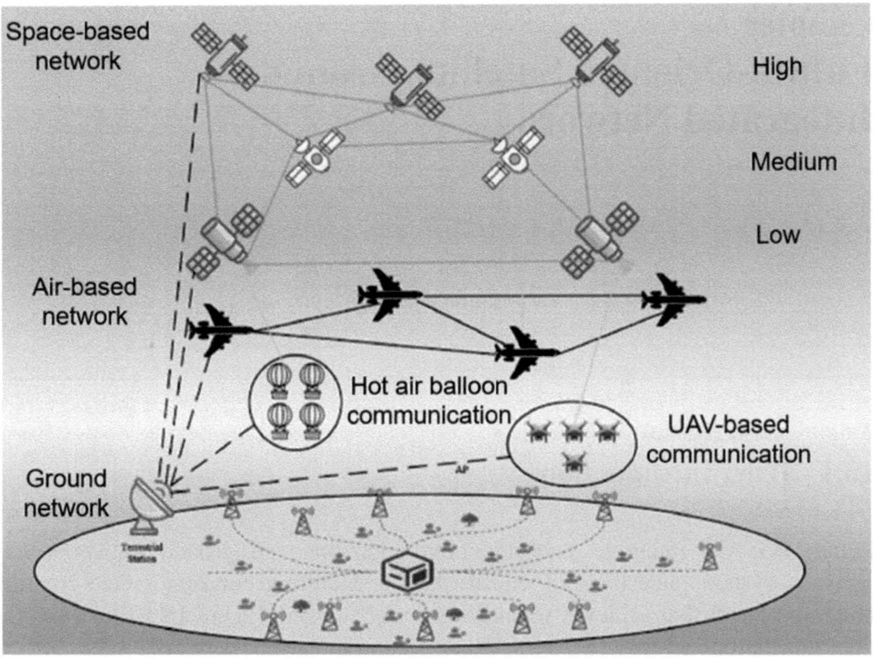

Fig. 18.1 Communication architecture of a satellite-ground integrated network

and the Space Communications and Navigation (SCaN) of National Aeronautics and Space Administration (NASA) [1].

The satellite-ground integrated intelligent network features wide coverage and flexible deployment, and is highly dynamic and protected against ground disasters. It provides unified service support and scheduling for users, enabling users to access the network anytime and anywhere and obtain seamless communication services. It can play an important role in various service scenarios, such as smart factory, telemedicine, smart home, autonomous driving, and smart agriculture, to meet the future-oriented requirements of intelligent connection of everything.

18.2 Integration Trend and Research Direction

In general, satellite-ground integration will be implemented from the aspects of standard, terminal, network, spectrum, and resource management, accompanied by research into new network architecture, unified adjustable air interface configuration, orthogonal multiple access (OMA) and non-orthogonal multiple access (NOMA), spectrum sharing, and interference management. Future research will focus on the following directions.

18.2.1 Intelligent Networking Architecture

A satellite-ground integrated network is a task-driven weak-connection network. It features network fragmentation and dynamic topology. As such, there is a high probability that satellites responsible for connections on a space-based network are unavailable in certain areas in which case low-earth orbit satellites need to be deployed independently. Therefore, it is necessary to study intelligent, dynamic, and adaptive networking methods that can adapt to the dynamic spatiotemporal changes on the satellite-ground integrated network, and are scalable to support more complex service scenarios in the future. In addition, mobility management is more complex on a satellite-ground integrated network. The fast time-varying feature of the network causes inaccurate channel estimation; frequent horizontal and vertical handovers of devices inside and outside the heterogeneous network result in high space–time overheads; routing optimization efficiency is low due to the difficulty in user positioning. Therefore, a satellite-ground integrated network desperately needs innovative mobile management solutions. So far, this integrated network has given rise to a more diverse range of services with differing QoS requirements, but the relationships between the QoS requirements and multi-dimensional resources are not yet clear. The network's heterogeneity feature also leads to different communication modes, media, devices, and application scenarios for each subnet. Given this, precise service-oriented unified resource management and reasonable resource allocation can also be incorporated into future research.

18.2.2 Intelligent Sensing

Sensing tasks of satellite-ground integrated networks are beyond the capability of traditional sensors, and require new sensors, which should be redesigned from the physical, control, and application layers based on the space-ground integrated sensing structure. The spectrum sensing and power control at the physical layer are crucial to improving the spectral efficiency and maximizing the power gains of the sensors, and they also pose design challenges. In addition, due to the complex heterogeneity of the satellite-ground integrated network, a new control-layer protocol should be formulated to accurately forward the sensing data to the application layer while ensuring system robustness. Lastly, the application layer should clean and filter the sensing data to complete the sensing tasks.

18.2.3 Intelligent Access Control

The space-based network is a supplement to the ground-based network. The satellite-ground integrated network can provide far-reaching communication coverage for user equipment, and must be able to seamlessly integrate the space-based network and ground-based network. Therefore, the integrated network should be highly dynamic and intelligent in support of a vast range of services. However, satellites that provide random access on the satellite-ground integrated network have limited transmission coverage, and therefore the diverse access devices must be centrally controlled. Such devices complicate effective access control for data transmission, necessitating the design of proper access and handover standards for central controls on the access devices. The dynamic network environment of access devices and the multi-dimensional heterogeneity of resources and services mean that some parts of the network may have redundant access capabilities, whereas other parts are unable to provide efficient access. Therefore, resource allocation must consider the features of different parts. A time-varying resource diagram can be used to describe the evolution of satellite and spatial network resources, and these resources can be further integrated with available terrestrial resources to dynamically ensure access capabilities and data transmission services of the network.

18.2.4 Dynamic Topology

The satellite-ground integrated network topology is dynamic. First, nodes such as satellites and space vehicles will be added, disabled, or updated, and new connections will be established, or original links will be disconnected between different nodes. Second, different satellites will constantly orbit their designated paths. The network topology changes in real time as time passes and space-relative locations vary. At the same time, space vehicles, such as unmanned aerial vehicles (UAVs) are extremely mobile. When receiving instructions in the air, they need to constantly change their position, which often generates extremely irregular network topologies. The dynamic network topology poses significant challenges for managing the satellite-ground integrated network, because the software-defined network (SDN) controller must obtain node status information in real time, and continuously update and deliver new flow tables to ensure service link connectivity.

18.3 Key Technologies

18.3.1 Network Integration Technology

The limited ground network resources cannot fulfill the QoS requirements of large-scale services, and seamless coverage cannot be ensured in remote areas and harsh environments. For this reason, heterogeneous network integration technology becomes an inevitable development trend of future mobile communications, which requires existing wireless access technologies and wireless networks to be deployed to complement and converge with one another. An integration technology with low cost and high conversion efficiency should be used for heterogeneous networks due to significant differences in networking protocols between them and high requirements for interconnection and integration [2].

18.3.2 Mobility Management Technology

Mobility management consists of location management and handover management. Location management allows the system to obtain real-time user locations and periodically update the location data in the system information database. Handover management refers to the management of calls and session transmission of mobile terminal nodes or mobile network nodes during mobility. More specifically, handover management involves horizontal handovers between homogeneous networks and vertical handovers between heterogeneous networks. A satellite network features high-speed mobility. One network node is generally incapable for maintaining a communication link for a long period of time, and prompt node switching is required based on the handover algorithm. If a handover policy causes frequent handovers between the serving and target nodes, the ping-pong effect will lead to abnormally fast consumption of network resources and power, which increases the risk of handover failures. Not only that, the large number of mobile nodes on the network significantly increases the signaling overhead related to mobility. As such, the quality of the handover algorithm directly affects the reliability of communication services [2].

18.3.3 Terahertz

Expanding the communication bandwidth is one of the most effective ways to improve the rate. The terahertz frequency band ranges from 0.1 THz to 10 THz, which can deliver a rate as high as 20 Gbit/s, 10 times that of 5G mmWave under ideal conditions. We know that scattering occurs when the surface bulge size of an object is of the same order of magnitude as the wavelength. In mmWave, the wavelength is 5 mm in the 60 GHz band and is 1 mm in the 300 GHz band. This wavelength is equivalent to the surface bulge size of a normal object and will cause strong scattering to help cover shadow areas. In massive multiple-input multiple-output (MIMO) systems, scattering increases the number of multipaths, provides higher multiplexing capability, and improves spectral efficiency. However, a higher frequency indicates a shorter wavelength and leads to poorer signal scattering and larger loss. The loss increases with the transmission distance, which in turn shrinks the coverage area of the base station. Fortunately, the terahertz technology of 6G makes developing a massive MIMO transceiver equipped with tens of thousands of antennas possible, and the beamforming technology can be used to effectively resist severe path loss of signals and further expand the communication range. Terahertz technology can provide ubiquitous connections and can be applied to satellite-ground integrated communications to provide global application services.

18.3.4 Access Technology

The height that a satellite orbits earth significantly expands the communication distance than that of a terrestrial 5G system. The high orbit increases the satellite-ground transmission delay, rendering the existing terrestrial 5G random access solution unsuitable for the satellite scenario. On one hand, the random access signal format needs to be modified and designed to resolve long delay in the satellite scenario. On the other hand, the long delay of satellite-ground links significantly increases the time required for completing an uplink access process between a terminal and a satellite, which poses challenges to some applications with low access delay requirements. OMA delivers limited system capacity because physical resources occupied by different access users are orthogonal to each other. NOMA has attracted wide attention as one of core technologies of the 5G air interface. Unlike OMA, NOMA allows concurrent transmission of multiple users in the same time domain, frequency domain, and code domain, so more user information can be carried over limited physical resources and then separated at the receive end by using a multi-user detection algorithm. This increases the system capacity and significantly improves the user access capability [3].

18.3.5 Synchronization Technology

The main problem with synchronization channel is how to achieve time and frequency synchronization between UEs and satellites, and how to obtain the cell identity information and system broadcast information, so as to prepare UEs for accessing the network. The downlink synchronization of a UE can be divided into three steps: satellite signal search, cell ID search and synchronization, and cell broadcast information acquisition. Blind search and satellite signal-assisted search should be considered. Blind search may be used for geostationary earth orbit (GEO) signals, but it is slow at searching. Low-earth orbit (LEO) signals usually have accurate ephemeris information and require a much shorter search time. After tracking a satellite, a UE will first detect the primary synchronization signal (PSS) to enable coarse synchronization in time and frequency, and then the UE will use the secondary synchronization signal (SSS) to achieve minor synchronization in time and frequency to obtain the cell ID information. Moreover, the UE demodulates the physical broadcast channel (PBCH) and obtains the master information block (MIB), the transmission time of the synchronization signal and PBCH block (SSB) for subsequent random access processing, the Remaining Minimum SI (RMSI), and other system parameters [4].

References

1. Qu ZC (2020) Architecture and key technologies for space-terrestrial integrated LEO satellite-based internet of things system. Nanjing University of Posts and Telecommunications, Nanjing
2. Li X (2021) Research on performance analysis and optimization technology of space-terrestrial integrated network. Beijing University of Posts and Telecommunications, Beijing
3. Nie YQ (2021) Research on the fast random access technology of satellite-5G system. Beijing University of Posts and Telecommunications, Beijing
4. Chen SZ, Sun SH, Kang SL (2020) System integration of terrestrial mobile communication and satellite communication—the trends, challenges and key technologies in B5G and 6G. China Commun 17(12):156–171

Chapter 19
Typical Application Scenarios of 6G in Industries

19.1 Metaverse

Metaverse first appeared in academia around 2000 as the development of VR technology and computer graphics started to take off. The outbreak of the pandemic in 2020 has impacted people's work and life and led to an increased tendency towards online activities, with relevant key technologies, tools, and devices gradually mature. Against this backdrop, discussions have been reignited about the metaverse that has science fiction roots.

The metaverse is the convergence of virtuality and reality, representing a simulated world parallel to and interconnected with the real world. As early as 1992, science-fiction writer Neal Stephenson coined the term "metaverse" in his novel *Snow Crash* and predicted various human activities in the future metaverse before the advent of the mobile Internet. Then, movies *The Matrix* in 1999 and *Ready Player One* in 2018 brought the imagination of the metaverse to the big screen. The metaverse not only exists in movies and novels, but has also entered our lives. In the Pokémon Go game launched by Nintendo in 2016, players can use their mobile phones to capture virtual monsters based on their real-world locations. Likewise, in July 2020, users took on virtual roles and hosted the academic conference Animal Crossing Artificial Intelligence (ACAI) workshop in the virtual world game Animal Crossing: New Horizons.

The current development path of the metaverse is similar to that of consumer-level VR/AR. It is now mainly used for entertainment and gaming and will see applications in other vertical industries in the future. The metaverse cannot be a completely closed universe, or a completely flat, open universe. It is similar to our real universe where open and closed systems coexist, big and small universes are nested, a small universe has the potential to expand, and a big universe has the opportunity to collide and integrate with others.

According to David Baszucki, CEO of Roblox, one of the most successful players in the metaverse world, the main characteristics of the metaverse include Identity, Economy, Immersive, Anywhere, Low Friction, Variety, Civility, and Friends. The

P. Sun, *A Guidebook for 5GtoB and 6G Vision for Deep Convergence*,
Management for Professionals, https://doi.org/10.1007/978-981-99-4024-0_19

virtual world in the metaverse mimics the real world. Users have no sense of space or distance between the real and virtual worlds, but they can switch between them at any time and even exchange material information.

The functionality and development of the metaverse depend on high-end technologies in many fields, such as digital twin, holographic projection, semantic communication, and on-purpose network. Among them, game and interaction technologies must be collaboratively developed to realize explosive user growth in the metaverse. Game technologies provide extremely rich content for the metaverse, and cutting-edge interaction technologies support efficient computing and transmission of massive data in the metaverse. Semantic communication technologies can extract customized semantic information to break through the limitation of traditional syntax bit transmission, and significantly compress communication traffic volume to ensure efficient transmission of massive data. On-purpose networks can analyze, model, and predict the intent of intelligent machines, and use native intelligence of intelligent agents to support on-demand networking and communication between intelligent agents in the metaverse. The cloud-based integrated intelligent network will become the infrastructure at the bottom of the metaverse. By providing large-scale access with high rate, low latency, high computing power, and strong AI, the network will deliver a real-time and smooth immersive experience for metaverse users. Cloud computing and edge computing enable metaverse users to use lightweight and cost-effective terminals, such as VR/AR/MR glasses with high definition and high frame rates.

19.2 Human–Computer Interaction

Human–computer interactions are multi-disciplinary by nature, involving fields as diverse as cognitive science, human engineering, psychology, computer science, behavioral science, design engineering, communication and media studies, and linguistics. Human–computer interaction aims to associate neural signals with electronic signals through electrodes, so that human minds and computers can communicate with each other. 6G networks will accelerate new research trends, such as emotional interaction [1] and brain-computer interaction [2], in place of traditional intelligent interaction devices with intelligent agents, which possess sensing and cognitive capabilities, thereby revolutionizing human interactions.

By introducing the *X* interaction concept, future mobile communication networks will facilitate coordination between humans and their physical and virtual worlds, accelerate human intelligence, and make machine intelligence conscious. This enables humans to participate in the network and allows for interconnection between humans and everything. Common fields for human–computer interaction include VR, AR, MR, and XR, all of which create a new sense of immersion, thanks to the seamless switching between real and virtual worlds.

The future XR system will support complex user-environment interactions in a proactive intelligent manner, including those related to voice, gesture, head, and eye movement. Unlike the current syntactic communication system that struggles to ensure a smooth wireless experience with XR due to limited transmission capabilities, semantic communication technology can significantly reduce the communication traffic volume through communication compression and semantic extraction. This technology can help build a national software and hardware ecosystem centered on the cloud rendering technology, drive the large-scale development of the entire XR industry chain, incubate several innovative cloud XR backbone enterprises, strengthen the cloud XR commercial platform as well as cloud XR key technologies and standards, and enable converged applications of cloud XR in some vertical industries to bring brand-new interaction experience to users.

19.3 Future Hyperconnected Cities

Rapid global urbanization in human society, together with increasingly challenging city development problems such as uneven allocation of public security, education, and healthcare resources, traffic congestion, and low energy efficiency, highlights the need for smart cities. As such, increasingly more governments around the world are using ICT technologies to sense, analyze, and integrate key information of core city operation systems to realize smart city management and operation. China's smart city construction and development kicked off in 2009, and the core principles of this plan were determined in 2014, which were: broadband information network, intelligent infrastructure, digital planning and design, convenient public services, precise social governance, and modern industry development. After 2016, China proposed the new concept of data-driven smart city aiming to enhance sensing, collaboration, insight, and innovation, make breakthroughs in city governance models, industry models, service models, and development concepts, and unlock the real value of smart cities.

In recent years, as AI, 5G, big data, IoT, cloud, GIS, and blockchain technologies gradually become more mainstream, Digital China has become an increasingly definitive strategic direction. Future cities will be smart in all aspects, presenting a future hyperconnected city form [3, 4, 5]. A future hyperconnected city will be a complex, huge system that has fragmented scenarios and needs to apply intelligence at specific locations in specific periods on demand. New technological innovations such as AI, intent sensing, and semantic communication will lead city development and transformation and build a new mechanism for long-term and healthy city growth to gradually encourage city modernization.

However, requirements on intelligence are scenario-based, and intelligence in different scenarios requires different computing power. Weak computing power will harm intelligence, while strong computing power will waste resources. Therefore, computing power should be properly adapted to intelligence, ensuring optimal computing power while saving costs.

"Cloud + Edge + Intelligence" will become the main pattern of a future hyperconnected city. In this kind of structure, a command center is established on the cloud and some "smart" tasks are delivered to the edge for quick sensing, response, and governance, thereby avoiding latency-triggered slow decision-making. A future hyperconnected city has the following six characteristics:

1. Crypto Native, co-construction and sharing of personal data centers, and large-scale, virtualized, and integrated data center services will become the main features. The service connotation will be extended from cabinet leasing, line bandwidth sharing, and hosting and maintenance to data storage and computing capability virtualization as well as comprehensive device maintenance and management supported by semantic representation measures.
2. Video experience will be supreme with ubiquitous large screen/multi-screen, HD/ultra-high-definition (UHD), and VR/AR facilities in cities. Semantic compression in semantic communication will help implement key core technologies in key fields, such as high-performance storage chips, UHD image sensors, high-speed data transmission interfaces, video/audio encoding and decoding, audio processing algorithms, and UHD broadcast-level channel cameras.
3. Real industries will be digitalized and industrial IoT will be decentralized, facilitating interconnections. On the basis of the dissipative structure theory, semantic intelligence will guide the phase transition and emergence of industrial Internet, promote the evolution to simple networks, and break the bottleneck of E2E real-time and reliable network interconnection.
4. Platform-based attempts will be made, mainly by creating digital IDs for initial setup of social networks across interstellar civilizations and science fiction cultures.
5. Various cloud devices will become new breakthrough points for distributed business organizations, cryptography economy, and certificate transactions. Online and offline devices will be streamlined, and services will be gradually virtualized.
6. Public management and social governance of future cities will be chain native. Next-generation information communication technologies, such as semantic sensing/transmission/computing will mine, analyze, match, and store government big data. Middle platform technologies and visualization technologies will enhance decision-making and auxiliary support for real-time dynamic data, gradually support the transformation of government decision-making from subjective "experience-based decision-making" to objective "scientific decision-making", and improve the scientific, intelligent, and refined urban governance through smart governance.

The digital transformation of future hyperconnected cities will be based on cloud native and chain native technologies. Technologies such as semantic representation measures, semantic sensing, semantic transmission, and semantic computing will be used to analyze the orderly structure and basic rules of future hyperconnected cities from the system perspective, break down the "simplified" structure of the network from the theoretical level to promote the construction of a digital-twin and interconnected super future city.

19.4 Smart Factory

Smart factories are an extension of digital factories that use data as the axis to stimulate enterprise intelligence. IoT and device monitoring technologies are used to enhance security management and device security maintenance, improve the controllability of the production process, and enhance remote control and auxiliary assembly, to quickly and accurately adjust the production schedule [6]. Figure 19.1 illustrates typical applications of smart factories where automated devices with different latency and reliability requirements are distributed. Due to a surge of smart devices in factories and the popularization of IoT applications, providing ultra-reliable low-latency communication (URLLC) services for URLLC terminals without compromising performance of other devices so as to maximize system performance has become an urgent problem.

As one of the key technologies in the 6G communication scenario, URLLC imposes strict requirements on statistical service quality of system applications involving massive terminals, and needs to implement E2E reliability and low latency features. This is rather difficult because high communication quality is required in the event of numerous communication terminals. In a smart factory, conflicts and collisions are likely to occur when a large number of users concurrently request network access. This not only causes system overload and user access failures, but also increases the access delay, reduces reliability, and causes system breakdown, severely affecting factory production.

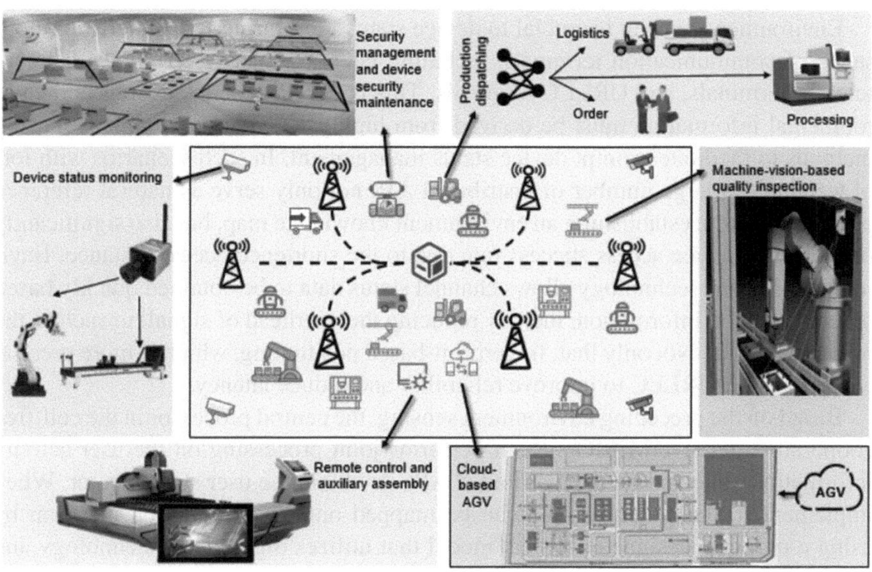

Fig. 19.1 Typical applications of smart factories

To accommodate a large number of automated devices with varying latency and reliability requirements in a smart factory, the cell-free cooperative massive MIMO system can be used to streamline smart factory operations because this system allows an extensive number of APs distributed vastly to provide URLLC services for massive terminals.

In the cell-free cooperative massive MIMO system, environment sensing technology can provide environment information and device positioning methods for security management and device security maintenance of smart factories. Dynamic deep reinforcement learning technology can effectively optimize production dispatching. Active user detection technology can accurately monitor real-time device status. NOMA and edge computing technologies can be used to enhance real-time performance and reliability of machine-vision-based quality inspection. Meanwhile, user clustering and AP association can assist machine remote control and assembly.

19.4.1 Device Status Monitoring and Security Maintenance

In a smart factory filled with terminals, there are certain moments at which most devices are silent and only some devices need to access the network. To control system running costs, the active status of devices needs to be monitored and security maintenance needs to be performed in real time to implement high-speed and reliable access.

Environment sensing is crucial to device status monitoring when there are thousands of communication terminals, including latency-sensitive terminals, latency-tolerant terminals, and URLLC terminals. This is because their locations and environmental information must be derived from highly-efficient environment sensing methods to facilitate prompt device status management. In such scenarios with lots of terminals, a large number of distributed APs not only serve as natural reference points to assist in establishing an environment knowledge map, but also significantly improve the device access success rate due to the shortened access distance. Environment sensing technology allows channel status data to be obtained quickly based on user location information, thereby reducing the overhead of signaling such as the reference signal. Not only that, fingerprint-based positioning, which is more precise, can be used in URLLC to improve reliability and reduce latency.

Based on the preceding environment sensing, the central processor in the cell-free cooperative massive MIMO system performs joint processing on the user activity information collected by the distributed APs, to obtain the user status vector. When implemented, the collected data can be mapped onto the simulation platform by using a properly designed physical model that utilizes digital twin technology and sensors. Through simulation, multiple signal collection sensors are properly deployed in a limited area to form a multi-access-point communication scenario, and apply the proposed active device detection scheme. After multiple debugging attempts, the verified algorithm solution can be migrated to the real environment for joint commissioning. Moreover, real-time communication needs to be maintained between IoT devices and between IoT and control systems. Through environment sensing, once an IoT device is faulty, it can be precisely located based on its channel status information by means of fingerprint-based positioning, for even stronger device security.

19.4.2 Production Dispatching

It is of paramount importance to efficiently adapt limited resources to a large number of terminals in smart factories in URLLC scenarios, where QoS requirements must be met on the basis of improved resource utilization. Factory terminal clustering technology can perform priority-based clustering on devices at the function level in line with service requirements. For example, by using latency as an indicator, devices requiring low latency will be allocated high-priority time–frequency resources. This can avoid some unnecessary conflicts between URLLC devices in scenarios with massive terminals, reduce transmission latency, and improve access reliability. Device-AP association technology can implement scalability to reduce the computing complexity and fronthaul capacity overload of APs in scenarios with a large number of terminals, as well as improve the processing rate and transmission efficiency, and further reduce the processing delay and transmission delay to meet QoS requirements.

To meet URLLC requirements in scenarios with massive terminals, NOMA allows multiple devices to share the same time–frequency resource by means of multiplexing in the power domain or code domain, resulting in reduced transmission delay and queuing delay. With user clustering and AP association, multi-dimensional resources such as non-orthogonal time–frequency-space and power resources can be allocated in a better way.

Changes to the market environment and increased complexity of industrial production may create unpredictable dynamic events that will disrupt the original production dispatching, which in turn hinders factory production. To cope with dynamic events in the production and manufacturing environment, multiple types of optimization objectives and performance evaluation indicators, such as order changes, workpiece processing, and device status, need to be considered. Various scheduling objectives may be correlated with, independent of, or even conflict with each other. Therefore, a multi-objective dynamic optimization algorithm needs to be used for targeted optimization. APs or AP clusters in the cell-free cooperative massive MIMO system can be used as intelligent agents to collect sensor delivered information in real time and to create scheduling models and optimize scheduling solutions in response to dynamic events based on deep reinforcement learning.

19.4.3 Machine-Vision-based Quality Inspection

In a smart factory, machine-vision-based inspection strains network bandwidth and real-time performance. On one hand, NOMA may be used to improve spectral efficiency by power-domain multiplexing and increase the overall transmission rate of the system. On the other hand, edge computing can quickly process computing-intensive tasks offloaded from mobile devices and effectively reduce the computing latency and processing latency of AI, thereby shortening the E2E latency to an ultra-low one in milliseconds. Meanwhile, multiple edge nodes can collaborate with each other to not only streamline problem solving, but also reduce energy consumption and improve data transmission reliability, implementing machine-vision-based quality inspection of URLLC with massive terminals. However, due to the limited computing and storage capabilities of edge devices, frequent data offload for balancing the overall system load will result in unnecessary transmission delay and energy consumption, and computing resource allocation on the mobile edge computing server also influences the task execution delay. When using edge computing to reduce E2E latency, how to effectively manage communication and computing resources, dynamically deploy computing and storage capabilities on a large scale, and ensure efficient collaboration and seamless interconnection of the cloud and devices remains a major challenge in resource allocation.

The local and edge servers can store the images and recognition results provided by the image collection system, and periodically upload the results to the AI platform for iterative model optimization, whereby the model accuracy can be improved with less misjudgment and manual re-check, improving inspection reliability. The architecture of the cell-free cooperative massive MIMO system is more conducive to information transmission between local and edge servers.

19.4.4 Remote Control and Auxiliary Assembly

In a smart factory, a large number of machine-type devices perform different work processes. These devices are classified into four classes based on their latency and reliability requirements and then clustered based on intelligent algorithms to deliver high reliability and low latency. These classes are: latency-reliability-sensitive, latency-sensitive, reliability-sensitive, and common communication devices. To reduce association complexity, nearby APs can be separately selected for each cluster of devices based on large-scale information to form cluster-specific AP sets. Hybrid resource scheduling, such as orthogonal reference signals for devices in a cluster, orthogonal reference signal multiplexing for devices in different clusters, grant-free two-step access for delay-sensitive devices, OMA for reliability-sensitive devices, and four-step NOMA for common devices, can be employed to meet diversified requirements.

In terms of auxiliary assembly, edge devices and sensors can collect environmental information of factory production lines and operation systems and send the information to edge servers for analysis and processing. Edge intelligent collaborative computing technology can implement efficient and real-time information collection and processing, helping onsite personnel implement standard assembly actions of complex or refined devices.

19.5 Smart Agriculture

Smart agriculture applies IoT technologies to traditional agriculture and fully utilizes computer network, IoT, and wireless communication technologies to realize intelligent agricultural plant protection, remote control, and intelligent agricultural logistics.

Smart agriculture integrates the Internet, cloud-edge-network-device computing, and IoT technologies. Thanks to various sensor nodes, UAV-based 3D detection, and the cell-free cooperative massive MIMO system, information covering temperature, humidity, and images can be quickly exchanged locally and across the Internet, supporting intelligent decision-making for agricultural production and transportation.

19.5.1 Smart Plant Protection

Figure 19.2 illustrates UAV-based smart plant protection in agriculture. The optimal operation route of UAVs can be planned according to seeding and fertilization requirements, and UAVs connected into networks can be remotely controlled for precise seeding and fertilization, driving the modernization and intelligentization of the agriculture industry. The big data strategy and digital village strategy were introduced at the 18th CPC National Congress with the aim of vigorously promoting a number of initiatives including "Internet + " modern agriculture. With this in mind, remote sensing and big data technologies of 6G networked UAVs can dynamically gather and monitor information regarding regional crop growth and land fertility, to maximize crop procurement. Crop growth can be modelled intelligently to accurately predict key information such as crop yield, nutrients, diseases, and pests. This continuously deepens big data construction in the agriculture sector and rural areas, allowing the penetration of digital technologies such as intelligent sensing, intelligent analysis, and intelligent control at a faster pace and providing a broad space for the digitalization of agricultural and rural production and management services. The basic construction of the space-air-ground integrated observation network, together with the improved data collection system, basic data resource system, and cloud platform for the agriculture sector and rural areas, can accelerate the integration of digital technologies into the agricultural industry system, production system, and operation

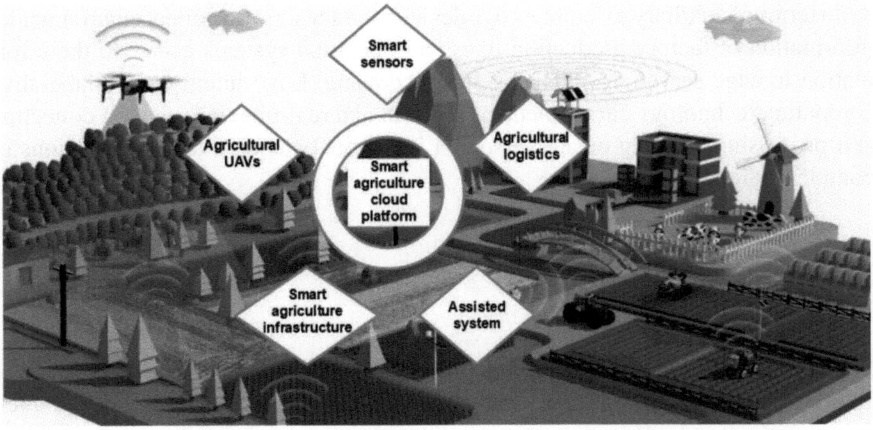

Fig. 19.2 UAV-based smart plant protection

system. This way, the digitalization of agricultural production and operation can make significant strides, with the digital level of managed services as well as the proportion of agricultural digital economy undergoing big improvements, while the rural digital governance system experiences considerable reform.

19.5.2 Smart Logistics

Smart logistics uses intelligent technologies and means such as intelligent hardware, IoT, and big data to strengthen analysis, decision-making, and intelligent execution of the logistics system. It empowers the rapid, efficient, and smooth operation of information flows and material flows, so as to cut social costs, integrate social resources, and improve production efficiency.

6G-based logistics will be smarter than traditional logistics not just because of new technologies and devices such as IoT, AI, and electronic labels during the logistics process, but also due to the integration of new concepts and models such as collaboration and sharing, instant response, real-time visualization, and flexible customization of the entire supply chain. In the transportation and distribution phases, smart logistics will accelerate the application of intelligent scenarios such as automatic transportation and unmanned driving in the logistics field, which consequently cuts labor costs, while in the packaging and transportation phases, smart robots will be more widely used. High-tech logistics scenarios, such as UAVs, unmanned warehouses, and unmanned stores are transforming the way the logistics industry operates. Contactless point-to-point take-out delivery services will also be possible.

6G technologies will further enable smart logistics and digital warehousing, promoting the construction of digital automatic logistics networks. In the future smart logistics, 6G technologies will be coupled with automation tools to fulfill capabilities within smaller timeslots, streamlining the entire logistics system.

The maturity of UAV technologies has seen that civil UAVs begin to penetrate the national economic industry and are more closely related to overcome typical problems in the agriculture industry. Considering agricultural market requirements and technology maturity, logistics distribution has a huge demand for UAVs. However, UAV applications are still in the early development phase, and UAV flight involves air control. With regulatory policies yet to be determined, UAVs have not been put into large-scale commercial use in the logistics field. Many logistics, express, and e-commerce enterprises from home and abroad are testing a large number of logistics-oriented UAV applications in an attempt to add UAVs as common logistics distribution tools. The next three years will be a key phase for the government and industry to coordinate and promote the implementation of regulatory policies. The release of commercial license for UAVs will lead to the gradual growth of UAV applications in the logistics industry. As such devices inherently operate in the air, UAVs can be equipped with HD cameras and megaphones to facilitate real-time display and unified dispatching in the farm command center. In terms of logistics transportation, a UAV-based urban air freight network can be built, which has significant advantages over ground transportation. On one hand, it is more convenient and faster. In some traffic-paralyzed road sections, urban congested areas, and some remote areas, ground transportation is not always smooth, and can in fact lead to long delivery delays, high delivery costs, and even significant losses in case of emergencies. This is where UAV-based delivery stands out. On the other hand, the rational use of idle low-altitude resources can effectively relieve the burden of ground traffic while saving land resources and infrastructure investment. High-rise buildings in some cities will be equipped with increasingly more helipads, as is the case in some developed countries and regions. In some towns and villages, small runways can also be easily set up for UAVs to take off and land. What's more, UAV-based transportation outperforms common air transportation and navigation transportation in that it can achieve lower costs and flexible dispatching as well as supplement existing air transportation capacity. Recent years have seen a year-on-year increase in the demands for air freight services, and the number of licensed pilots and supporting resources have become insufficient. The high labor cost of hiring pilots and crew also restricts the growth of the air freight industry. UAV-based freight has relatively low costs, and its unmanned driving feature enables the intensive development of all elements in airport construction and operation management and relieves the restrictions caused by staff management. There are certain demands for fast air transportation between fourth-tier or five-tier cities and between cities in a region. However, such demands usually involve a short distance and a small load, in which case developing a branch of UAV-based freight would be more feasible, because traditional air freight struggles to achieve economic benefits of its capacity in terms of take-off and landing, height, flight distance, volume, and load. UAV-based transportation is also more suitable for agricultural products in some remote mountainous areas and dangerous areas with rivers and seas, where land transportation and water transportation are extremely challenging.

19.6 Smart Mining

Smart mining is digital and information-based construction that guarantees mine production as well as occupational health and safety, and provides technical and logistic support. In a complex mine environment with high security requirements, the unique distributed access points and collaborative processing mechanism of the cell-free cooperative massive MIMO system can handle the challenges of performing stereoscopic inspection-based geographical mapping and emergency rescue in mines, as well as the high requirements on real-time performance. It facilitates the construction of smart mines. In stereoscopic inspection-based geographic mapping, UAVs collect stereoscopic data through stereoscopic inspection, access the network through adjacent access points in the cell-free cooperative massive MIMO system, and facilitate data processing in the computing center. In disaster relief, UAVs, satellites, and the ground network form an integrated space-air-ground architecture, which helps unleash the advantages of the cell-free cooperative massive MIMO system.

Geological disasters can occur in mines, endangering property and life. Such disasters include a collapse, landslide, debris flow, ground collapse, ground crack, and ground subsidence, which may be instigated by natural or human phenomena. China suffers from various types of natural disasters that are high in frequency. A number of factors, such as economic development and population growth, also affect and aggravate the frequency and scope of natural disasters to some extent. Especially in recent years, frequent geological disasters lead to substantial economic losses for the country and people. The pressure of disaster relief can be enormously mitigated as a result of fast-developing UAV-based communication. When a natural disaster occurs, UAVs can arrive at the site faster than humans would to inspect the area under remote control, capture and upload real-time onsite HD images over the 6G space-ground integrated network, and help rescue personnel quickly identify key factors of the disaster and facilitate disaster rescue decision-making, disaster assessment, and post-disaster reconstruction planning. As some serious emergencies require assistance from other departments, such as the police and the fire department, the three-dimensional live-view platform can also effectively help external rescue parties to quickly understand the onsite environment, make rescue plans, plan rescue routes, and promptly detect the nearest available facilities, achieving efficient communication between personnel onsite and not onsite.

Emergency satellite communication generally utilizes satellites and terrestrial stations to deal with mine disasters. Terrestrial stations include a fixed station—a communication central station—and a vehicle-mounted station. The fixed station is connected to the command center to summarize information and release commands, while the vehicle-mounted station is flexible and can arrive at the site promptly when needed and tackle emergencies efficiently. In the event of an emergency, a temporary command center can be set up onsite and transmit signals to the rear command center through satellites to provide data support for emergency rescue decision-making. Figure 19.3 illustrates the architecture of the satellite-communication-based emergency rescue network.

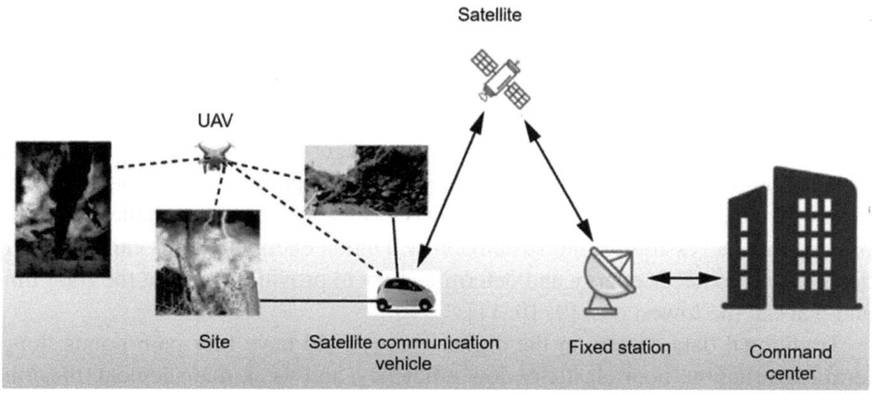

Fig. 19.3 Satellite-communication-based emergency rescue network architecture

Satellite communication has the following features: (a) Long communication distance that allows for long-distance far-reaching communication; (b) multiple access connections whereby satellite communication within the coverage can be well implemented through terrestrial stations; (c) flexibility that enables satellite communication devices to be loaded on multiple types of carriers, such as individuals, cars, and hulls, breaking through space restrictions, especially in areas where communication conditions are unsatisfactory or networks are underdeveloped; (d) wide bandwidth of a single satellite, which expands as technologies advance; (e) line stability that ensures good communication quality because of few influencing factors; (f) low cost of emergency communication. Another striking feature is that the distance of satellite communication is not directly related to the communication cost, making satellite communication ideal for long-distance emergency communication [7].

Emergency satellite communication not only utilizes terrestrial stations and other satellite communication resources, but also relies heavily on dynamic bandwidth allocation and signal modulation, demodulation, and coding technologies, to transmit images, videos, and audio data of disaster areas in real time, helping the command center acquire an understanding about the accident status quickly and respond as soon as possible [8].

19.7 Digital Energy

In recent years, the world energy market has shown increasing signs of instability. Environmental issues, particularly air pollution, have become a serious threat to us all and a significant challenge to how energy enterprises operate. A green, low-carbon, and sustainable energy supply has become an important factor affecting a country's economic development, and digitalization is regarded as an important way to achieve energy transformation. Digital energy is an in-depth integration of IoT technologies into the energy industry. By using big data and AI technologies and connecting energy facilities to the IoT, digital energy streamlines the connection between the physical and digital worlds, adds bit streams to watt streams to manage watts by bits and organize energy flow, extends energy categories and breaks through boundaries, amplifies facility utility, and optimizes inter-category collaboration, building a more efficient, clean, and economical modern energy system with enhanced security, productivity, accessibility, and sustainability. Digital energy systems can accurately determine who needs energy and determine how to provide energy at the right time and place at the lowest cost [9, 10, 11].

Traditional data centers in the digital energy field have four pain points: long-term construction, poor elasticity, low efficiency, and weak management, meaning they fail to meet the requirements of the cloud era. Considering the shifting business trend from support system to production system, the need for more advanced data centers is becoming more urgent. Semantic communication uses semantic information for representation measures, extracts semantic information from massive data in multi-environment industry scenarios, and establishes a unified semantic-information comprehensive representation theory. It improves the edge network communication capacity and computing capability, assists in providing such energy solutions as all-scenario intelligent sites and edge-to-cloud intelligent data centers for numerous application scenarios in the ICT industry, and supports smooth evolution of ICT networks to 6G-based all-cloud networks. For example, in intelligent power supply and distribution scenarios, large-, medium-, and small-sized data centers can be built through semantic extraction and trusted computing, semantic communication networks can be deployed based on smart modules to implement full connection and wide coverage, and the modular design, all-in-one integration, and one-stop on-demand deployment can achieve on-demand expansion, module-level security configuration, and flexible adaptation to multiple power supply and distribution scenarios [12]. In intelligent geological survey scenarios, semantic communication allows data nodes to complete upload, storage, computing and prediction, and 3D visualization of a large amount of seismic data anytime and anywhere. Semantic communication also supports fast self-networking in emergency. For power rescue and relief scenarios in disasters such as earthquakes, rains, and snows, semantic emergency communication systems' self-networking and large-bandwidth backhaul capabilities support HD video trunking communication onsite as well as commanding and decision-making.

Furthermore, semantic communication is mainly established based on context knowledge of both the transmitter and receiver and is more confidential than traditional communications. The same sentence may contain significant information for the target receivers who have the background knowledge, but may be meaningless for unauthorized listeners. This reduces the burden of data privacy management in future data-enabled digital energy. Besides mining data value, data can be encrypted during storage, transfer, and processing, implementing secure data transfer. Through trusted computing, industry information and related information can be automatically identified, and networks in common areas and key areas can be logically isolated to limit the spread of insecure factors, so that the data center architecture can be modularized and intelligentized, leading energy digitalization and building a green and intelligent world.

19.8 Smart Grid

In recent years, the development of renewable clean energy has become a focal point of the green development trend. However, new clean energy fluctuates uncontrollably, and its implementation is demanding on the power grid. The emergence of distributed energy systems enables common users to produce and store energy, breaks the bottlenecks of traditional centralized power supply, maximizes economic benefits, and transforms power grids into "microgrids". The distributed energy system manages multiple energy sources and devices with different interfaces and types, and therefore requires more refined control to save energy from energy production to energy consumption.

A large number of distributed and intermittent energy systems in different forms have penetrated into power grids. The requirements of energy development are now beyond the reach of traditional passive, unidirectional power grids. High-quality power services such as power distribution automation, emergency communication, and precise load control, also drive the upgrade of the power grid architecture and resource management mode. This transforms the grid into an active, bidirectional smart grid. Two-way grid and communication flows can improve the reliability, security and efficiency of power system. As power grids continuously connect to increasingly diversified devices and systems, smart grids must be capable to provide access to numerous terminals and bearers with low latency and high reliability, as well as have the scalability for openness and sharing, security management and control capabilities, and efficient and flexible operation management capabilities.

A smart grid covers power generation, transmission, conversion, distribution, and consumption. Due to the diversity of electric power services, a powerful communications network is needed for transmitting massive data generated in a smart grid. Smart grids bring new challenges to communications networks, which must feature good isolation, high reliability, ultra-low latency, and low costs, in order to meet different grid users' needs.

In the future, with more distributed point-to-point connections in smart grids, functions of the main grid system will be gradually moved closer to users, leading to an increased demand for local control and linkage control with the main grid system, millisecond-level latency, and 99.999% reliability. The types of collection services in grids will increase sharply, and the number of connections will expect a 50- to 100-fold growth to reach billions, while access density will rise to 6000/km^2. Mobile application services in grids will deliver a single-terminal throughput of 50 Mbit/s with dozens of milliseconds of latency and develop towards 4 K/8 K HD and VR/AR services with the single-terminal access bandwidth of 50 Mbit/s. There will be an increase of 5–10 times in the communication bandwidth, 15–24 times in the data collection frequency, and 10–13 times in the number of connected users.

To achieve strict and diversified QoS for future smart grids, future 6G communications should be designed to provide diversified QoS provisioning technologies that can effectively utilize limited or confined resources. Network slicing is a new communication function for implementing personalized services and can construct an E2E logical network to provide one or more network services on demand based on specified network functions and specific network access technologies. Through flexible network customization, network slicing can provide an optimal network resource allocation solution. Therefore, it is possible and feasible to apply network slicing to grid services so that network slices can be created to fulfill different service requirements in a smart grid. Figure 19.4 illustrates the application of network slicing in a smart grid. In a cell-free cooperative massive MIMO system, distributed APs in the physical architecture can facilitate grid control through point-to-point connections by moving functions of the main grid system closer to users, and can help achieve ultra-large-scale coverage and coordinate multiple types of devices. At the logical layer, SDN- and NFV-based network slicing [13] can optimize the joint allocation of multiple resources such as grid and communication resources to provide customized services.

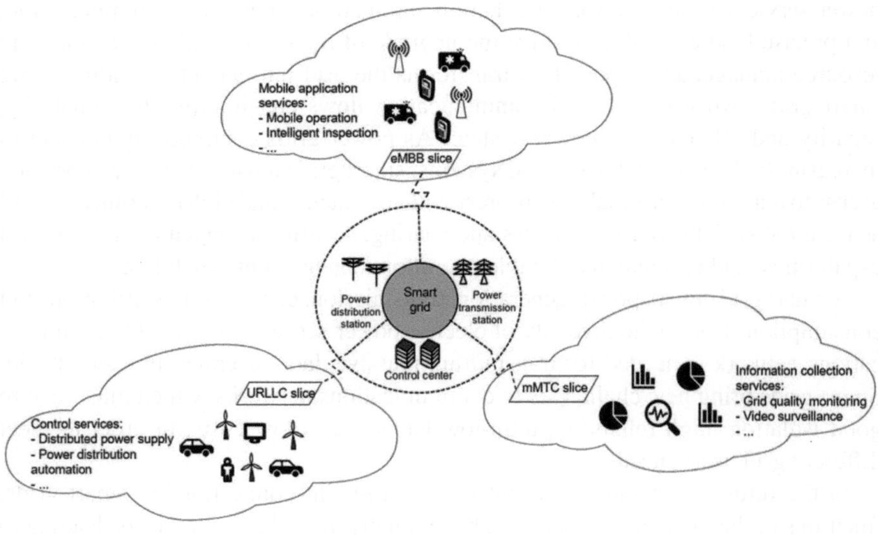

Fig. 19.4 Application of network slicing in a smart grid

19.9 Smart Healthcare

The future healthcare industry will fully integrate cutting-edge technologies such as cloud computing, big data, VR/AR, and AI to make healthcare services truly digital and intelligent, and promote growth of the healthcare sector. China proposes eight directions for future development in the smart healthcare field, including emergency treatment, remote diagnosis, remote treatment, hospital management, intensive care, traditional Chinese medicine diagnosis and treatment, health management, and intelligent disease control. Therefore, future smart healthcare will not be limited to the use of auxiliary information devices such as electronic medical records and self-service machines in hospitals. Instead, driven by technologies such as IoT and AI, smart healthcare will provide digital, networked, and intelligent medical facilities and solutions for more medical scenarios. After the pandemic outbreak, various IT and communication technologies have been integrated into the traditional healthcare field, providing doctors and healthcare personnel with a large amount of scientific evidence that can be searched, analyzed, and quoted to support diagnosis, assisting in the fight against the pandemic. For example, the intelligent screening, diagnosis, and grading system oriented to the coronavirus diagnosis and treatment process studied by scholars from Beijing University of Posts and Telecommunications was promptly deployed in multiple hospitals in multiple regions, including Hubei. The system can intelligently diagnose coronavirus infection in just seconds with an average accuracy of 90%. It has passed the clinical verification of multi-center, real-world scenarios and played a crucial role in preventing and controlling the pandemic worldwide [14].

The development of smart healthcare depends on cutting-edge technologies such as IoT, AI, and semantic communication. AI technologies, represented by Federated Learning, can implement secure and efficient flow of disease-related data among patients, medical devices, medical personnel, and medical institutions [15]. Medical personnel can use information such as images provided by medical instruments and devices to learn about patients' physical condition and make diagnosis. What's more, medical personnel can communicate with each other in the same department, across departments, across medical institutions, or across regions. Technologies such as semantic communication and natural language processing make it possible to automatically analyze medical records described in natural languages, helping build a medical semantic knowledge base that fully leverages medical data and solves the existing problem of low data quality in smart healthcare. The semantic knowledge base on the cloud can promote cross-platform integration of medical data, service process integration between hospitals, as well as sharing and exchanging of medical information and resources.

The future smart healthcare network will focus on the full-process informatization of patients before, during, and after medical treatment, and build a telemedicine platform, an image-assisted diagnosis cloud platform, and a regional information integration platform [16]. Also, the construction of future smart healthcare and smart hospitals will benefit from the higher rate, lower latency, and smart interaction platform that integrates mobility and big data analysis of 6G. This simplifies access to smart healthcare services and fulfills expectations of future healthcare. By making full use of cutting-edge technologies such as semantic communication and AI, future smart healthcare can support more application scenarios such as mobile first aid, AI-assisted diagnosis and treatment, holographic projection, telemedicine, remote emergency, remote clinic, smart operating room, smart ward, and smart hospital guide.

19.10 Smart Transportation

Smart transportation makes full use of information technologies such as mobile Internet, IoT, and edge computing, and tools such as AI, environment sensing, system methods, and knowledge mining to promote the transformation and upgrade of transportation towards a more efficient, economical, environmentally friendly, and secure direction.

19.10.1 Smart Vehicle

With a new round of deep technological revolution and industry transformation, Smart Vehicle has become the strategic direction of the global automotive industry. A Smart Vehicle system is a comprehensive system equipped with advanced vehicle-mounted sensors, controllers, and executors and is integrated with next-generation information communication technologies to provide environment sensing, decision-making and planning, and multi-level assisted driving [17]. Computers, modern sensing, information convergence, communications, AI, and automatic control technologies are used to implement intelligent information exchange and sharing between vehicles and X (such as a pedestrian, vehicle, infrastructure, or cloud). Complex environment sensing, intelligent decision-making, and collaborative control functions empower safe, efficient, comfortable, and energy-saving driving, delivering the next-generation automobile that can operate properly with zero human intervention.

There are two Smart Vehicle systems: One is Autonomous Vehicle, which senses the surrounding environment through various sensors and relies on vehicle-mounted controllers for decision-making and control. The decisions and control policies will be transferred to the bottom layer for execution, implementing autonomous driving. The other is Connected Vehicle, in which vehicle-to-everything (V2X) communication is used for vehicles to obtain external environment information and perform decision-making and control [18, 19]. As AI and communications intelligence develops, the two systems are continuously integrated, from assisted autonomous driving to networked autonomous driving. The ultimate vision is to achieve full automation [20]. *Intelligent Connected Vehicle Technology Roadmap 2.0* released at the 2020 World Intelligent Connected Vehicle Conference proposes the future development goals of Smart Vehicle.

Figure 19.5 illustrates the development objectives of Smart Vehicle, which fully integrates the Internet and information industry and will be more intelligent, networked, and platform-based. Vehicles will gradually transform from pure transportation tools to smart mobile space and application terminals, becoming an important carrier of emerging businesses. However, the 5G-based next-generation V2X system is a complex cyber-physical converged system, whereby cloud nodes of vehicles are connected through digital communication networks. Large-scale application and deployment of the system cannot be completed overnight due to the following four challenges: (a) The system is complex and involves a large number of tasks; (b) The channel is not ideal and difficult to control; (c) The structure is variant and difficult to optimize; (d) The existing methods are isolated from each other and cannot be combined from both macroscopic and microcosmic perspectives. For example, remote driving requires the reliability illustrated in Fig. 19.5 with the user-plane latency being 3 ms, which is challenging to existing communication technologies. This necessitates the use of semantic communication to promote autonomous driving [21].

The real-time semantic processing framework in semantic communication fully utilizes the maximum computing power of the heterogeneous computing platform to meet the computing requirements of multiple services and environments. It can quickly feed back vehicle status information, predict the surrounding environment, and assist in driving decision-making, improving driving safety and convenience as well as delivering brand new user experience. What's more, semantic communication

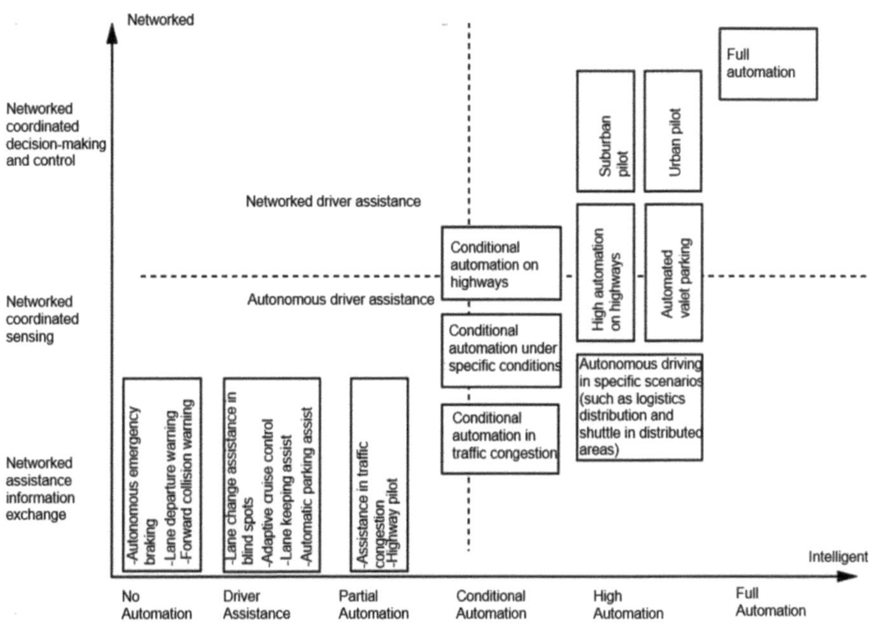

Fig. 19.5 Development goals of Smart Vehicle

will further improve the reliability of vehicle-oriented semantic networks. Semantic compression and smart sensing will facilitate closer inter-vehicle collaboration and road environment modeling, while intelligent vehicle-cloud nodes and simplified V2X networks will motivate the evolution of multi-dimensional pedestrian-vehicle–road collaboration, helping build a smart transportation system, making it feasible to implement fully autonomous driving.

Based on semantic communication technologies, the future Smart Vehicle network will be a large-scale communication system that combines the intra-vehicle network, inter-vehicle network, and vehicle-cloud network and complies with established communication protocols and data exchange standards to implement wireless communication and information exchange between vehicles and *X* (such as a pedestrian, vehicle, infrastructure, or cloud), realizing intelligent traffic management, dynamic information service, and intelligent vehicle control.

19.10.2 High-speed Railway

1. Current Situation
   As science, technology, and economy develop rapidly, railway communications, particularly high-speed railway communications, have garnered the attention of academia and industries due to high mobility, security, environmental protection, comfort, transparency, predictability, and reliability. To meet the requirements of future intelligent railway communications, the railway transportation industry needs to develop new communication network architectures and key technologies to provide high-quality and reliable communication services for passengers. 6G is envisioned to effectively improve stability in high-speed railway scenarios, and the future 6G-based high-speed railway (HSR) wireless networks will consider spectrum expansion by using mmWave frequency bands. To improve coverage in mmWave, which features much higher path loss than low frequency bands, the beamforming technology can be used in the cell-free cooperative massive MIMO system to concentrate the energy of expected signals on the target direction to ensure the radiation range, so the beam direction of a base station can be adapted properly to the speed and direction of a high-speed train. Fig. 19.6 illustrates a mmWave cell-free cooperative massive MIMO system used for high-speed railway communication.
2. Wireless Environment Map
   The wireless environment map is initially applied to the WLAN and can integrate multi-dimensional information such as time, frequency, and location and display the information visually. At present, this concept has been adopted in technical documents of international standardization organizations such as the International Telecommunication Union (ITU) and the Institute of Electrical and Electronics Engineers (IEEE), and has been applied to many scientific research projects at home and abroad. Due to the periodicity and regularity of high-speed railway movement tracks, beamforming in high-speed railway communication

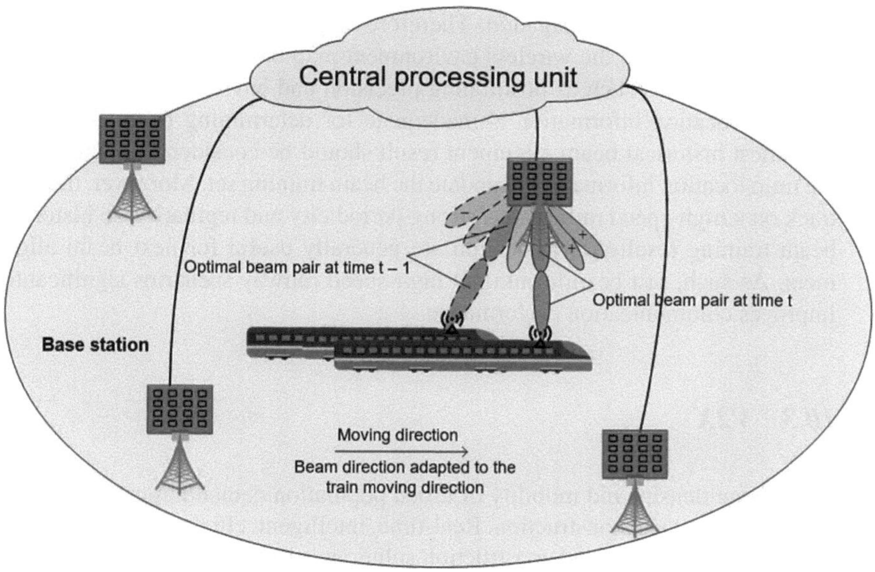

Fig. 19.6 mmWave cell-free cooperative massive MIMO system for high-speed railway communication

can rely on the wireless environment map, which can be used to save terrain information, channel model, wireless equipment information, propagation loss, large-scale fading, channel characteristics, electromagnetic interference, and communication quality test results along the railway. Such information facilitates compensation for Doppler shifts, resource allocation, and implementation of other adaptive techniques, thereby ensuring efficient, continuous, secure, and stable communication.

3. Beam Alignment in High-Speed Railways
   To provide reliable connections for train passengers, a high-speed train uses a two-layer network to separate indoor and outdoor channels. That is, APs are deployed in the train compartments to receive all communication data in the train, and mobile repeaters are deployed outside on top of the train to transmit the corresponding data to outdoor base stations. This transmission architecture avoids direct connections between train passengers and roadside base stations, reduces the high penetration loss caused by signals entering and leaving compartments, and strengthens the communication quality.

   In radio communications, channel quality is highly dependent on the propagation environment, such as the physical location of a user. At present, railways are generally constructed on viaducts, especially in China (for example, a field test illustrates that 86% of high-speed trains between Beijing and Tianjin run on viaducts), featuring simple scattering environment and approximately line of sight (LOS) wireless channel. After beamforming technology is used, non-line-of-sight (NLOS) propagation with relatively low signal strength can be ignored compared with LOS propagation. Therefore, approximate beam directions can first be derived from the wireless environment map based on the train location. Due to limitations such as positioning precision and environment information, only the location information is inadequate for determining beam directions. The latest historical beam alignment result should be considered together with the train location information to update the beam training set. Moreover, the train track on a high-speed railway has strong periodicity and regularity, so historical beam training results for a location are generally useful for next beam alignment. As such, fast beamforming in high-speed railway scenarios significantly improves communication performance.

19.10.3 V2X

The ever-rising density and mobility of urban population demand upgrade of urban traffic management and construction. Real-time, intelligent, efficient, and networked urban traffic management and construction solutions will be indispensable to future smart cities, where smart V2X will definitely play an important role. Figure 19.7 illustrates the V2X application based on the cell-free cooperative massive MIMO system. While sensing functions such as real-time vehicle detection, positioning, identification, and imaging can be implemented on future 6G networks by using communication signal information, this system can obtain surrounding road environment information through the sensing functions, and generate ultra-high-resolution images by means of advanced algorithms, edge computing, and AI technologies, to deliver security and reliability for infinite connections between humans, machines, and things anytime and anywhere. This will open up an array of brand-new services and experience for improving people's lives.

Connectivity of everything is the engine of V2X, which can be broadly classified into vehicle to vehicle (V2V), vehicle to infrastructure (V2I), vehicle to pedestrian (V2P), and vehicle to data center. Vehicles are expected to exchange information with all influencing entities to reduce accidents, mitigate traffic congestion, reduce environmental pollution, and provide other information services. Currently, two V2X technologies dominate the market. One is the dedicated short-range communications (DSRC) technology from the IEEE. This technology was first proposed in the United States and has been developed and tested for many years since the standard was released in 2010. Its wireless access part consists of multiple IEEE protocols, which are collectively referred to as wireless access protocols for vehicular environments. The other is C-V2X from the 3GPP organization. This technology was launched after DSRC and is evolved based on cellular network communication technologies such as 4G/5G. It includes two types of communication interfaces: One is the interface for short-distance direct communication among vehicles, pedestrians, and roads, and the other is the interface between terminals and base stations to implement reliable communication over a longer distance and a larger coverage. Integration of 5G into V2X can provide ultra-large coverage for urban traffic based on the cell-free

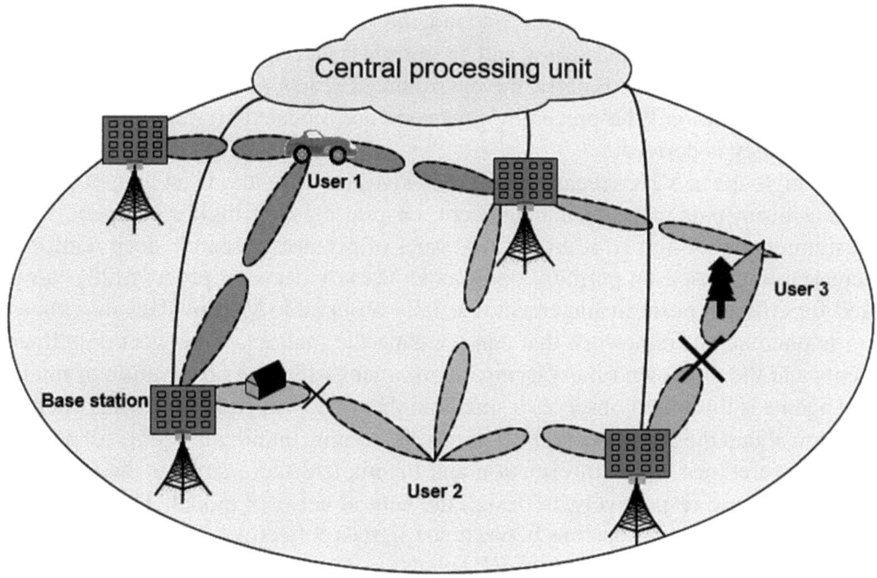

Fig. 19.7 V2X application based on the cell-free cooperative massive MIMO system

cooperative massive MIMO system, driving smarter V2X applications and deeper connectivity of everything.

Future 6G technologies will expect to build a new type of communication network featuring smart interaction, inclusive intelligence, communication sensing, and all-domain coverage. The existing sub-6 GHz communication struggles to meet future requirements of 6G on ultimate performance and stringent technical specifications. This opens the door for mmWave communication in 30 GHz to 300 GHz, which gradually attracts more attention. Major advantages of mmWave communication include rich spectrum resources, strong beam steering characteristics, and high data transmission rates. However, severe path loss and poor penetration capability are also challenges facing communication in high frequency bands due to short wavelengths. In the cell-free cooperative massive MIMO system architecture, the environment sensing-based intelligent beam management technology can sense dynamic environment information to implement beam scanning, tracking, and alignment between distributed APs and mobile terminals. On the basis of enhanced communication coverage and controlled energy consumption, mmWave communication propels its strengths and mitigates its weaknesses, making it suitable for smart V2X in cities and providing technical assurance for the implementation of smart V2X.

When it comes to beam management in high-speed V2X scenarios, deep reinforcement learning can be an effective tool. This form of learning is an E2E sensing and control system with strong commonality and its learning process may be described as follows: First, intelligent agents interact with the environment at each episode to obtain a high-dimensional observation, and use deep learning for observation sensing to obtain features of a specific state. Second, the value function of each action is evaluated based on the expected return, and the current state is mapped to a corresponding action by using a policy. Finally, the environment reacts to this action and obtains the next observation. The preceding processes are repeated continuously until the optimal policy is derived.

In a high-speed V2X scenario, the exhaustive search method can help obtain the optimal beam pair in beam management, despite it being highly complex, time-consuming, and costly. To address such issues of exhaustive search, deep reinforcement learning based on partially observable Markov decision process [22] can be used for efficient beam management. Partially observable Markov decision process is a mathematical framework that can simulate the interaction between intelligent agents and the unknown time-varying environment when the observation of intelligent agents is limited. It observes a small number of scores and predicts unobserved scores to obtain the score set of all RF beams. A learning intelligent agent will provide observed reference signal information and incomplete score table as the input and expected output, respectively, to design the neutral network model. The purpose is to learn the hidden relationships between the signals collectively received by all APs and the rates of different groups of RF beams.

After the neural network of deep reinforcement learning is trained to obtain the weight matrix, intelligent agents can predict the optimal RF beam pair in the operation phase by multiplying the weight matrix and the channel vector of the corresponding location of each instance (together) to determine the beam direction. In other words, beam control means setting the beam direction based on the changes in the channel vector. The channel vector of each AP is affected by high mobility and may relate to the channel vectors of other APs. The weight matrix may be considered as a serial connection of different vectors, where each vector corresponds to a different channel vector. Such beam control renders lower complexity because the received reference signals are mapped directly to hybrid beamforming vectors, without the need of heavy computation or lots of signaling data.

Based on current development of V2X and technical support of future 6G communications, smart V2X with device-pipe-cloud synergy can further improve the lives of everyone. "Device" simply refers to terminals, and in the future, 6G technologies are expected to implement connectivity of new-type smart terminals, ensuring intelligent network access of a large number of vehicles in a city. Meanwhile, in the cell-free cooperative massive MIMO system architecture, distributed APs can provide massive connections to smart devices. For example, traffic information in each corner of a city's roads can be collected by combining static monitoring of traffic cameras and dynamic inspection of intelligent UAVs, to realize connectivity of everything. After this, "pipe" emphasizes the process. In urban transportation, pedestrians and vehicles are usually moving. Traditional communication technologies have to consume a large amount of energy to ensure quality of service, whereas highly beam-steering and intelligent mmWave beam management technologies can provide efficient and cost-effective solutions to guarantee accurate beam tracking and beam alignment even in high-mobility movement scenarios. And last but not least, "cloud" stresses data processing. The central data processing center in the cell-free cooperative massive MIMO system can be integrated with technologies such as big data and cloud platform to centrally process an array of information collected by distributed APs and display visual analysis results to commanders, and issue instructions to complete the communication process. Not only that, the multi-access edge computing (MEC) technology handles the problem that smart terminals cannot process a large amount of data. The core idea of this technology is to deploy edge computing nodes on distributed APs to offload a large amount of data to be processed from vehicles. This significantly improves the QoS of smart V2X.

New-type smart V2X technologies can disruptively change how people live and move around. For example, with real-time road condition prediction, vehicles can upload road conditions to the traffic management department at any time, streamlining vehicle flow control and route planning on the cloud to prevent traffic congestion. With intelligent parking management, drivers can be notified of the nearest entrance of a parking lot and the number of vacant parking spots before they arrive at their destination and complete parking under the guidance of the intelligent system. With secure driving of V2X, vehicles can move around safely by constantly analyzing surrounding environment through their own sensors and connect to various traffic lights and other control signals in cities to automatically warn and avoid dangers. 6G will slash network latency, provide larger coverage and higher transmission rate, and ensure the intelligence, efficiency, and network connection of the V2X industry.

19.10.4 Traffic Inspection

In future intelligent transportation, 6G will build a new network featuring intelligent interconnection and collaboration between humans, machines, and things in the real world, assisting all-round, large-capacity, efficient, and cost-effective traffic inspection tasks in cities. Driven by the innovation of multiple new disciplines, 6G will be heavily integrated with information technologies such as advanced computing, AI, big data, and IoT. The cell-free cooperative massive MIMO system architecture will allow a large number of new-type 6G smart terminals to be deployed and streamlines real-time communication between them. Based on the notion of distributed information collection and centralized information processing, the overall intelligent traffic management will combine static monitoring and dynamic inspection. Figure 19.8 illustrates smart security protection in a city.

Traditional urban traffic inspection mainly involves roadside camera surveillance and stationed traffic police. Such inspection methods struggle to meet the requirements of current traffic conditions as the urban population increases sharply. This highlights the need for new 6G technologies to strengthen urban traffic management. Urban traffic inspection comprises three parts, which are traffic information collection, information processing, and control instruction delivery. (1) Traffic information collection: With intelligent agents in the cell-free cooperative massive MIMO system, traffic information collection can be completed by combining static monitoring of

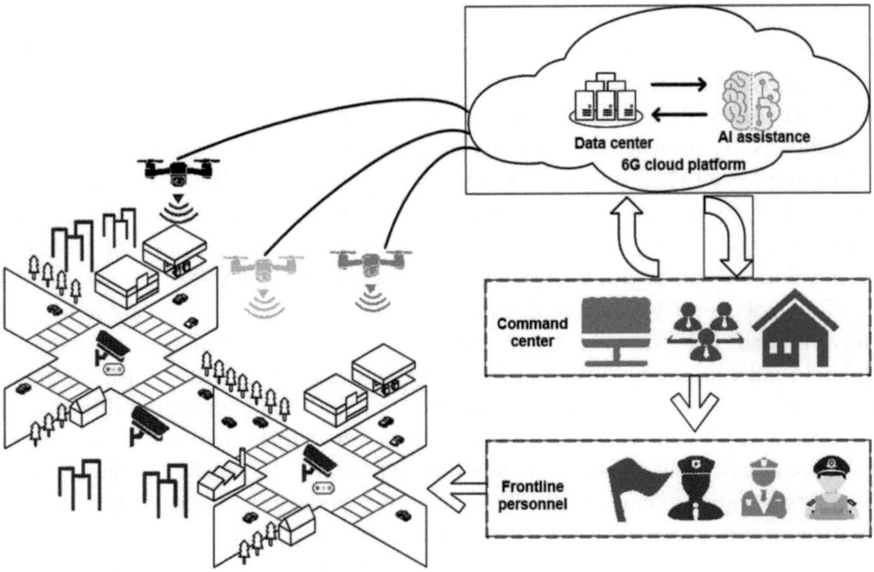

Fig. 19.8 Smart security protection in a city

existing traffic cameras and dynamic inspection of intelligent UAVs. UAVs can be an ideal solution for traffic dispersion when manual dispersion becomes slow or road conditions cannot be accurately determined. UAVs can also collect traffic violation evidence to compensate blind spots in ground camera surveillance, conduct large-scale surveys of the accident site while saving time and labor, and perform traffic planning while working with HD cameras to implement all-round urban traffic surveillance. Cell-free cooperative massive MIMO technology allows multiple UAVs to work concurrently on different tasks, while intelligent beam management technology for environment sensing means APs can effectively control all UAVs. (2) Information processing: After traffic information collection, new-type 6G smart terminals working in the cell-free cooperative massive MIMO architecture can quickly and accurately send large amount of traffic data back to the cloud computing center by using URLLC technologies. What's more, new 6G technologies can be integrated with technologies such as big data and cloud platform to realize accurate and effective processing of large-scale urban traffic information, while the cloud computing center can utilize AI algorithms to forecast future urban traffic conditions to prevent traffic accidents from occurring. (3) Control instruction delivery: Once data analysis is complete, the command center categorizes and issues management and control instructions to frontline personnel separately so that they can perform their own responsibilities and handle urban traffic assignments within their jurisdiction, implementing effective staff management. In this phase, the command center can achieve instant communication with individuals, ensuring that commands can be received immediately and problems can be solved promptly. So far, the overall urban traffic inspection layout of live video-center-front line has been formed.

References

1. Li MJ (2021) Research on personalized affective interaction based on multimodal semantic analysis. University of Science and Technology Beijing, Beijing
2. Tiwari N, Edla DR, Dodia S et al (2018) Brain computer interface: a comprehensive survey. Biol Inspired Cogn Arch 26:118–129
3. Li WJ, Long Y (2020) Technologies and cities: general smart city technologies promote city resilience. Shanghai City Plan 2:64–71
4. Batty M (2018) Artificial intelligence and smart cities. Environ Plan B: Urban Anal City Sci 45(1):3–6
5. Sholla S, Naaz R, Chishti MA (2018) Semantic smart city: context aware application architecture. Proceedings of 2018 Second International Conference on Electronics, Communication and Aerospace Technology (ICECA). Piscataway: IEEE Press, 721–724
6. Ren M (2021) Digital intelligence. Four typical application scenarios of smart factory (in Chinese). China Informatiz 1:47–49
7. Xu ZH (2019) Discussion on the application and development of satellite communications in emergency communications (in Chinese). China New Commun 21(17):2–3
8. Li N, Feng HR (2019) Discussion on the application and development of satellite communications in emergency communications (in Chinese). Inf Technol Informatiz 11:146–148
9. Huawei Technologies Co., Ltd (2021) Digital energy 2030 (in Chinese)
10. Digital Energy Industry Think Tank (2021) White paper on ten trends of digital energy (in Chinese)
11. Masanet E, Shehabi A, Lei NA et al (2020) Recalibrating global data center energy-use estimates. Science 367(6481):984–986
12. State Grid (2021) Energy digital transformation white paper (in Chinese)
13. Shang FJ, Li X, Zhai D et al (2021) October 9) Two-phase resource allocation technology for network slices in smart grid. Journal of Computer Applications 41(7):2033–2038
14. Wang G, Liu X, Shen J et al (2021) A deep-learning pipeline for the diagnosis and discrimination of viral, non-viral and COVID-19 pneumonia from chest X-ray images". Nat Biomed Eng 5(6):509–521
15. Yin KJ (2021) Design and implementation of internet of things management platform for intelligent medical treatment. Beijing University of Posts and Telecommunications, Beijing
16. Internet Healthcare Industry Alliance (2019) 5G smart healthcare white paper (in Chinese)
17. Dai YF, Li Keqiang (2015) Development status and trend analysis of intelligent connected vehicles (in Chinese). Automot Manuf 18:14–17
18. Chen S, Hu J, Shi Y et al (2018) Technologies, standards and applications of LTE-V2X for vehicular networks. Telecommun Sci 34(4):1–11
19. Wang JQ, Huang H, Zhi P et al (2019) Review of development and key technologies in automatic driving. Appl Electron Tech 45(6):28–36
20. China Academy of Information and Communications Technology (CAICT) (2020) Connected vehicle white paper (connected autonomous driving volume) (in Chinese)
21. Lu YQ, Asghar MR (2020) Semantic communications between distributed cyber-physical systems towards collaborative automation for smart manufacturing. J Manuf Syst 55:348–359
22. Fozi M, Sharafat AR, Bennis M (2022) Fast MIMO beamforming via deep reinforcement learning for high mobility mmWave connectivity. IEEE J Sel Areas Commun 40(1):127–142

MIX
Papier aus verantwortungsvollen Quellen
Paper from responsible sources
FSC® C105338

If you have any concerns about our products,
you can contact us on
ProductSafety@springernature.com

In case Publisher is established outside the EU,
the EU authorized representative is:
Springer Nature Customer Service Center GmbH
Europaplatz 3, 69115 Heidelberg, Germany

Printed by Libri Plureos GmbH
in Hamburg, Germany